Lecture Notes in Networks and Systems 659

The series "Lecture Notes in Networks and Systems" publishes the latest developments in Networks and Systems—quickly, informally and with high quality. Original research reported in proceedings and post-proceedings represents the core of LNNS.

Volumes published in LNNS embrace all aspects and subfields of, as well as new challenges in, Networks and Systems.

The series contains proceedings and edited volumes in systems and networks, spanning the areas of Cyber-Physical Systems, Autonomous Systems, Sensor Networks, Control Systems, Energy Systems, Automotive Systems, Biological Systems, Vehicular Networking and Connected Vehicles, Aerospace Systems, Automation, Manufacturing, Smart Grids, Nonlinear Systems, Power Systems, Robotics, Social Systems, Economic Systems and other. Of particular value to both the contributors and the readership are the short publication timeframe and the world-wide distribution and exposure which enable both a wide and rapid dissemination of research output.

The series covers the theory, applications, and perspectives on the state of the art and future developments relevant to systems and networks, decision making, control, complex processes and related areas, as embedded in the fields of interdisciplinary and applied sciences, engineering, computer science, physics, economics, social, and life sciences, as well as the paradigms and methodologies behind them.

Indexed by SCOPUS, INSPEC, WTI Frankfurt eG, zbMATH, SCImago.

All books published in the series are submitted for consideration in Web of Science.

For proposals from Asia please contact Aninda Bose (aninda.bose@springer.com).

Nenad Filipovic
Editor

Applied Artificial Intelligence: Medicine, Biology, Chemistry, Financial, Games, Engineering

Springer

Editor
Nenad Filipovic
Department of Applied Mechanics
and Automatic Control
University of Kragujevac
Kragujevac, Serbia

ISSN 2367-3370 ISSN 2367-3389 (electronic)
Lecture Notes in Networks and Systems
ISBN 978-3-031-29716-8 ISBN 978-3-031-29717-5 (eBook)
https://doi.org/10.1007/978-3-031-29717-5

This Springer imprint is published by the registered company Springer Nature Switzerland AG
The registered company address is: Gewerbestrasse 11, 6330 Cham, Switzerland

Preface

The book covers knowledge and results in theory, methodology and applications of Artificial Intelligence and Machine Learning in academia and industry. Nowadays, artificial intelligence has been used in every company where intelligence elements are embedded inside sensors, devices, machines, computers and networks. The chapters in this book integrated approach toward global exchange of information on technological advances, scientific innovations and the effectiveness of various regulatory programs toward AI application in medicine, biology, chemistry, financial, games, law and engineering. Readers can find AI application in industrial workplace safety, manufacturing systems, medical imaging, biomedical engineering application, different computational paradigm, COVID-19, liver tracking, drug delivery system and cost-effectiveness analysis. Real examples from academia and industry give beyond state of the art for application of AI and ML in different areas. These chapters are extended papers from the First Serbian International Conference on Applied Artificial Intelligence (SICAAI), which was held in Kragujevac, Serbia, on May 19–20, 2022 (www.aai2022.kg.ac.rs).

Nenad Filipovic

Contents

Advances in the Use of Artificial Intelligence and Sensor Technologies for Managing Industrial Workplace Safety 1
Arso M. Vukićević and Miloš Petrović

Implementation of Deep Learning to Prevent Peak-Driven Power Outages Within Manufacturing Systems 29
Milovan M. Medojević and Marko M. Vasiljević Toskić

Reproductive Autonomy Conformity Assessment of Purposed AI System 45
Dragan Dakić

Baselines for Automatic Medical Image Reporting 58
Franco Alberto Cardillo

HR Analytics: Serbian Perspective 75
Dragan V. Vukmirović, Željko Z. Bolbotinović, Tijana Z. Comić, and Nebojsa D. Stanojević

Ontology-Based Analysis of Job Offers for Medical Practitioners in Poland 90
Paweł Lula and Marcela Zembura

Synergizing Four Different Computing Paradigms for Machine Learning and Big Data Analytics 103
Veljko Milutinović and Jakob Salom

Pose Estimation and Joint Angle Detection Using Mediapipe Machine Learning Solution 109
Katarina Mitrović and Danijela Milošević

Application of AI in Histopathological Image Analysis 121
Jelena Štifanic, Daniel Štifanić, Ana Zulijani, and Zlatan Car

The Projects Evaluation and Selection by Using MCDM and Intuitionistic Fuzzy Sets 132
Aleksandar Aleksić, Snežana Nestić, and Danijela Tadić

Application of MCDM DIBR-Rough Mabac Model for Selection of Drone for Use in Natural Disaster Caused by Flood 151
Duško Z. Tešić, Darko I. Božanić, and Boža D. Miljković

Improving the Low Accuracy of Traditional Earthquake Loss Assessment Systems 170
Zoran Stojadinović

SecondOponionNet: A Novel Neural Network Architecture to Detect Coronary Atherosclerosis in Coronary CT Angiography 186
Mahmoud Elsayed and Nenad Filipović

Ontology-Based Exploratory Text Analysis as a Tool for Identification of Research Trends in Polish Universities of Economics 198
Edyta Bielińska-Dusza, Monika Hamerska, Magdalena Kotowicz, and Paweł Lula

Improved Three-Dimensional Reconstruction of Patient-Specific Carotid Bifurcation Using Deep Learning Based Segmentation of Ultrasound Images 223
Milos Anić and Tijana Đukić

Seat-to-Head Transfer Functions Prediction Using Artificial Neural Networks 249
Slavica Mačužić Saveljić

A Review of the Application of Artificial Intelligence in Medicine: From Data to Personalised Models 271
Anđela Blagojević and Tijana Geroski

Digital Platform as the Communication Channel for Challenges in Artificial Intelligence 306
Jelena Živković and Đorđe Ilić

Mathematical Modeling of COVID-19 Spread Using Genetic Programming Algorithm 320
Leo Benolić, Zlatan Car, and Nenad Filipović

Liver Tracking for Intraoperative Augmented Reality Navigation System 332
Lazar Dašić

Intelligent Drug Delivery Systems 342
Ana Mirić and Nevena Milivojević

Cost Effectiveness Analysis of Real and in Silico Clinical Trials for Stent Deployment Based on Decision Tree 367
Marija Gačić

Author Index 383

Advances in the Use of Artificial Intelligence and Sensor Technologies for Managing Industrial Workplace Safety

Arso M. Vukićević[1(✉)] and Miloš Petrović[2]

[1] Faculty of Engineering, University of Kragujevac, 6 Sestre Janjić Street, 34000 Kragujevac, Serbia
arso_kg@yahoo.com
[2] School of Electrical Engineering, University of Belgrade, 73 Bulevar Kralja Aleksandra, 11000 Belgrade, Serbia
petrovic.milos@etf.bg.ac.rs

Abstract. With technological progress, workplace safety standards have increased, so a growing scientific community focused on the application of technologies for improving workers' safety and well-being. In this study, we review recent advances in applying cloud technologies, artificial intelligence, and numerous sensors to assess various problems in safety science, ranging from reporting and management roles to improving the ergonomics of physical tasks. Particularly, we review studies focused on applying or combining cloud technologies, artificial intelligence, sensors, and robotics for studying or improving industrial workplaces. The emphasis was on topics covered with our recent project AI4WorkplaceSafety (http://ai4workplacesafety.com), where we were focused on: 1) developing a lightweight framework for easing the collection and management of safety reports (related to unsafe acts and unsafe conditions). 2) Automating of PPE compliance using computer vision, which represents specific cases of unsafe acts. 3) Assessing and detecting unsafe acts related to pushing and pulling (typical examples are workplaces in warehouses and transportation). 4) Finally, we briefly presented a modular and adaptive laboratory model (industrial workstation) design for a human–robot collaborative assembly task. It is concluded that ongoing technological progress and related multidisciplinary studies on this topic are expected to result in a better understanding and prevention of workplace injuries.

Keywords: Artificial intelligence, workplace, safety, engineering.

1 Introduciton

Industry 4.0 (I4.0) is a term used to indicate the global industrial transformation driven by rapid technological advances. According to the official Global Industry Classification Standard (GICS), there are 11 sectors, 24 groups, 69 industries, and 158 subindustries [1]. Considering such diversity, it is nowadays more precise to talk about the I4.0 branches; such are: Quality 4.0 [2], Maintenance 4.0 [3], and Safety 4.0 [4]. So far, in many industrial branches, the major goal has been set towards automation which has

N. Filipovic (Ed.): AAI 2022, LNNS 659, pp. 1–28, 2023.
https://doi.org/10.1007/978-3-031-29717-5_1

brought tremendous progress in many manufacturing sectors (automotive, electronics manufacturing, welding, etc.). However, it is shown that in practice there are still many workplaces that cannot be adequately or fully automatized. Although there is a growing trend of supplementing laborious workplaces with (co)robots that can collaborate with human operators [5] - authorities agree that the further evolution of technology and industry will remain even more human-centered [6]. Thus, there is an increasing need for technologies that could help in improving the well-being of human operators in an industrial environment.

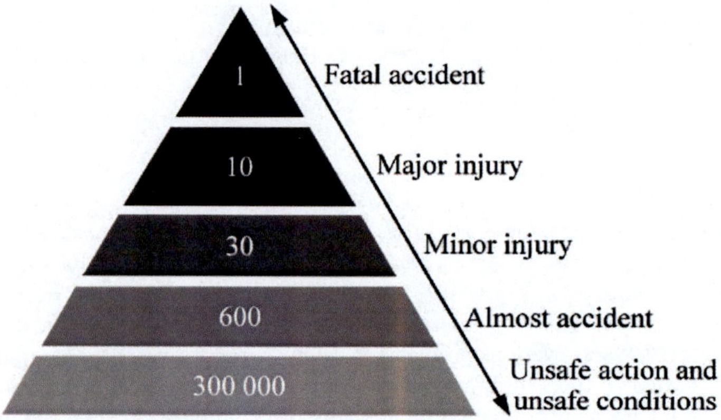

Fig. 1. Workflow of the digitalized management of unsafe conditions and unsafe events

1.1 Workplace Safety Management in SMEs

Risk and safety management is a broad topic [7], so the focus of this chapter will be restricted to occupational safety and health (OSH). Briefly, the OSH scientists and professionals aim to improve the safety, health, and welfare of people at work, with the end goal to the number of production injuries and accidents down to zero. To reach this goal, companies are focusing on proactive identification of the accidents' precursors to prevent accidents. According to Heinrich's pyramid (Fig. 1), proactive identification of unsafe conditions (UC) and unsafe acts (UA) have the biggest impact on safety [8]. Although there are recommendations set by regulatory bodies and international standards, traditional management of workplace safety has shown to be a slow, subjective, and complex task when it comes to industrial practice. In the rest of this chapter, the emphasis is put on reviewing research studies that cover SMEs' needs because of the fact that they generate most of the GDP and employment opportunities in developed countries. Moreover, it is more likely that compact solutions proposed in the literature will be first applied in SMEs on a smaller scale before being incorporated into enterprises' ICT systems. From the SMEs' viewpoint, enterprise solutions frequently are too expensive, especially when it comes to the incorporation of additional and/or nonstandard features specific to their type and size of business [9]. In these terms, cloud technologies and compact web frameworks have shown the biggest potential to improve safety management, along with computer vision techniques - as the recognition of UC/UA is a visual task.

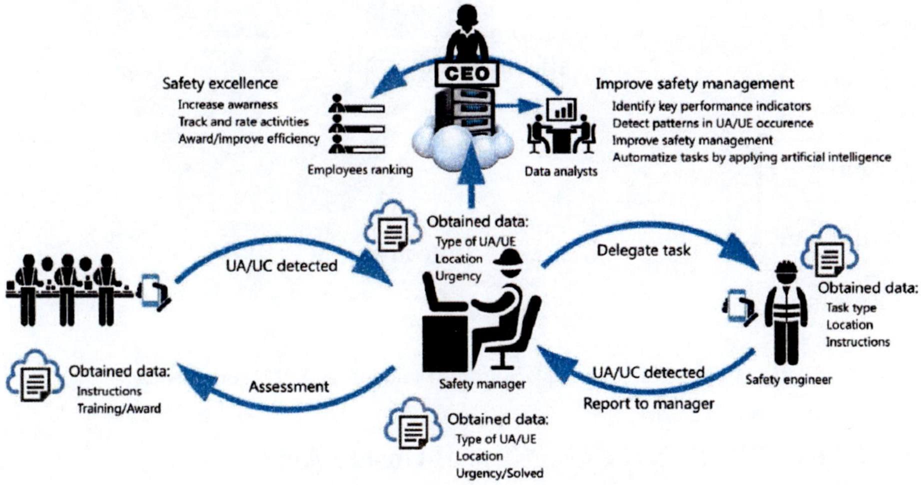

Fig. 2. Workflow of the digitalized management of unsafe conditions and unsafe events [10]

1.2 The Importance of Timely and Objective Identification of UC/UA

With the progress of the Safety 4.0 paradigm, traditional paper and manual reporting are being replaced with cloud-based applications and services run on smartphones and edge devices. One such solution is the SafE-Tag, a minimalistic framework released with the aim to enhance the collection of safety reports and delegation of corresponding tasks - with the end goal of encouraging employees to proactively contribute and learn about safety [10]. The graphical illustration of the concept proposed in the same study is given in Fig. 2. The composing parts of the proposed solution are a) central cloud server and b) remote mobile device - so that employees are allowed to collect and report UC and UA, as well as to receive and respond to assigned tasks. Along with an efficient collection of safety reports, the long-term benefit of digitalized safety reporting is to enable in-depth analysis of safety performances by employing business intelligence.

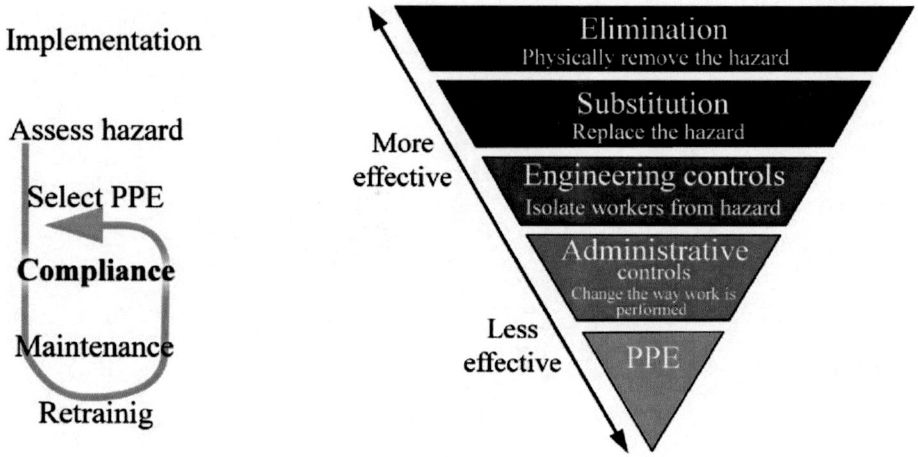

Fig. 3. Relation of workplace safety standards and PPE compliance

2 Misuse of PPE as the Use Case of Unsafe Acts

The Occupational Safety and Health Administration (OSHA) has proposed the five levels of OSHA controls (Fig. 3): 1) elimination, 2) substitution, 3) engineering, 4) administrative, and 5) use of personal protective equipment (PPE) [11]. In this sense, the use of PPE may be considered a first-line barrier between employees and hazards when applied. Despite the availability of PPEs, and corresponding PPE standardization and use guidance, the industrial practice has shown that misuse of PPE still represents a serious problem. Briefly, reports indicate that PPE misuse causes a number of injuries and large losses to national economies [12]. This is explained by supervisors' inability to timely and objectively notice PPE non-use in large manufacturing halls where the number of workers fluctuates [10]. Although PPEs are commonly stratified into four levels (A-D) [13], in related studies PPEs are commonly split according to physiological functions that they aim to protect. Initially proposed approaches are variants of radiomics-based detection of helmets [14]. Recent studies are based on the use of convolutional neural networks [15]. In terms of deep learning architectures used, the most frequently used detectors are YOLOv3 [16, 17], Fast R-CNN [18]. In a recent study, Nagrath et al. demonstrated the application of combining SSD and MobileNetV2 classifier for Covid19 mask detection [19].

There is a high variability of PPEs concerning appearance and design. Therefore, compliance of a head-mounted PPE is a very specific case. Moreover, sometimes it is required for an employee to wear multiple PPEs simultaneously (e.g., hard hats, safety glasses, safety masks, and earmuffs). Compared to previous studies that were running multiple classifiers or multi-class classifiers for the head- mounted PPE and mainly focused on hard hats and face masks, our goal was to assess the usage of object detectors in a more efficient approach and perform comprehensive validation by accounting for more PPE types that are relevant for wider industrial application of the computer vision-based compliance of PPE (Fig. 4).

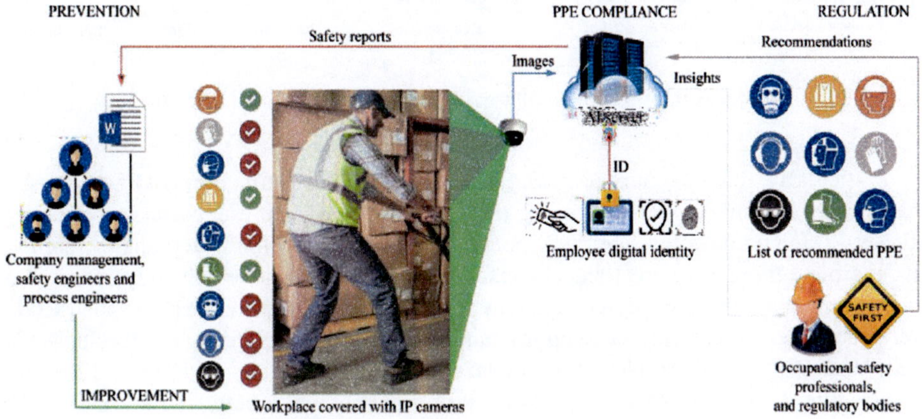

Fig. 4. The concept of AI-driven PPE compliance [20]

In this chapter, we review our recent study which proposed a generic procedure composed of four steps (Fig. 5): 1) employee detection/identification of an in the workspace (Fig. 5a); 2) pose estimation for detecting body landmark points (Fig. 5b); 3) use of the pose landmark points to define regions of interest (ROI); and 4) classification of ROIs (Fig. 5c).

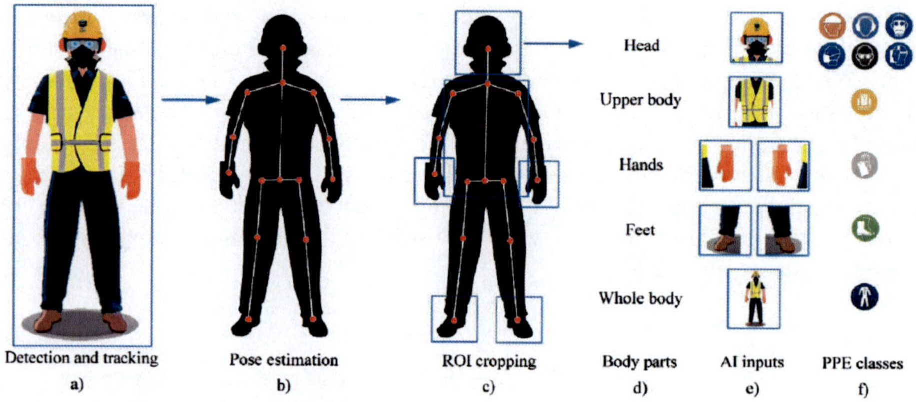

Fig. 5. Workflow of the proposed pose-aware PPE compliance [20]

Particularly, the procedure used HigherHRNet for pose estimation [21], which estimates body landmark points by using the high-resolution representation provided by HRNet [22]. The detected landmark points were used for defining regions of interest (ROI) around five body parts (head, hands, upper body, legs, and whole body). Since the PPE compliance was considered as a classification problem, previously cropped ROIs were subjected for the various deep learning classification architectures MobileNetV2 [23], VGG19 [24], Dense-Net [25], Squeeze-Net [26], Inception_v3 [27], and ResNet [28] - while the MobileNetV2 was the most optimal choice. Briefly, the MobileNetV2 is based on an inverted residual structure, where the input and output of the residual block are thin bottleneck layers, while the intermediate expansion layer uses lightweight depthwise convolutions to filter features as a source of non-linearity [23]. The authors performed the transfer learning by using the model pre-trained on the ImageNet data set [29]. The training was performed using the Adam optimization algorithm [30] with the cross-entropy loss function and the following online augmentations: random rotation (±30°), random flip, random crop, and Gaussian noise. The data set used in this study was developed by combining web-mined images and public PPE datasets (from the Roboflow hardhat train data set and the Pictor PPE data set). The metrics selected for the evaluation and comparison of developed models were accuracy, precision, recall, and f1 score. Considering the current privacy regulations and costs/complexity of using AI, the solution is recommended for the use in controlled conditions, such as: 1) self-check points (when users are asked to confirm their identity by using e.g., RFID card, while AI is used solely for the PPE compliance but not for the purpose of identification and tracking), and on 2) monitoring of particular workplaces/machines with high risk from injuries (so that AI could ensure timely detection and mitigation of occurred risks). In Fig. 6, we showed a couple of use cases [20].

Fig. 6. Sample results of PPE compliance [20]

3 The Use of AI for Assessing the Safety of Pushing and Pulling Activities

The focus of this section is on managing workplace safety in workplaces that involve tasks of pushing and pulling (P&P), such as warehouses and transportation. Non-ergonomic

P&P causes musculoskeletal disorders (MSD), including pain in the back, arms, neck, etc. For employees, in addition to job loss (or forced retraining), MSD also has negative long-term consequences in the form of permanent disabilities and inability to perform everyday activities.

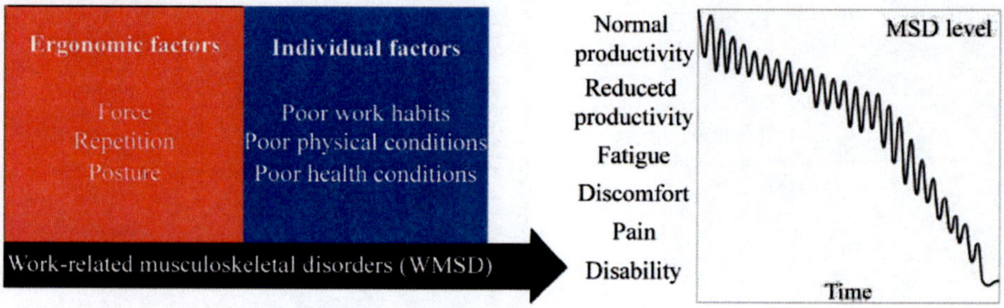

Fig. 7. Progress of the musculoskeletal disorders caused by repetitive non-ergonomics acts at a workplace

Industrial practice recognizes two categories of a cargo P&P: 1) with wheel-based tools (handcarts, forklifts, etc.) and 2) without wheels-based tools (rolling, sliding, and pulling) [31]. This chapter is focused on the first group; however, the proposed methods are applicable to the second group as well. Risks related to the P&P can be divided into ergonomic (workplace-related) and individual (poor health habits or poor physical conditions) [32]. The three key ergonomics risks, which are of interest for this study, are: 1) High frequency-repetition; 2) Excessive effort-overload; and 3) Incorrect (unnatural) body posture. When a worker is exposed to these risks over time, fatigue of the human body accumulates – and when the fatigue level overcomes the ability of the body to recover, regenerate and adapt – MSDs and injuries occur (Fig. 7). The risk assessment of P&P involves consideration of: 1) handcarts type and conditions; 2) weight of cargo; 2) operator's body posture; 3) P&P path shape; 4) distance to be covered; 5) condition of the floor; and 6) presence of obstacles. As may be noted, operators' habits do not directly determine most of these factors, or they do not change significantly over time. For example, the condition of floors and equipment is in the charge of maintenance engineers, process engineers are responsible for choosing the optimal route and mode of transport, management and procurement are responsible for the optimal choice of the type of handcrafts, etc. However, the position of the body during the P&P is extremely subjective and it is not easy to be improved or monitored during working hours. The current bottleneck of workplace safety practice is the assumption that a supervisor will notice non- ergonomic handling of handcarts and timely warns operators – which is very difficult to be managed manually. Accordingly, this study aimed to enhance this risk assessment task and facilitate safety engineers' precise preventive actions.

Fig. 8. Experimental setup for P&P task

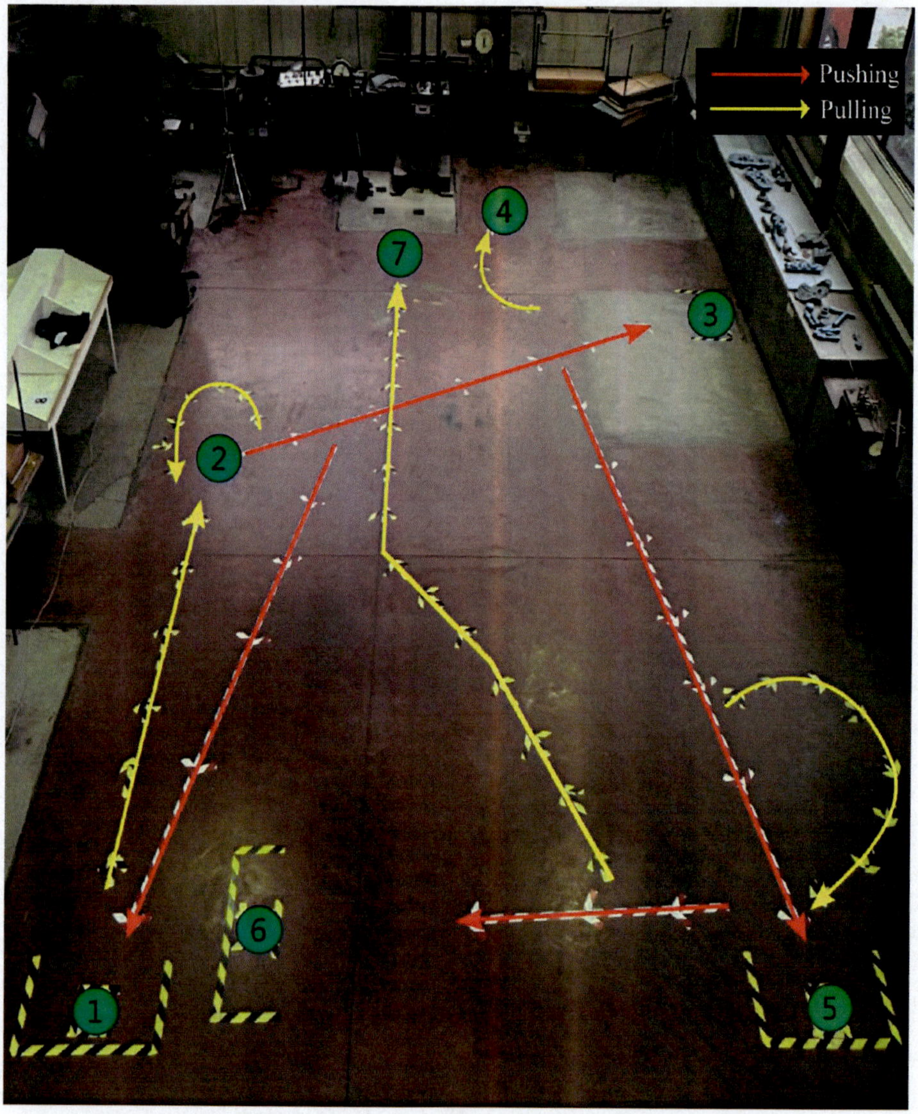

Fig. 9. Experiment environment and P&P path

Figure 8 and Fig. 9 show the experiment environment and pushing and pulling path used in the study [33] and our ongoing study. As may be noted, it is composed of complex turnovers and push/pull maneuverings. In the recent study [33], we aimed to use force IoT sensors to measure P&P force for various participants. Sample force diagrams are shown in Fig. 10, where different colors were used to separate left- and right-hand forces. From this sample diagram, there are considerable differences in signals measured from the left and right hand, as well as for vertical and horizontal components of forces.

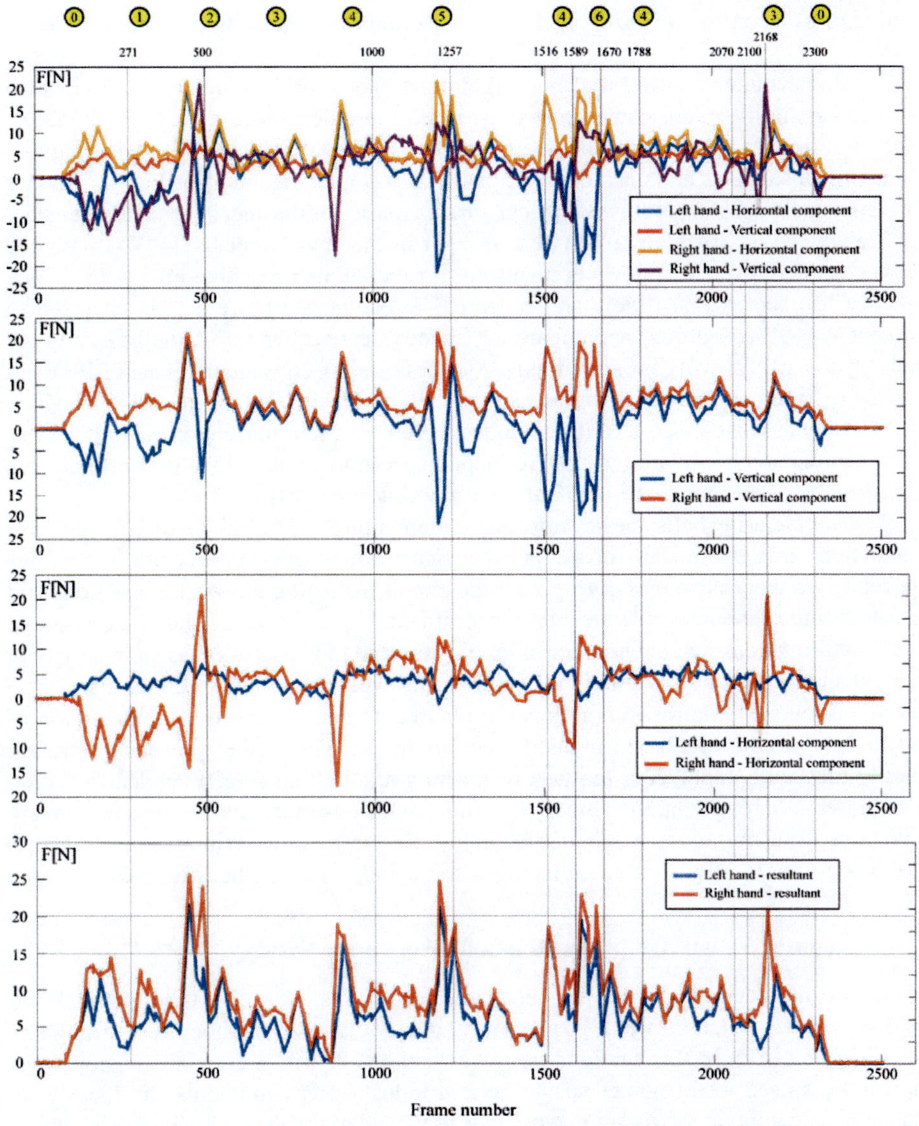

Fig. 10. Pushing and pulling forces for experiments performed in study [33]

3.1 Workplace Musculoskeletal Disorders and Injuries

There is an increasing need to improve the interface between human operators and new technologies while ensuring the implementation of the highest workplace safety standards and well-being of human operators in an industrial environment. Besides, new workplace safety standards declared zero injuries as an ultimate goal. To achieve this challenge, safety science and ergonomics aim to design and improve workplaces by minimizing discomfort, exertion, and stress and eliminating hazards and risks of injuries [34].

Previous studies showed that non-ergonomic execution of repetitive and physical tasks is among the major causes of work-related musculoskeletal disorders (WMSDs) [35]. It is important to emphasize the difference between difficulties in detecting and managing unsafe acts and unsafe conditions. For example, the misplaced tools, missing PPE, and unclear floors, represent typical unsafe conditions that can be instantly detected and mitigated [20, 36]. Contrary to that, unsafe acts that may be related to WMSDs need to be considered repetitive events resulting in accumulated negative effects [37]. The practice has shown that timely and objective detection of unsafe acts is essential to prevent WMSDs, and their accompanying negative consequences (disabilities and the inability to perform everyday activities) [38]. The costs and consequences of WMSDs are studied by international organizations such as World Health Organization [39] - which reports indicate that ~ 126.6 million adults in the US have a musculoskeletal disorder; while similar reports related to the EU population indicate that 33% of workers have unnatural body postures for > 25% of their working time [40].

In manufacturing halls, the key effort in implementing and follow-up of safety recommendations are performed by onsite safety managers and safety supervisors. Their roles are related to workplace monitoring with the aim of managing the worker's actions and to detects their distinctions from safety recommendations. However, the practice (large manufacturing halls and the number of employees that move across) has shown that the manual supervision of workers is ineffective and expensive. As a solution, a series of initiatives tend to propose computerized tools to automate or improve the detection of unsafe acts in both in-lab and industrial environments. The studies presented here are mainly focused on analyzing the task of pushing and pulling (P&P) handcarts, which was chosen as a representative, highly dynamic task whose variants are present in many industries (transportation, warehouses, healthcare, etc.). Another interesting task that will be covered is collaborative polishing with the help of a collaborative robot.

3.2 Computer Vision, Deep Learning and Workplace Safety

Detection and recognition of objects and (unsafe) actions is a well-studied topic in the field of computer vision, which has recently rapidly evolved with the breakthrough of deep learning [41]. In this section, we review studies in which computer-vision techniques were used to recognize unsafe acts in industrial environments. In a study by Han et al., a computer vision framework was proposed to identify critical unsafe behavior in construction, specifically ladder climbing [42]. Three actions were considered ascending, descending, and reaching far to a side (unsafe act). The detection of unsafe actions was performed by combining the results of both 2D pose estimation and 3D reconstruction and using the motion templates and skeleton models. The number of correctly detected actions was further enhanced with a more detailed human skeletal model (with more joints) applying the same methodology [43]. Another approach for safety assessment of ladder climbing tasks was presented, considering dynamic behavior as a static posture and using a mathematical model of the human skeleton to identify unsafe behaviors based on value ranges of joint parameters [44, 45]. Classification of postures regarding human back, arms, and legs (and their three levels - straight, bent, and bent heavily) were employed to ensure ergonomic posture recognition [46]. An RGB camera was used to capture skeleton motions, view-invariant features in 2D skeleton motions were selected, and the function that approximates the relationship between real-world 3D angles/lengths and the corresponding projected 2D angles/lengths was defined. Risk assessment for several outdoor jobs was performed using OpenPose [47] outputs and computing Rapid Upper Limb Assessment (RULA) scores from snapshots and digital videos. Monitoring construction workers is not a new concept, and procedures for detection that localize construction workers in video frames and initialize tracking have been developed [48]. Some authors suggested that applications of deep learning, even though more complex, could provide satisfactory results in the field of safety management. Seo et al. offered a comprehensive review of systems for safety monitoring on construction sites, categorized previous studies into groups, and emphasized research challenges [49]. A new hybrid deep learning model (CNN + LSTM) for automatic recognition of workers' unsafe actions was developed [50]. The approach was experimentally validated in several scenarios on the task of ladder climbing, where a combination of CNN and LSTM adequately examined spatial and temporal information. An improved CNN that integrates red–green–blue, optical flow, and gray image streams for activity assessment in construction are proposed [51]. It was tested on a dataset of real-world construction videos containing actions of walking, transporting, and steel banding. To prevent workers from falling from heights in construction, Fang et al. developed an algorithm using a faster region-based CNN for detecting the presence of workers and a deep CNN for determining if they are wearing a safety harness [52]. An interesting framework for risk management of railway stations generalizable to a wide range of locations and some additional types of risks was presented [53]. CNN was applied as a supervised machine learning model to automatically extract and classify risky behaviors (fall, slip, and trip) in the stations.

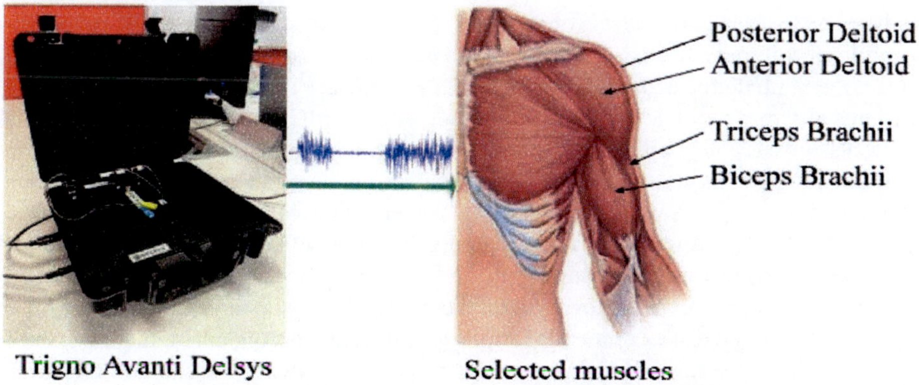

Fig. 11. EMG measuring equipment and selected arm muscles

3.3 The Use of Sensors for Analyzing Workplace Safety

As an alternative to computer vision, there are numerous methods for safety management and recognition of human activities using sensor data. Yan et al. proposed a wearable Inertial Measurement Units (WIMU), a based warning system for construction workers that guarantees self-awareness and self- management of risk factors that lead to WMSDs of the lower back and neck without disturbing their operations [54]. A smartphone application processes real- time data (quaternion data transferred into angles of flexion, extension, lateral bending, and rotation) captured by the IMU sensors fastened to the back of a safety helmet and the upper part of the back. Yang et al. presented a computationally efficient method for activity recognition as a lightweight classification using activity theory for representing everyday human activities, radiofrequency identification (RFID) sensor data, and penalized Naive Bayes classifier [55]. Hofmann et al. used ordinary smartphone sensor data and LSTM for human activity recognition and detection of wasteful motion in production processes [56]. The activities considered were walking, standing, sitting, and jogging, and the reported accuracies for each activity were above 98%. Ordóñez et al. proposed a deep framework for activity recognition - DeepConvLSTM (convolutional and LSTM recurrent units) suitable for homogeneous sensor modalities and multimodal wearable sensors [57].

EMG has the potential to guide our understanding of motor control and provide knowledge of the underlying physiological processes determining force control. It opens the possibility of acquiring insight into muscle activity (load) and better interpreting overexertion, thus preventing the threat of WMSDs and enhancing industrial workplace safety. Even though this idea is not new [58, 59], the scientific fields of biomechanics and biomedical engineering still need to be further investigated, and more effort needs to be put into analyzing industrial task execution. Detailed instructions for EMG measurement methodology have already been presented and widely used [60, 61].

EMG sensors were utilized, as a primary tool, in many recent studies that tried to analyze and assess the risk levels of WMSDs. An extensive study (more than 100 workers with and without a history of chronic pain) was conducted, testing lumbar paraspinal muscles as a predictor of low-back pain (LBP) risk [62]. The same participants were reevaluated two years later, and by examining some EMG variables, it was possible to successfully identify a subgroup of subjects with a higher risk of back pain. In-lab experiment on the risk assessment of non-fatal, cumulative musculoskeletal low back disorders among roofers was presented [63]. The effect of working on uneven rooftops, different working postures (stoop and kneeling), facing direction, and working frequency was evaluated using EMG measurements and 3D human motion data. An analysis of experienced and inexperienced rodworkers using EMG sensors and Xsens MTx Xbus system was conducted to examine the factors that affect the risk of developing lower back musculoskeletal disorders [64]. Working strategies of the two groups were compared, with the accent on levels of back moments L4/L5 and the time spent in an upright and flexed posture. A novel wearable wireless system capable of real- time assessment of the muscular efforts and postures of the human upper limb for WMSD diagnosis was proposed [65]. This real-time system that combines IMU and EMG sensors was tested on the task of repetitive object lifting and dropping, and the risk was estimated based on Rapid Upper Limb Assessment (RULA) and the Strain Index (SI). A biomechanical analysis was conducted in the solid waste collection industry investigating five occupational LBP risk factors for three techniques of waste collection, throwing and three garbage bag masses [66]. LBP risk factors were computed using a full-body musculoskeletal model in OpenSim, where muscle activity was estimated in two ways: using EMG electrodes (more accurate) and the conventional static optimization method.

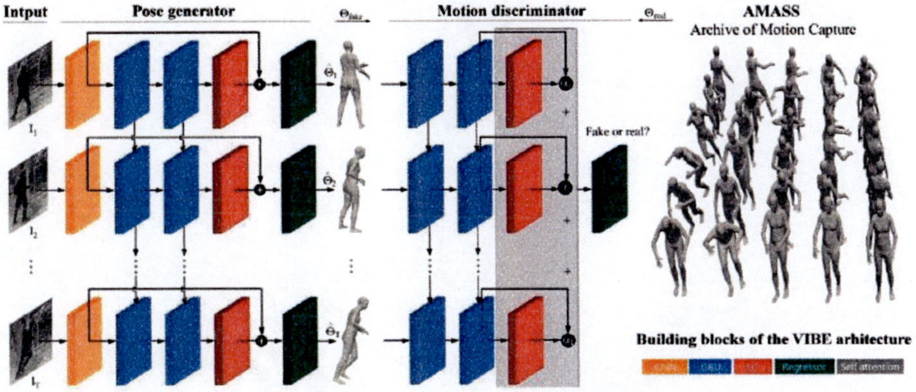

Fig. 12. VIBE architecture for 3D pose estimation from monocular images [33, 67]

3.4 The Use of 3D Pose Estimation and Human Body Models

The experiments were recorded using four DAHUA IPC-HFW2831TP-ZS 8MP WDR IR Bullet IP cameras with a DAHUA PFS3010-8ET-96 8port Fast Ethernet PoE switch. The host PC had an 1151 Intel Core i3-8100 3.6-GHz 6-MB BOX CPU. In our experiments, the Video Inference for Body Pose and Shape Estimation (VIBE) architecture [67] is used to solve the 3D pose reconstruction problem in an adversarial manner. VIBE is a video pose and shape estimation method. The first step of VIBE is a pose generator that extracts image features from video input using a pretrained CNN. Temporal encoder - bidirectional Gated Recurrent Units (GRU) processes these features to make use of the sequential nature of human motion, thus incorporating information from past and future frames which is beneficial when the body of the person is occluded or its pose is ambiguous in a particular frame. Then, the regressor predicts the parameters of the Skinned Multi-Person Linear (SMPL) body model [68] for the whole input sequence at each time instance to obtain realistic and kinematically plausible 3D body shapes and poses (motions). SMPL parametric model delivers a detailed 3D mesh of a human body composed of Quad4 elements with N = 6890 vertices and K = 23 joints. The SMPL model is composed of 82 parameters Θ, which are divided into: 1) pose parameters $\theta \in R^{72}$ (rotation of the 23 body landmark points), and 2) shape parameters $\beta \in R^{10}$ (the first 10 coefficients of a PCA space).

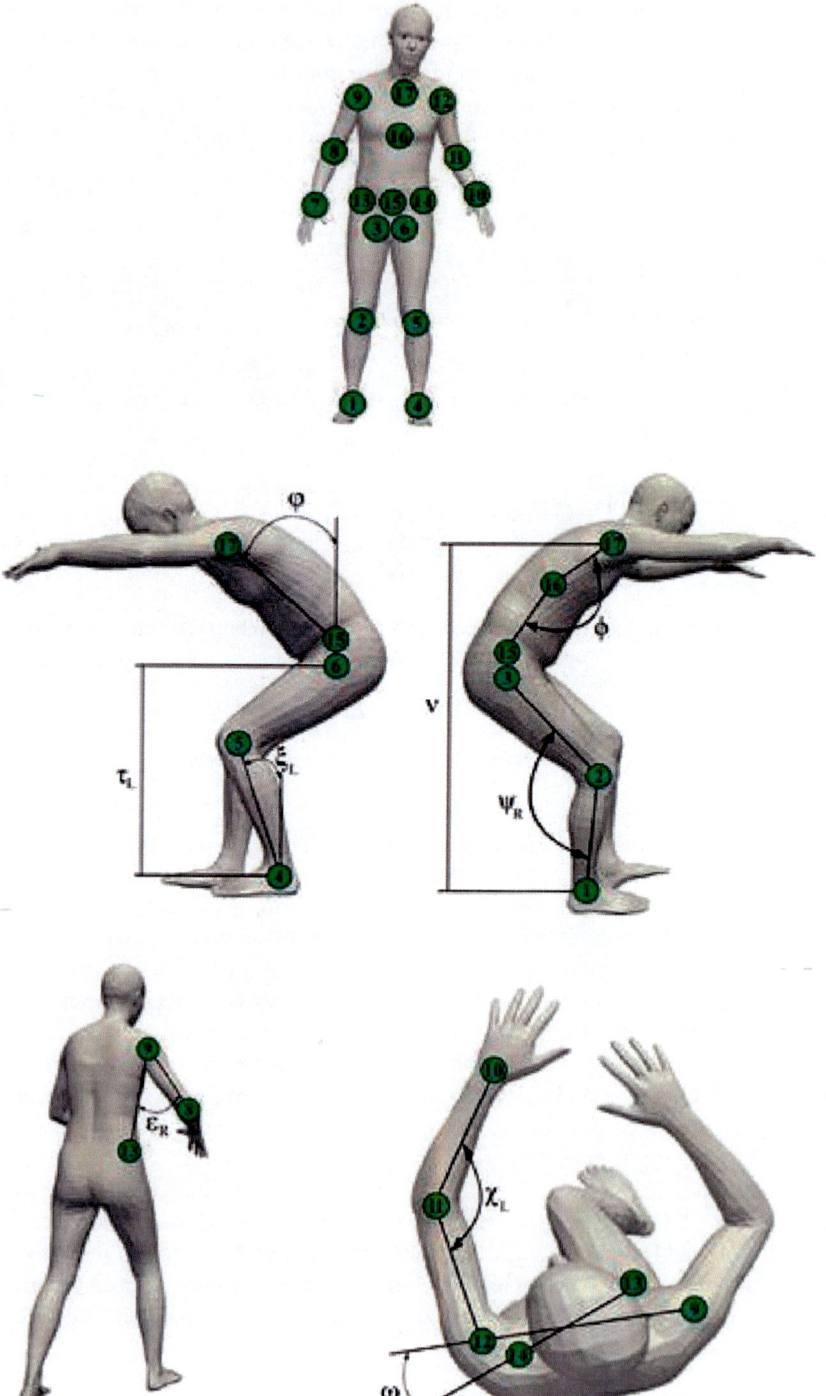

Fig. 13. SMPL model with ergonomic parameters

Accordingly, the SMPL model is a differentiable function, $M(\theta, \beta) \in R^{6890 \times 3}$. The motion discriminator has the task of deciding whether the generated sequence of poses corresponds to a realistic or fake sequence. It uses a stack of GRU layers to process poses sequentially. Then the self-attention mechanism dynamically aggregates and amplifies the contribution of important frames. The motion discriminator is trained on AMASS database (~11,000 movements of ~ 300 subjects) [69], and it takes predicted pose sequences along with pose sequences from AMASS. If the motion discriminator cannot spot the difference between predicted poses and the poses generated from AMASS, then the predicted motion is realistic (adversarial approach).

From the reconstructed SMPL pose, we chose 17 landmark points and computed a series of 13 parameters (divided into three groups) that were used to assess P&P ergonomics (Fig. 13) [33]:

- Leg parameters - the angle of the left knee $\psi L(\sphericalangle 1, 2, 3)$, the angle of the right knee $\psi R(\sphericalangle 4, 5, 6)$, the angle of the left lower leg and the vertical axis $\xi L(\measuredangle 1, 2, Z)$, and the angle of the right lower leg and the vertical axis $\xi R(\measuredangle 4, 5, Z)$;
- Spine parameters - the angle of spine $\phi(\measuredangle 15, 16, 17)$, the angle of the spine and the vertical axis $\varphi(\measuredangle 17, 15, Z)$, the vertical distance between landmark points 1 and 3 $\tau|1, 3|Z$, the vertical distance between points 1 and 17 $\upsilon|1, 17|Z$, and the angle of torsion between the shoulder and the hips ω $(\measuredangle 13, 14, 12, 9)$;
- Arm parameters - the angle of the left elbow $\chi L(\measuredangle 10, 11, 12)$, the angle of the right elbow $\chi R(\measuredangle 7, 8, 9)$, the angle between the left upper arm and the torso $\varepsilon L(\measuredangle 13, 9, 8)$, and the angle between the right upper arm and the torso $\varepsilon R(\measuredangle 14, 12, 11)$.

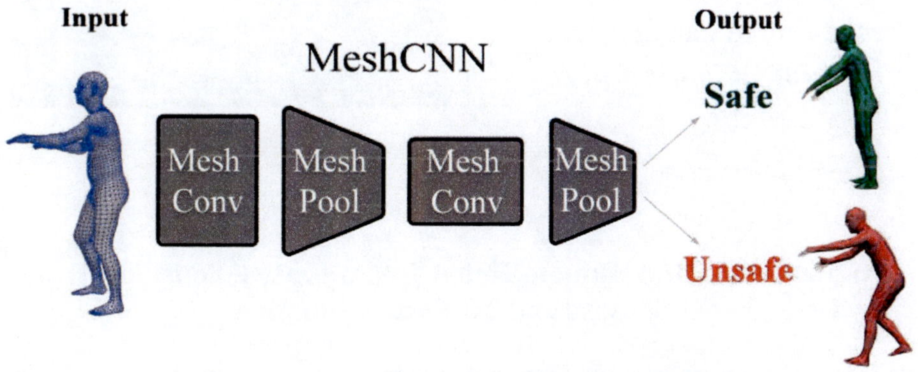

Fig. 14. Concept for MeshCNN pose classification

The polygonal meshes provide an efficient, non-uniform representation that approximates surfaces via 2D polygons in 3D space, explicitly capturing both shape surface and topology [70]. MeshCNN is a deep convolutional neural network designed explicitly for triangular meshes [71]. It comprises customized convolution and pooling operations tailored to operate with the 3D mesh edges. Convolutions process edges accounting for mesh geodesic connections, while the pooling layers preserve the surface topology through the edge collapse. MeshCNN spatially adapts and learns which edges to collapse, unlike classic edge collapse, which removes edges to minimize geometric distortion. Through the successive call of mesh convolution and pooling, the network iteratively generates valid mesh connectivity by learning to preserve and expand important features and discard and collapse redundant ones. The idea is to use 3D pose estimation algorithms, extract a realistic 3D model of a human body, and perform a classification to safe and unsafe acts using irregular structures directly as inputs of a deep neural network (Fig. 14).

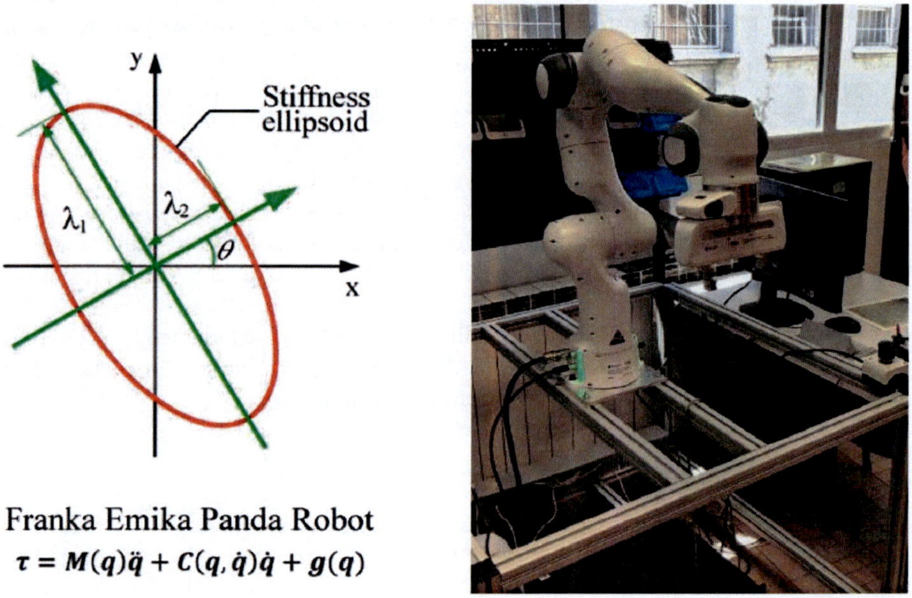

Fig. 15. Collaborative robot and its laboratory setup

4 Assessment of the Human–Robot Collaborative Polishing Task by Using EMG Sensors and 3D Pose Estimation

Our recent study presented a method to improve human–robot collaboration in the industrial setting [72]. The proposed method can be a tool to enhance ergonomics during complex dynamic interactions between a human and a robot, and it can enable the worker to be replaced by a collaborative robot capable of achieving workers' level of performance. The inspiration came from the vision of the factories of the future, where humans and robots will work alongside. This goal is still far away, and more analyses are required from the perspective of collaborative robot control, motion planning, safety, and ergonomics. Previous studies focused on numerous aspects of human–robot collaboration, and a unique framework is proposed for robot adaptation to human motor fatigue in collaborative industrial tasks [73]. KUKA Lightweight Robot was equipped with a Pisa/IIT Softhand and controlled in hybrid force/impedance mode, and EMG measurements were providing information about muscle activity. In order to improve ergonomics, several configurations for collaborative power tooling tasks were tested using an MVN Biomech suit (Xsens Technologies BV) and a Kistler force plate [74]. Conclusions about preferable human poses were obtained based on the analysis of overloading joint torques and muscle activities. The same group of authors introduced the joint compressive forces to enhance the previously proposed model more precisely [75]. Finally, to account for multiple potential contributors to WMSDs, the set of ergonomic indexes are defined, and more extensive experiments were conducted in a laboratory setting [76]. None of the aforementioned solutions incorporated knowledge from the field of computer vision nor performed 3D pose estimation using conventional cameras.

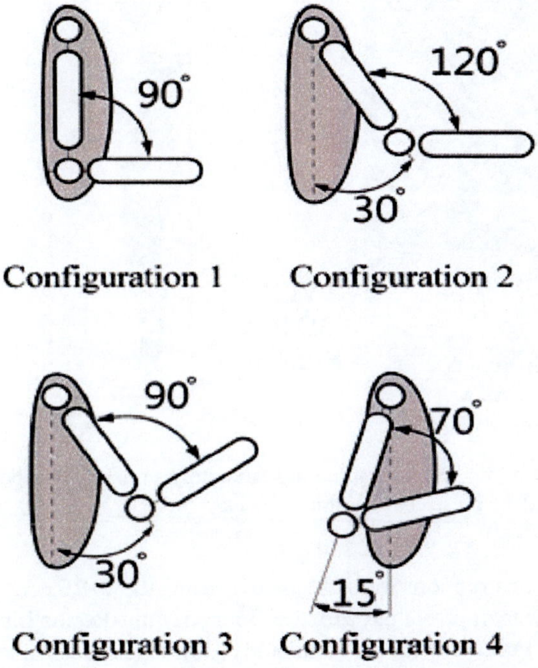

Fig. 16. Four different task configurations adopted by human co-worker

In our previous work, we conducted a laboratory study for human ergonomics monitoring and improvement in a human–robot collaborative polishing task. The data regarding the human whole-body motion, the force exerted on the working piece, and the human muscle activities were recorded through the experimental sessions to investigate the trend of human muscular activity and its correlation with body posture configurations and acting force during the collaborative task. Each subject was instructed to adopt four different body postures (Fig. 16) and to perform the polishing task (using a 1.2 kg polisher) in each configuration for 2 min, exerting a constant force of 10 N and

20 N, respectively. For the analysis of human muscle activity, dominant muscles for the polishing task were selected: Biceps Brachii (BB), Triceps Brachii (TB), Anterior Deltoid (AD), and Posterior Deltoid (PD), and four EMG surface electrodes (Trigno Avanti sensors by Delsys) were placed on the subject's skin (Fig. 11). The EMG signals were processed following the same methodology described in detail in our previous work [77]. The robot (Franka Emika Panda Robot (Fig. 15)) was controlled in impedance mode, and its task was to bring the board to the human and place it in different positions and orientations in the workspace for each experimental session. The board was provided with a force torque sensor to measure the interaction force between the tool and the working piece.

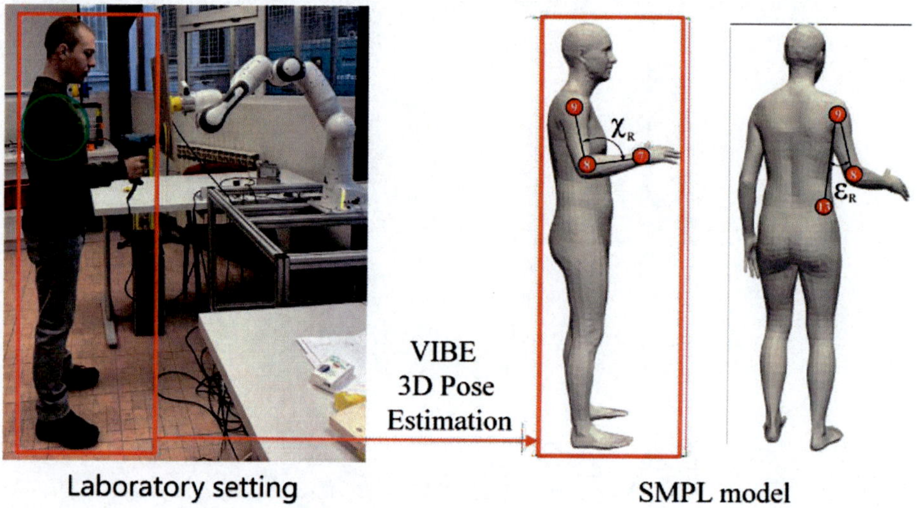

Fig. 17. The concept of the proposed solution - laboratory setting with robot and EMG sensors setup; and SMPL model with selected postural angles

The 3D pose reconstructions were obtained by using the VIBE deep learning architecture, and the poses were represented with the Skinned Multi-Person Linear (SMPL) parametric body model. Following the previous study [33], we extracted a series of ergonomic parameters and selected two postural angles that are important collaborative polishing task - χR and εR, defined on the SMPL model with key points 7–8-9 and 8–9-13, respectively (Fig. 17). Furthermore, human arm manipulability $w = \sqrt{det(J(q)J(q)^T)}$ [78] was taken into account.

The results indicated that for this collaborative task, muscles with dominant activity are the Anterior Deltoid and Biceps Brachii, while the activity of Triceps Brachii and Posterior Deltoid is almost insignificant (below 10% of maximal voluntary contraction for all configurations). Configurations 2 and 3 have overall higher muscle activity than configurations 1 and 4. A similar result can be found in previous work on this topic [74]. With the increase of exerted force, a notable increase, especially in Anterior Deltoid and to some degree in Biceps Brachii activity is observed. Configuration 2 provides the lowest arm manipulability capacity, which affects task productivity and makes it even less preferable than configuration 3. To conclude, preferable poses that are safe and ergonomic and should be adopted by a human coworker in the collaborative polishing task are configuration 1 and configuration 4. The results of 3D pose estimation, more precisely the estimation of two considered postural angles, suggest that there is a potential for successful pose differentiation that opens the possibility of performing accurate pose classification of these four configurations. Pose classification can lead to the adaptation of the robot behavior to assist and guide the human partner to conduct the collaborative task with less effort and in a more ergonomic way.

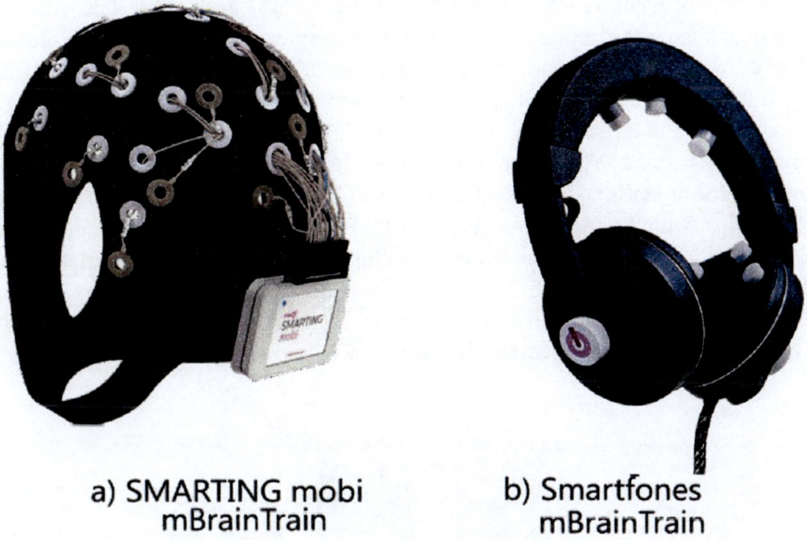

Fig. 18. mBrainTrain EEG measurement devices [81]

5 The Use of EEG for Workplace Safety Assessment

In the late 1920 s, Hans Berger, a German psychiatrist, invented the electroencephalogram (EEG) to assess the electrical activity of the cerebral cortex. Briefly, EEG is a recording method of the macroscopic electrical activity of the surface layer of the brain using small metal discs (electrodes) attached to the scalp [79]. Recently, a lot of effort was put into investigating the use of EEG in the industry (as it can reveal other useful information, e.g., emotions [80]) - opening a new scientific discipline called neuroergonomics. Monitoring workers' attention and brain cognitive activity during repetitive tasks has been studied using EEG. The potential of EEG to measure the cognitive workload of human operators in the chemical process control room was evaluated [82], and a multi-feature EEG- based workload metric with detailed insight into the evolution of the operator's mental models during training was developed [83]. Deep learning algorithms were used to investigate goal-directed context-dependent and context-independent behavioral patterns as neurological terms and goal-directed decision-making based on a correlation of brain regions' activity [84]. Detection of cognitive overload in highly controlled, specially designed tasks was also explored [85]. In order to develop future fatigue countermeasures and reduce fatigue-related accidents, EEG frequency bands and four different algorithms were used to detect fatigue [86].

5.1 Development of Modular and Adaptive Laboratory Set-Up for Neuroergonomic and Human–Robot Interaction Research

Our recent study presented a modular and adaptive laboratory model (industrial workstation) design for a human–robot collaborative assembly task [87]. The goal was to analyze the task from the perspective of neuroergonomics and human- robot interaction: to explore current industrial workers' problems, including performance, well-being, and injuries, and to create a solution that meets the operator's anatomical, physiological, and biomechanical characteristics. The workstation (Fig. 19) was comprised of several specific components: Poka–Yoke system (six lines that supply six different components), collaborative robot station, EEG system (EASYCAP GmbH, Wörthsee, Germany and SMARTING, mBrainTrain, Serbia (Fig. 18a)), EMG sensors (biosignalsplux muscleBAN), and touchscreen PC screen. Initial conclusions were that it is possible to improve the physical, cognitive, and organizational aspects, thus increasing workers' productivity and efficiency through transforming standard workplaces into the workplaces of the future by applying ergonomic research laboratory experimental set-up.

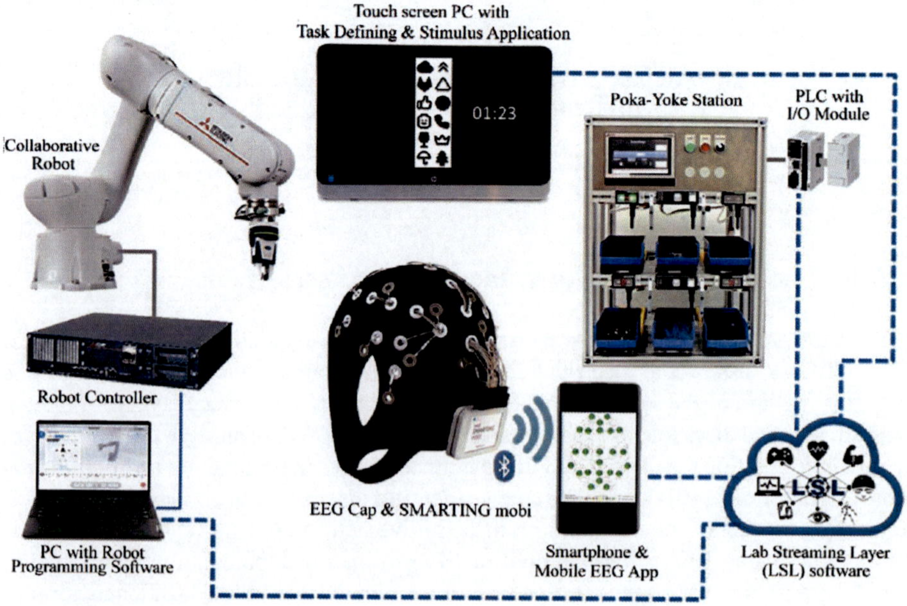

Fig. 19. Lab Streaming Layer integration of key measurement set-up elements [87]

6 Influence of Operators' Psychological and Physiological Characteristics on Workplace Safety

An interesting perspective for research when it comes to workplace safety is examining the correlation between the psychological and physiological status of the subject and its working performance. P&P is highly associated with risks from musculoskeletal injuries, and there is interest in the effects of various aspects of the P&P task on loads experienced by the body, especially the spine [88]. It is reasonable to assume that the presence of pain syndromes of the upper extremities or spine and the reduced mobility of the spine could affect the way of performing the task. Workplace activities require continuous attention in order to perform them safely. A person's psychological/emotional status affects attention, and it is expected that individuals with a more negative psychological status (anxiety, stress, depression, or apathy) have a higher risk of workplace injury. In our recent study, we analyzed handcart pushing and pulling using IoT force and EMG sensors and the relationship with operators' psychological status and pain syndromes [89]. None of the previous studies have considered the association between psychological status and patterns of activities related to specific tasks and the actual performance of individuals with different states of musculoskeletal health. The study included 20 male individuals divided into two subgroups, Group 1 (without high psychological scores of stress, depression, anxiety, and apathy; without pain syndromes of the upper extremity; and without limited mobility of the thoracolumbar spine) and Group 2 (with high scores of stress, depression, anxiety, or apathy; with pain syndromes of the upper extremity; and with limited mobility of the thoracolumbar spine). A detailed quantitative analysis of the acquired signal was performed - the computation of 10 force parameters (for both left and right side), one EMG parameter (for three different muscles, both left and right side), and three time-domain parameters.

7 Conclusions

With the ongoing progress, there is a trend of replacing or supplementing human operators with technologies varying from software tools to cobots. However, it is widely accepted that further technological evolution will remain even more human-centered, which consequently increases the need to improve human operators' well-being and safety. Here, we reviewed our and related studies on this topic that range from the application of cloud technologies to artificial intelligence, sensors, and robotics. Within the first topic, we highlighted the SafE- Tag, which represents a web/mobile framework that eases the collection of safety reports (related to unsafe acts and unsafe conditions) and the delegation of corresponding tasks. Furthermore, we reviewed studies focused on the digitalization of PPE compliance - as a specific case of unsafe acts. The emphasis was on approaches related to our recent study that proposed a generic procedure for PPE compliance (by combining pose estimation, ROI identification, and classification). Finally, we reviewed studies related to assessing and detecting unsafe actions - with an emphasis on industrial tasks that involve pushing and pulling (typical examples are workplaces in warehouses and transportation). Studying the safety of repetitive physical tasks is important to prevent musculoskeletal disorders (MSD), such as the back, arms, neck, etc., that have negative long-term effects. We primarily focused on reviewing studies that were focused on utilizing computer vision and pose estimation algorithms to asses safety or detect unsafe acts. Additionally, we reviewed related studies that propose or combine various sensor systems for the recognition of human activities as an alternative to computer vision. We emphasized our findings in our recent study that aimed to improve human–robot collaboration in the industrial setting [72]. Finally, we briefly presented a modular and adaptive laboratory model (industrial workstation) design for a human–robot collaborative assembly task which was introduced in our recent study to further enhance the assessment of industrial workers' performance and well-being [87]. Knowing that performance is affected by a person's psychological/emotional status, as well as physiological status, we also briefly discuss how individuals with a more negative psychological status (anxiety, stress, depression, or apathy) and with pain syndrome would be at a higher risk of workplace-related injury. It may be concluded that further technological progress and multidisciplinary studies on this topic will result in a better understanding and prevention of workplace injuries by their objective and timely detection.

Acknowledgements. This study was supported by the Science Fund of the Republic of Serbia, project ID 6524219 - AI4WorkplaceSafety.

References

1. The Global Industry Classification Standard. https://www.msci.com/our-solutions/indexes/gics. Accessed 30 Aug 2022
2. Gunasekaran, A., Subramanian, N., Ngai, W.T.E.: Quality management in the 21st century enterprises: research pathway towards Industry 4.0. Int. J. Prod. Econ. **207**, 125–129 (2019)
3. Bengtsson, M., Lundström, G.: On the importance of combining "the new" with "the old"–one important prerequisite for maintenance in Industry 4.0. Procedia Manuf. **25**, 118–125 (2018)
4. Badri, A., Boudreau-Trudel, B., Souissi, A.S.: Occupational health and safety in the industry 4.0 era: a cause for major concern? Saf. Sci. **109**, 403–411 (2018)
5. Galin, R.R., Meshcheryakov, R.V.: Human-robot interaction efficiency and human-robot collaboration. In: Kravets, A.G. (ed.) Robotics: Industry 4.0 Issues & New Intelligent Control Paradigms. SSDC, vol. 272, pp. 55–63. Springer, Cham (2020). https://doi.org/10.1007/978-3-030-37841-7_5
6. Zarte, M., Pechmann, A., Nunes, I.L.: Principles for human-centered system design in industry 4.0 – a systematic literature review. In: Nunes, I.L. (ed.) AHFE 2020. AISC, vol. 1207, pp. 140–147. Springer, Cham (2020). https://doi.org/10.1007/978-3-030-51369-6_19
7. Hollnagel, E.: Safety–I and Safety–II: The Past and Future of Safety Management. CRC Press, Boca Raton (2018)
8. Heinrich, H.W.: Industrial Accident Investigation – A Scientific Approach. McGraw-Hill Book Company, New York (1941)
9. Micheli, G.J.L., et al.: Barriers, drivers and impact of a simplified occupational safety and health management system in micro and small enterprises. In: Arezes, P. (ed.) AHFE 2018. AISC, vol. 791, pp. 81–90. Springer, Cham (2019). https://doi.org/10.1007/978-3-319-94589-7_8
10. Vukićević, A.M., Djapan, M., Stefanović, M., Mačužić, I.: SafE-Tag mobile: a novel javascript framework for real-time management of unsafe conditions and unsafe acts in SMEs. Saf. Sci. **120**, 507–516 (2019)

11. GUIDELINES FOR PERSONAL PROTECTIVE EQUIPMENT (PPE), Environmental Health and Safety, University of Washington, February 2022. https://www.ehs.washington.edu/system/files/resources/ppeguidelines.pdf. Accessed 30 Aug 2022
12. Wong, T.K.M., Man, S.S., Chan, A.H.S.: Critical factors for the use or non-use of personal protective equipment amongst construction workers. Saf. Sci. **126**, 104663 (2020)
13. United States Environmental Protection Agency (USEPA) (2021). https://www.epa.gov/emergency-response/personal-protective-equipment. Accessed 30 Aug 2022
14. Rubaiyat,A.H., et al.: Automatic detection of helmet uses for construction safety. In: 2016 IEEE/WIC/ACM International Conference on Web Intelligence Workshops (WIW), pp. 135–142. IEEE, October 2016
15. Wu, J., Cai, N., Chen, W., Wang, H., Wang, G.: Automatic detection of hardhats worn by construction personnel: a deep learning approach and benchmark dataset. Autom. Constr. **106**, 102894 (2019)
16. Delhi, V.S.K., Sankarlal, R., Thomas, A.: Detection of personal protective equipment (PPE) compliance on construction site using computer vision based deep learning techniques. Front. Built Environ. **6**, 136 (2020)
17. Tran, Q.H., Le, T.L., Hoang, S.H.: A fully automated vision-based system for real-time personal protective detection and monitoring. KICS Korea-Vietnam Int. Jt Work Commun. Inf. Sci. **2019**, 6 (2019)
18. Zhafran,F., Ningrum, E.S., Tamara, M.N., Kusumawati, E.: Computer vision system based for personal protective equipment detection, by using convolutional neural network. In: 2019 International Electronics Symposium (IES), pp. 516–521. IEEE, September 2019
19. Nagrath, P., Jain, R., Madan, A., Arora, R., Kataria, P., Hemanth, J.: SSDMNV2: a real time DNN-based face mask detection system using single shot multibox detector and MobileNetV2. Sustain. Cities Soc. **66**, 102692 (2021)
20. Vukicevic, A.M., Djapan, M., Isailović, V., Milašinović, D., Savković, M., Milošević, P.: Generic compliance of industrial PPE by using deep learning techniques. Saf. Sci. **148**, 105646 (2022)
21. Cheng, A., Xiao, B., Wang, J., Shi, H., Huang, T.S., Zhang, L.: HigherHRNet: scale-aware representation learning for bottom-up human pose estimation. In: Proceedings of the IEEE/CVF Conference on Computer Vision and Pattern Recognition, pp. 5386–5395 (2020)
22. Sun, K., Xiao, B., Liu, D., Wang, J.: Deep high-resolution representation learning for human pose estimation. In: Proceedings of the IEEE/CVF Conference on Computer Vision and Pattern Recognition, pp. 5693–5703 (2019)
23. Sandler, M., Howard, A., Zhu, M., Zhmoginov, A., Chen, L.C.: Mobilenetv2: inverted residuals and linear bottlenecks. In: Proceedings of the IEEE Conference on Computer Vision and Pattern Recognition, pp. 4510–4520 (2018)
24. Simonyan, K., Zisserman, A.: Very deep convolutional networks for large-scale image recognition. arXiv preprint: arXiv:1409.1556 (2014)
25. Huang, G., Liu, Z., Van Der Maaten, L., Weinberger, K.Q.: Densely connected convolutional networks. In: Proceedings of the IEEE Conference on Computer Vision and Pattern Recognition, pp. 4700–4708 (2017)
26. Iandola,F. N., Han, S., Moskewicz, M.W., Ashraf, K., Dally, W.J., Keutzer, K.: SqueezeNet: AlexNet-level accuracy with 50x fewer parameters and< 0.5 MB model size. arXiv preprint: arXiv:1602.07360 (2016)
27. Szegedy, C., Vanhoucke, V., Ioffe, S., Shlens, J., Wojna, Z.: Rethinking the inception architecture for computer vision. In: Proceedings of the IEEE Conference on Computer Vision and Pattern Recognition, pp. 2818–2826 (2016)
28. He, K., Zhang, X., Ren, S., Sun, J.: Deep residual learning for image recognition. In: Proceedings of the IEEE Conference on Computer Vision and Pattern Recognition, pp. 770–778 (2016)

29. Russakovsky, O., et al.: Imagenet large scale visual recognition challenge. Int. J. Comput. Vis. **115**(3), 211–252 (2015)
30. Kingma, D. P., Ba, J.: Adam: A method for stochastic optimization. arXiv preprint: arXiv: 1412.6980 (2014)
31. A Swedish Work Environment authority, Ergonomics for the Prevention of Musculoskeletal Disorders, Stockholm, Sweden, vol. 2 (2012)
32. Karwowski, W.: Human factors and ergonomics. In: Handbook of Standards and Guidelines in Ergonomics and Human Factors. Lawrence Erlbaum Associates Publishers, Mahwah (2006)
33. Vukićević, A.M., Mačužić, I., Mijailović, N., Peulić, A., Radović, M.: Assessment of the handcart pushing and pulling safety by using deep learning 3D pose estimation and IoT force sensors. Expert Syst. Appl. **183**, 115371 (2021)
34. Yaris, A., Ditchburn, G., Curtis, G.J., Brook, L.: Combining physical and psychosocial safety: a comprehensive workplace safety model. Saf. Sci. **132**, 104949 (2020)
35. Antwi-Afari, M.F., Li, H., Edwards, D.J., Pärn, E.A., Seo, J., Wong, A.Y.L.: Biomechanical analysis of risk factors for work-related musculoskeletal disorders during repetitive lifting task in construction workers. Autom. Constr. **83**, 41–47 (2017)
36. Isailović, V., et al.: Compliance of head-mounted personal protective equipment by using YOLOv5 object detector. In: 2021 International Conference on Electrical, Computer and Energy Technologies (ICECET), pp. 1–5. IEEE, December 2021
37. Anderson, S.P., Oakman, J.: Allied health professionals and work-related musculoskeletal disorders: a systematic review. Saf. Health Work **7**(4), 259–267 (2016)
38. Summary - Work-related musculoskeletal disorders: prevalence, costs and demographics in the EU. https://osha.europa.eu/en/publications/summary-msds-facts-and-figures-overview-prevalence-costs-and-demographics-msds-europe. Accessed 30 Aug 2022
39. Cieza, A., Causey, K., Kamenov, K., Hanson, S.W., Chatterji, S., Vos, T.: Global estimates of the need for rehabilitation based on the global burden of disease study 2019: a systematic analysis for the global burden of disease study 2019. Lancet **396**(10267), 2006–2017 (2020)
40. Occhipinti, E., Colombini, D.: A toolkit for the analysis of biomechanical overload and prevention of WMSDs: Criteria, procedures and tool selection in a step-by-step approach. Int. J. Ind. Ergon. **52**, 18–28 (2016)
41. Goodfellow, I., Bengio, Y., Courville, A.: Deep Learning. MIT press, Cambridge (2016)
42. Han, S., Lee, S.: A vision-based motion capture and recognition framework for behavior-based safety management. Autom. Constr **35**, 131–141 (2013)
43. Han, S., Lee, S., Peña-Mora, F.: Vision-based detection of unsafe actions of a construction worker: case study of ladder climbing. J. Comput. Civil Eng. **27**(6), 635–644 (2013)
44. Yu, Y., Guo, H., Ding, Q., Li, H., Skitmore, M.: An experimental study of real-time identification of construction workers' unsafe behaviors. Autom. Constr. **82**, 193–206 (2017)
45. Guo, H., Yu, Y., Ding, Q., Skitmore, M.: Image-and-skeleton-based parameterized approach to real-time identification of construction workers' unsafe behaviors. J. Constr. Eng. Manag **144**(6), 04018042 (2018)
46. Yan, X., Li, H., Wang, C., Seo, J., Zhang, H., Wang, H.: Development of ergonomic posture recognition technique based on 2D ordinary camera for construction hazard prevention through view-invariant features in 2D skeleton motion. Adv. Eng. Inform. **34**, 152–163 (2017)
47. Cao, Z., Simon, T., Wei, S.E., Sheikh, Y.: Realtime multi-person 2D pose estimation using part affinity fields. In: Proceedigs of the IEEE Conference on Computer Vision and Pattern Recognition, pp. 7291–7299 (2017)
48. Park, M.W., Brilakis, I.: Construction worker detection in video frames for initializing vision trackers. Autom. Constr. **28**, 15–25 (2012)
49. Seo, J., Han, S., Lee, S., Kim, H.: Computer vision techniques for construction safety and health monitoring. Adv. Eng. Inform. **29**(2), 239–251 (2015)

50. Ding, L., Fang, W., Luo, H., Love, P.E., Zhong, B., Ouyang, X.: A deep hybrid learning model to detect unsafe behavior: Integrating convolution neural networks and long short-term memory. Autom. Constr. **86**, 118–124 (2018)
51. Luo, H., Xiong, C., Fang, W., Love, P.E., Zhang, B., Ouyang, X.: Convolutional neural networks: computer vision-based workforce activity assessment in construction. Autom. Constr. **94**, 282–289 (2018)
52. Fang, W., Ding, L., Luo, H., Love, P.E.: Falls from heights: a computer vision-based approach for safety harness detection. Autom. Constr. **91**, 53–61 (2018)
53. Alawad, H., Kaewunruen, S., An, M.: A deep learning approach towards railway safety risk assessment. IEEE Access **8**, 102811–102832 (2020)
54. Yan, X., Li, H., Li, A.R., Zhang, H.: Wearable IMU-based real-time motion warning system for construction workers' musculoskeletal disorders prevention. Autom. Constr. **74**, 2–11 (2017)
55. Yang, J., Lee, J., Choi, J.: Activity recognition based on RFID object usage for smart mobile devices. J. Comput. Sci. Technol. **26**(2), 239–246 (2011)
56. Hofmann, C., Patschkowski, C., Haefner, B., Lanza, G.: Machine learning based activity recognition to identify wasteful activities in production. Procedia Manuf. **45**, 171–176 (2020)
57. Ordóñez, F.J., Roggen, D.: Deep convolutional and lstm recurrent neural networks for multimodal wearable activity recognition. Sensors **16**(1), 115 (2016)
58. Habes, D.J.: Use of EMG in a kinesiological study in industry. Appl. Ergon. **15**(4), 297–301 (1984)
59. Marras, W.S.: Industrial electromyography (EMG). Int. J. Ind. Ergon. **6**(1), 89–93 (1990)
60. Day, S.: Important factors in surface EMG measurement. Bortec Biomedical Ltd Publishers, pp. 1–17 (2002)
61. Konrad, P.: The ABC of EMG: a practical introduction to Kinesiological electromyography (2005)
62. Heydari, A., Nargol, A.V., Jones, A.P.: Humphrey AR, Greenough CG, EMG analysis of lumbar paraspinal muscles as a predictor of the risk of low-back pain. Eur. Spine J. **19**(7), 1145–1152 (2010)
63. Wang, D., Hu, B., Dai, F., Ning, X.: Sensor-based factorial experimental study on low back disorder risk factors among roofers (2015)
64. Salas, E.A., Vi, P., Reider, V.L., Moore, A.E.: Factors affecting the risk of developing lower back musculoskeletal disorders (MSDs) in experienced and inexperienced rodworkers. Appl. Ergon. **52**, 62–68 (2016)
65. Peppoloni, L., Filippeschi, A., Ruffaldi, E., Avizzano, C.A.: A novel wearable system for the online assessment of risk for biomechanical load in repetitive efforts. Int. J. Ind. Ergon. **52**, 1–11 (2016)
66. Molinaro, D.D., King, A.S., Young, A.J.: Biomechanical analysis of common solid waste collection throwing techniques using OpenSim and an EMG-assisted solver. J. Biomech. **104**, 109704 (2020)
67. Kocabas, M., Athanasiou, N., Black, M.J.: Vibe: video inference for human body pose and shape estimation. In: Proceedings of the IEEE/CVF Conference on Computer Vision and Pattern Recognition, pp. 5253–5263 (2020)
68. Loper, M., Mahmood, N., Romero, J., Pons-Moll, G., Black, M.J.: SMPL: a skinned multi-person linear model. ACM Trans. Graph. (TOG) **34**(6), 1–16 (2015)
69. Mahmood, N., Ghorbani, N., Troje, N.F., Pons-Moll, G., Black, M.J.: AMASS: archive of motion capture as surface shapes. In: Proceedings of the IEEE/CVF International Conference on Computer Vision, pp. 5442–5451 (2019)
70. Botsch, M., Kobbelt, L., Pauly, M., Alliez, P., Lévy, B.: Polygon Mesh Processing. CRC Press, Boca Raton (2010)

71. Hanocka, R., Hertz, A., Fish, N., Giryes, R., Fleishman, S., Cohen-Or, D.: Meshcnn: a network with an edge. ACM Trans. Graph. (TOG) **38**(4), 1–12 (2019)
72. Petrović, M., Vukićević, A.M., Lukić, B., Jovanović, K.: Assessment of the human-robot collaborative polishing task by using EMG sensors and 3D pose estimation. In: Müller, A., Brandstötter, M. (eds.) International Conference on Robotics in Alpe-Adria Danube Region, pp. 564–570, Springer, Cham (2022).https://doi.org/10.1007/978-3-031-04870-8_66
73. Peternel, L., Tsagarakis, N., Caldwell, D., Ajoudani, A.: Robot adaptation to human physical fatigue in human–robot co-manipulation. Auton. Robots **42**(5), 1011–1021 (2018)
74. Kim, W., Peternel, L., Lorenzini, M., Babič, J., Ajoudani, A.: A human-robot collaboration framework for improving ergonomics during dexterous operation of power tools. Robot. Comput.-Integr. Manuf. **68**, 102084 (2021)
75. Fortini, L., Lorenzini, M., Kim, W., De Momi, E., Ajoudani, A.: A real-time tool for human ergonomics assessment based on joint compressive forces. In: 2020 29th IEEE International Conference on Robot and Human Interactive Communication (RO-MAN), pp. 1164–1170, IEEE, August 2020
76. Lorenzini, M., Kim, W., Ajoudani, A.: An online multi-index approach to human ergonomics assessment in the workplace. IEEE Trans. Hum.-Mach. Syst. **52**, 812–823 (2022)
77. Radmilović, M., Urukalo, D., Petrović, M., Becanović, F., Jovanović, K.: Influence of muscle co-contraction indicators for different task conditions. In: ICEtran (2021)
78. Yoshikawa, T.: Manipulability of robotic mechanisms. Int. J. Robot. Res. **4**(2), 3–9 (1985)
79. Lee, J. H., et al.: Stress monitoring using multimodal bio-sensing headset. In: Extended Abstracts of the 2020 CHI Conference on Human Factors in Computing Systems, pp. 1–7, April 2020
80. Rahman, M.M., et al.: Recognition of human emotions using EEG signals: a review. Comput. Biol. Med. **136**, 104696 (2021)
81. mBraintrain SMARTING mobi and smartphones. https://mbraintrain.com. Accessed 30 Aug 2022
82. Iqbal, M.U., Srinivasan, B., Srinivasan, R.: Dynamic assessment of control room operator's cognitive workload using electroencephalography (EEG). Comput. Chem. Eng. **141**, 106726 (2020)
83. Iqbal, M.U., Shahab, M.A., Choudhary, M., Srinivasan, B., Srinivasan, R.: Electroencephalography (EEG) based cognitive measures for evaluating the effectiveness of operator training. Process Saf. Environ. Prot. **150**, 51–67 (2021)
84. Villalba-Diez, J., Zheng, X., Schmidt, D., Molina, M.: Characterization of industry 4.0 lean management problem-solving behavioral patterns using EEG sensors and deep learning. Sensors **19**(13), 2841 (2019)
85. Morton, J., et al.: Identifying predictive EEG features for cognitive overload detection in assembly workers in Industry 4.0. In: H-Workload 2019: 3rd International Symposium on Human Mental Workload: Models and Applications (Works in Progress), p. 1 (2019)
86. Jap, B.T., Lal, S., Fischer, P., Bekiaris, E.: Using EEG spectral components to assess algorithms for detecting fatigue. Expert Syst. Appl. **36**(2), 2352–2359 (2009)
87. Savković, M., Caiazzo, C., Djapan, M., Vukićević, A.M., Pušica, M., Mačužić, I.: Development of modular and adaptive laboratory set-up for neuroergonomic and human-robot interaction research. Front. Neurorobot. (2022)
88. Pinupong, C., Jalayondeja, W., Mekhora, K., Bhuanantanondh, P., Jalayondeja, C.: The Effects of Ramp Gradients and Pushing-Pulling Techniques on Lumbar Spinal Load in Healthy Workers. Saf. Health Work **11**(3), 307–313 (2020)
89. Petrović, M., et al.: Experimental analysis of handcart pushing and pulling safety in an industrial environment by using IoT force and EMG sensors: relationship with operators' psychological status and pain syndromes. Sensors **22**, 7467 (2022)

Implementation of Deep Learning to Prevent Peak-Driven Power Outages Within Manufacturing Systems

Milovan M. Medojević[1,2](✉) and Marko M. Vasiljević Toskić[3]

[1] The Institute for Artificial Intelligence Research and Development of Serbia, 1 Fruškogorska Street, 21000 Novi Sad, Serbia
milovan.medojevic@ivi.ac.rs

[2] EnergyPulse Doo, 14 Dragiše Brašovana Street, 21000 Novi Sad, Serbia

[3] Faculty of Technical Sciences, University of Novi Sad, 6 Trg Dositeja Obradovića, 21000 Novi Sad, Serbia
markovt@uns.ac.rs

Abstract. In this paper, a solution to effective energy consumption monitoring of fast-response energy systems in industrial environments was deployed, while the research focuses on the manner and intensity of energy use in the observed system as a consequence of nonlinearity in the performance of the dynamic system, to predict the near future relatively accurately. The paper addresses the quite common but still an inevitable case for the majority of manufacturing systems where constant jumps in peak loads on several machines simultaneously lead to the situation that the entire system remains without a power supply. This paper proposes a deep learning method, based on enhanced recurrent neural network (*RNN*), more precisely LSTM network (*Long Short-Term Memory*) to effectively predict future machine states in terms of energy consumption five steps ahead. The data sets were obtained for eight machines in one CNC metal-forming center on a monthly level at a one-second sampling rate by means of using a previously developed IoT device.

Keywords: Manufacturing · CNC machine tools · Energy consumption · IoT · Deep learning · LSTM

1 Introduction

The manufacturing sector has a critical role in the general economy's supply chain, as being recognized as an indispensable component that ensures the delivery of goods and services to other economic sectors. However, frequent power failures, driven by process machine operations as is the case within CNC machining processes, are sometimes inevitable. One of the reasons for that is that the switching capacities of machines are often many times higher in capacity than in the case of switches in electrical cabinets for powering the process itself. Due to this reason, frequent power outages can occur, bringing operations to a screeching halt and contributing to productivity, revenue, and

N. Filipovic (Ed.): AAI 2022, LNNS 659, pp. 29–44, 2023.
https://doi.org/10.1007/978-3-031-29717-5_2

material losses. Specifically, power outages force production systems to deal with production lines pushed to a sudden stagnation. This further manifests in the inability to produce and assemble goods, an increase in downtime that eventually affects supply chains to shut down altogether.

Traditionally, this problem was solved eventually by improving the management of electrical layout, relying on backup generators, frequent testing to prevent faulty behaviors, and so on. Moreover, with regular, time-based maintenance intervals, production machinery is often subjected to maintenance although no actual need for such activity exists, while components, tools, and accompanying elements are superseded as soon as their operation time expires or is close to expiry [1]. In many cases, those components could still be utilized. On the other hand, there are cases in which specific components reach an operating threshold before the regular time-based maintenance schedule and, before failure, perform faulty tasks that trigger the problem occuring in other fields, such as increasing the current intensity of the motors, which leads to power outages due to excessive simultaneous loading. In response to the previously mentioned, some studies recognized that near real-time processing of energy data could directly indicate certain anomalies in components, tools, and overall machine operation [2–4], while the exponential rise of IoT, Big Data and AI breakthroughs could support manufacturing systems to understand the vast amount of data fast and utilize the generated information to predict and prevent downtime.

1.1 The Role of IoT, Big Data, and AI in Managing Manufacturing Systems

In recent times, IoT has been characterized by the wealth of new data streams to simultaneously provide support to a variety of management activities whether it is about increasing productivity, energy, and resource efficiency, or even improving the maintenance processes within the manufacturing system [5]. Although there is a large amount of scientific research in this field [6], only some of those are briefly given below.

For example, [7] investigated distributed collaborative control for industrial automation with wireless sensor and actuator networks, while [8] were focused on coordination in wireless sensor-actuator networks and energy-aware, spatiotemporal correlation mechanism to perform efficient data collection in wireless sensor networks respectively. Moreover, [9] contributed to the application of wireless sensor and actuator networks to achieve intelligent microgrids with a promising approach toward a global smart grid deployment. In terms of IoT systems development, [10] proposed a system for automated multi-node environmental water parameter monitoring which can make a self-assessment of measurement quality. Similarly, [11] carried out the realization of a low-cost wireless weather station without any moving parts. In doing so, sophisticated methods characteristic of ultrasonic anemometers were applied. The proposed system considers logging data regarding dry bulb temperature, atmospheric pressure, humidity, as well as wind direction and speed, while the complete integration was performed on the ESP32 development board, which enabled them to transmit the acquired data to the WEB server and application simultaneously. On the other hand, the concept of IoT received a lot of attention in terms of arising security issues. [12] reports on access control in IoT environments, [13] on IoT data provenance implementation challenges, while [14] research focus was on the privacy of IoT-enabled smart systems. Also, [15] proposed a security monitoring system for IoT, while a general overview of IoT security was discussed by [16]. On top of the previously mentioned, [17] proffer a solution for effective energy consumption monitoring of fast-response energy systems in industrial environments. The research was oriented towards the design and development of an industrial IoT –based system for behavior profiling of non-linear dynamic production systems based on energy flow theory, while the developed solution was further used to generate the datasets considered within this study.

It is important to point out that a strong, interdependent relationship exists between any IoT system and AI solution deployment. This is mainly due to the well-established fact that the data is considered the food for AI. This implies that, although both, the code and data are recognized as foundations of AI systems, it is much more effective to focus on improving the data quality rather than spending significant efforts on the AI model itself. The reason for this is quite obvious and is hidden behind the term known as Big Data, or more precisely, data sets characterized by high volume, high velocity, and high variety which are necessary to be processed further to gain actionable information eventually.

When considering what makes Big Data Good Data, one should start by recognizing the importance of the 4th V component - veracity. For example, many manufacturing systems log and store a large amount of data (high volume) in various modes with different data points (high variety), and with a high sampling frequency (high velocity). However, if those data experience a lack of veracity, there is a high probability of being left with a large amount of data that cannot deliver any insight or intelligence. Therefore, the veracity component enables the manufacturing (or any other) system to have trust in the data that are being gathered, while logically, this could be achieved by careful design and development of custom-tailored IoT solutions. Consequently, IoT plays an important role in enabling such a harvest in which the data are reaped for the wealth of intelligence and then the 5th V component can be introduced, and that is value.

Since the data preparation typically represents 75% of the AI modeling efforts, to accelerate effectively toward reaching the top benefits it is necessary to put the data-centric approaches in the foreground instead of model-centric ones.

Although, a well-established system for good data provisioning is a huge step from the manufacturing management system point of view when it comes to AI solutions integration this is considered an obligatory prerequisite.

In general, AI could be considered as a technique that provides machines with the capability to emulate characteristically human behavior, especially in situations involving tasks related to predicting, classifying, learning, planning, reasoning, and perceiving, in general. Having in mind that the IoT is mostly to be credited for enabling data collection in real-time, the proliferation of such sensing technologies resulted in huge amounts of time-series data being produced by machines and processes within manufacturing facilities worldwide, which further caused an exponential growth in the research and development of AI techniques. Thus, a vast number of investigations and research studies related to diverse AI techniques deployment aimed to optimize various energy-related issues are being conducted more and more [18].

Hence, [19] considered ANN and LSTM for short-term EVs load forecasting since their large penetration leads to high uncertainty in the power demand of the power system. [20] elaborated on large timespan quasi-periodicity of load sequences based on ARIMA and LSTM. The data set used in their study contained more than 700 weeks of energy consumption data, while the generated model ensured superior performance compared with popular STLF models. Likewise, [21] recognized that the extensive use of distributed generations in smart grids fetched the additional need for the accuracy of STLF. The investigation considered the application of AM, RU, and Bi-LSTM while the validity was measured on the actual data sets from several countries. [22] proposes DBN-based prediction of the power system hourly load by deploying Copula models to compute the peak load indicative variables. [23] applied GA-PSO algorithm for optimization of BPNN parameters in their research considering the effect of interaction between forecasting and subsampling in a paper manufacturing system. [24] proposed a fuzzy-based ensemble model for week-ahead load forecasting by introducing hybrid DL neural networks to apply a fuzzy clustering idea to divide the input into clusters which were subjected to training via a neural network consisting of one RBF, one convolutional, one pooling, and two fully-connected layers. [25] considered advanced sequence prediction models LSTM and GRU to forecast the annual load of feeders' distribution to exploit the sequential information hidden in multi-year data.

Given the aforementioned, the exponential function of technological development became peaking steep with strong pressure to eliminate boundaries between production and management, providing that all critical systems are well-integrated to realize the growth opportunities that this new age of intelligent manufacturing brought upon us.

2 Research Approach, Aim, and Structure

Bearing in mind that energy represents a common denominator based on which it is possible to observe the hidden laws of interaction of practically all systems, the approach based on the identification of energy flows enables the determination of the states and behavior of CNC machine tools during operation within manufacturing systems. From this perspective, energy represents the inherent ability of the observed system to exert an external influence, or in other words to perform any type of given task. Thus, energy represents a state variable that is directly correlated with changes in work (which represents a process variable), over time. [17] indicates that changes in the use of machine energy in production systems are not constant over time, but dynamic due to the influence of the non-linearity of the production process and changes in the state of the machine over time. Complex machines consist of a large number of energy-using components that generate specific energy load profiles during operation [26]. A modern milling machine, for example, can contain a wide range of functions, including workpiece handling, lubrication, chip removal, tool change, tool anomaly detection, etc., all in addition to the machine tool's primary function which is material removal by cutting. Therefore, the variable determined in this way combines the effects of both, forces and velocities, and their product, power or intensity of energy change, characterized by the dynamic property, includes and reflects complete information about its balance and movement, and therefore exceeds the study of changes in force and motion separately [27].

With this in mind, the aim of this paper is to implement previously developed industrial IoT devices and to accurately quantify energy usage, the intensity of changes in power draw, to observe relevant peaks, etc., but also to generate a profile of the machine's behavior during operation (as a continuous series of recorded states) that provides insight into the process in which event forensic transparency enables visibility and identification of trends/patterns in the behavioral changes. Furthermore, the generated data sets are used to develop a referent behavior AI model. Eventually, both, the IoT devices and accompanying AI models, should be deployed throughout the manufacturing process/facility to detect anomalies. Here, anomalies are considered as deviations in states within the behavior profile recorded by IoT devices compared to the proposed AI referent model.

The structure of this research starts with the implementation of IoT nodes into the observed manufacturing systems. The data acquisition is performed for one month after which the generated data were further processed to be suitable for referent AI model development. Lastly, the necessary integrations were carried out to ensure automated operation.

2.1 Manufacturing System Overview

For analysis, the manufacturing system for metal forming and CNC processing was observed. This system represents a machine park specialized in the automotive and aerospace industries, with a focus on the production of machine parts, elements, components, assemblies, and subassemblies in flexible and scalable machining. Upon evaluation 8 machines (4 lathes, 3 mills, and 1 machine saw) were suggested for monitoring and data acquisition via IoT nodes.

2.2 IoT Device Overview

A previously developed device is called Current Profiler which stands for the hardware unit for non-invasive, continuous monitoring and acquisition of current intensity data for profiling system/process/machine behavior [17]. The main reason for developing this solution is to gather reliable data regarding the intensity of the electrical current. As stable voltage is one of the fundamental prerequisites for the operation of electrical machines in industrial environments, variations in the intensity of electric current reflect the behavior of the observed system through a series of recorded states, which at the same time provide exact data regarding energy consumption. In addition, a non-invasive current transformer (CT) was used for sensing, measuring, and logging the data regarding alternating currents. For communication purposes, the device is equipped with the ESP8266 Wi-Fi serial transceiver module, while in cases of any communication interruption the data are backed up and stored on an SD card.

Although the device represents a classical sensing node, the autonomy of the device is ensured by battery power. By default, the device prefers to ensure the required energy from the grid, while in the cases of a sudden power outage it shifts to a battery which enables time-based detection of interruption and re-establishment of the system power supply. In terms of technical functionalities, the device is capable to measure, data in real-time, on three channels, in the range from 0 to 100A. Also, the device measures and displays the operating temperature (–40 - 85 °C) which is primarily intended for monitoring optimal operating conditions in which the device is located. Furthermore, the device connects the measured values to the timestamp and records data to the SD card. The device power (3.3V or 5V) is provided using the USB port or an external power source via the DC connector. In terms of visualization, data could be displayed online as well as on a device in real-time. The sampling rate in this development stage is one sample per second, although it can be varied via software-defined telemetry services. Finally, the device is upgradeable and modular, so it can be expanded and connected to the necessary peripherals via the I2C or SPI protocol. Figure 1, illustrates the component layout on the integration PCB as well as the physical appearance of the observed device.

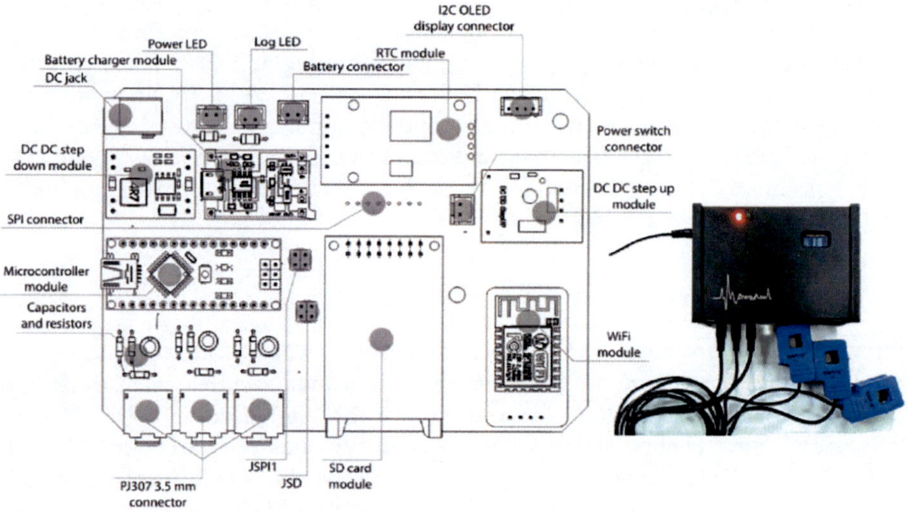

Fig. 1. IoT device for data acquisition

2.3 AI Model Development

When it comes to fast and accurate forecasting based on time series as a primary tool in the field of energy-based predictive maintenance, LSTM (*Long Short Term Memory networks*) models turned out to fit among the most suitable choices [28]. The main reason lies behind the fact that LSTMs are designed to avoid the long-term dependency problem while remembering information for long periods is their default behavior and not something they struggle to figure out [29].

LSTM network is a variant of RNN that overcomes the traditional RNN gradient demise problem [30], where the hidden layer is no longer a simple neural unit, but an LSTM unit with a unique memory property recognized as a state cell, which describes the current state of the LSTM unit [31] (Fig. 2). Here, c stands for the current state quantity, while h represents the current output of the LSTM unit. Reading and modifying the state units within the LSTM is provided by controlling the forget, input, and output gates [32], typically formed by sigmoid or tanh functions and Hadamard product operation.

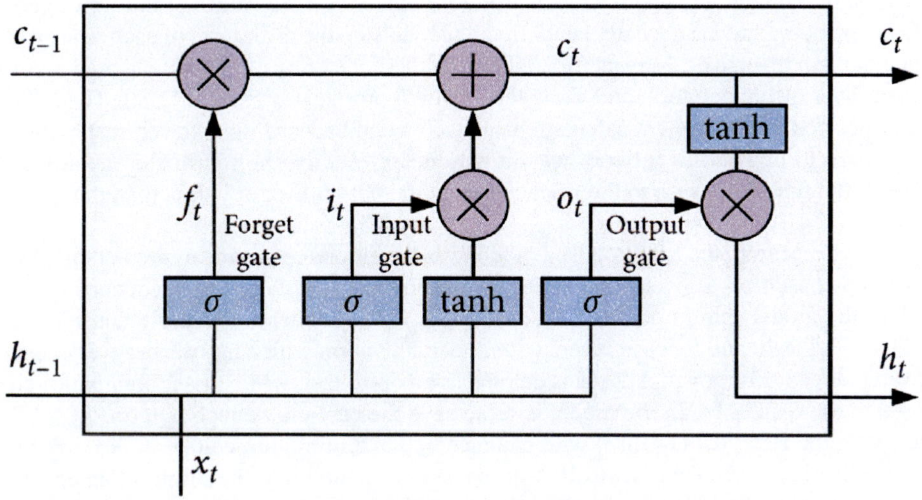

Fig. 2. Schematic diagram of LSTM cell structure [33]

The purpose of the forget gate is to determine c_{t-1} of the previous moment as well as how many components remain at the current time c_t. Similarly, the input gate determines the number of components saved to the cell state c_t. at current time x_t from the network, as well as the components quantity of the output gate control cell state c_t that is output to the current value h_t of the LSTM [33]. The set of equations regarding LSTM variables is given below:

$$f_t = \sigma\left(W_{fx}x_i + W_{fh}h_{t-1} + b_f\right),$$

$$i_t = \sigma(W_{ix}x_i + W_{ih}h_{t-1} + b_i),$$

$$g_t = \phi\left(W_{gx}x_t + W_{gh}h_{t-1} + b_g\right),$$

$$c_t = f_t \odot c_{t-1} + i_t \odot g_t,$$

$$o_t = \sigma(W_{ox}x_t + W_{oh}h_{t-1} + b_o),$$

$$h_t = o_t \odot \phi(c_t).$$

Here, W_{fx}, W_{fh}, W_{ix}, W_{ih}, W_{gx}, W_{gh}, W_{ox}, and W_{oh} are the weight matrices corresponding to the input of the network activation function, b_f, b_i, b_g, and b_o are the offset vectors, $\odot$ is the multiplication operator of the Hadamard product matrix, σ denotes the sigmoid activation function, while ϕ stands for the tanh activation function.

The LSTM training algorithm considers backpropagation through time (BPTT) where one could distinguish 4 crucial steps respectively [34]. The first step is a calculation of the output value of each neuron forward, where the values of the six vectors (f_t, i_t, g_t, c_t, o_t, and h_t) are calculated. In the second step the error term of each neuron is calculated in reverse on 2 levels (Spatial level, where the error term is propagated to the upper layer of the network, and Time level, which propagates back in time). Then, the third step considers gradient calculations of each weight according to the corresponding error term to update the network weight parameter. Finally, the fourth step iteratively repeats the first 3 steps up to the moment when the network error is less than the given value.

In this research, the LSTM-based model was developed to be able to predict the next five steps based on a given time series. More precisely, based on the sequence of 30 values the model should predict the subsequent 5. The model architecture considers 2 LSTM with Relu and 5 dense layers with linear activation which has illustratively been given in Fig. 3. Moreover, ADAM is defined as an optimizer, while losses are monitored through MSE and MAE. The model is set to save the best-only epoch in terms of MSE on validation (test) data (Vmse) with patience = 3 and minimum delta = 0.001.

Lastly, the batch size is set to 100, the same as the number of iterations of an epoch. The proposed model structure resulted as the most prominent one after several controlled experiment variations. More details about the data preparation and model development are provided in the Appendix[1].

[1] Appendix is available at: https://github.com/Backo-tech/Models/blob/main/5_steps_ahead_prediction_LSTM_model.py.

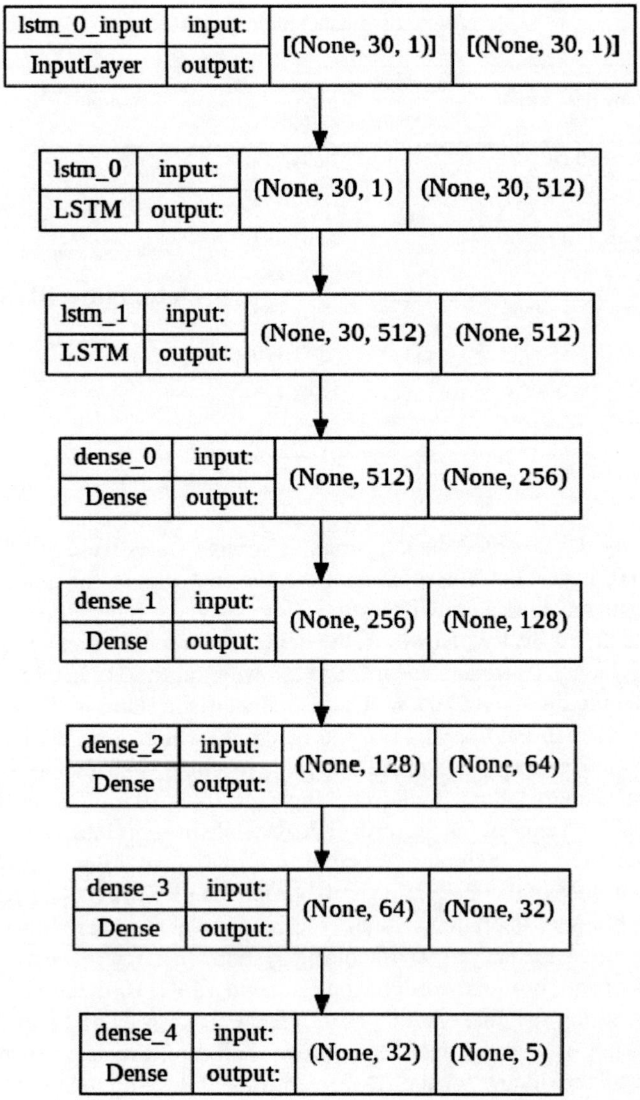

Fig. 3. Proposed LSTM model architecture

3 Obtained Results

In terms of data acquisition, the developed IoT device operated continuously without any malfunction. During the data acquisition period, about 17 million data samples were recorded. Table 1 provides an overview of data volume and structure.

Table 1. Acquired data volume and structure

No.	Machine designation	Machine type	No. samples	Sample structure
1	HAAS SL 20 HE	lathe	2.694.778	[Date, Time, S1, S2, S3, T] (date, time, current 1, current 2, current 3, temperature)
2	HAAS SL 20 THE (1)		2.480.446	
3	HAAS SL 20 THE (2)		2.587.392	
4	HAAS ST 20 Y		2.573.452	
5	SCHMID VMC-800P	mill	171.592	
6	SCHMID VMC-500P		2.571.390	
7	Pinnacle VMC1100S		1.342.193	
8	Kasto SBA-260AU	saw	2.393.416	

After plotting the recorded data, a series of strange states were spotted during the HAAS SL 20 HE operation. The profiles of the currents, that is, the power draw depict the periods of the machine's daily operation. Regularly, the typical power draw of this machine should be 10–15 kW. However, the device recorded a series of machine states that cannot be characterized as standard, nor allowed, on the 1st and 15th day of measurement in the amount of over 80 kW. These characteristic states indicate the existence of certain anomalies in the operation of the machine, which were discovered through machine energy consumption. Also, although this machine is characterized by an oversized switching capacity (short-circuit breaking capacity is 10 kA), the current intensity at full load should not exceed the prescribed 20A, while in short intervals it can function with a load of up to 35 A maximum. This indicates that, in addition to the possibility of failure, the occurrence of these situations can burden the network if these peaks coincide with other machines in operation, which results in sudden power outages within the process or sometimes the entire manufacturing system.

On the other hand, profile recordings on Pinnacle VMC1100S did not show significant anomalies during the observation period. However, characteristic peak loads were spotted but did not exceed the declared values. Therefore, the data set from Pinnacle VMC1100S[2] was considered for the AI referent model development in terms of training and validation (test), while the data set from HAAS SL 20 HE was used for model evaluation and verification in practice.

[2] The data set used for model development is available here.

Training and validation (test) were performed in an 80/20 ratio with defined callbacks in terms of *early_stop* and *save_checkpoint*. The training was completed after 19 epochs in which the total number of parameters amounts to 3.326.597, where all were trainable. Figure 4 provides a quick overview of the learning curve/loss optimization for the training and validation (test) data sets.

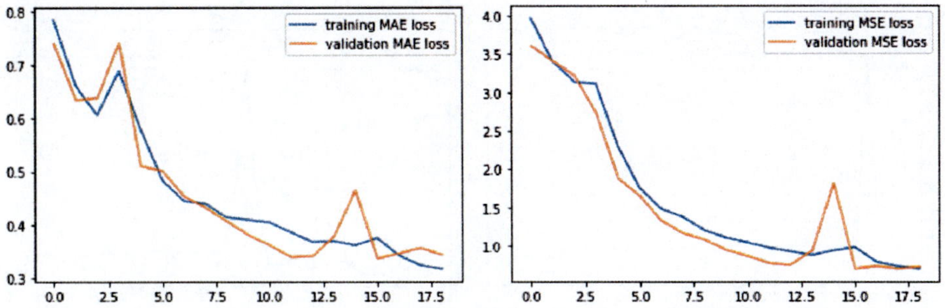

Fig. 4. The learning curve/loss optimization for the training and validation (test) data sets

Figure 4 shows the model training flow and serves as an indicator to decide if and when the optimal learning level has been reached. Based on the data presented, it can be seen that after 17 epochs, there is no significant drop in the error function (loss function), based on which it is possible to conclude that the current network architecture has reached its optimum. Training loss in terms of MSE reached 0.7094 (MAE = 0.3174), while validation (test) loss in MSE reached 0.7314 (MAE = 0.3433). Therefore, the generated model can predict the subsequent future states (based on the previous 30, it can predict 5 steps in the future), with acceptable results due to its complexity, while an increase in MSE is expected for each subsequent increase in the forecasting steps. Below, Fig. 5 shows real (red) and predicted values (green), while the input data sequence based on which the prediction is made is marked in blue. Also, the starting point in terms of time (starting sample in the observed data set) was randomly selected and indicated as P.

Similarly, Fig. 6 illustrates model performance comparison with real data from the evaluation data set in the form of exemplary starting positions.

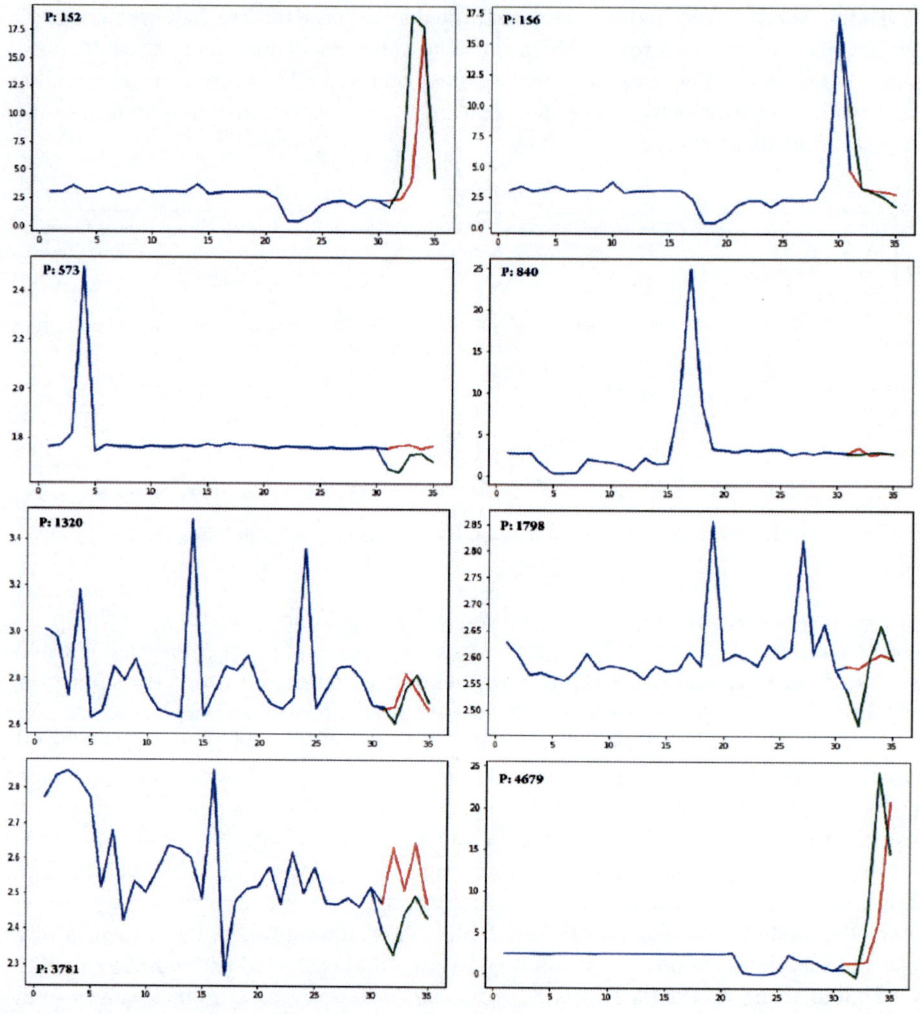

Fig. 5. Model performance comparison with real data from validation data set in the form of exemplary starting positions within the data set

Since the proposed model generalizes the observed problem well, the following integration solution toward deployment in the manufacturing system is conceptually proposed hereinafter.

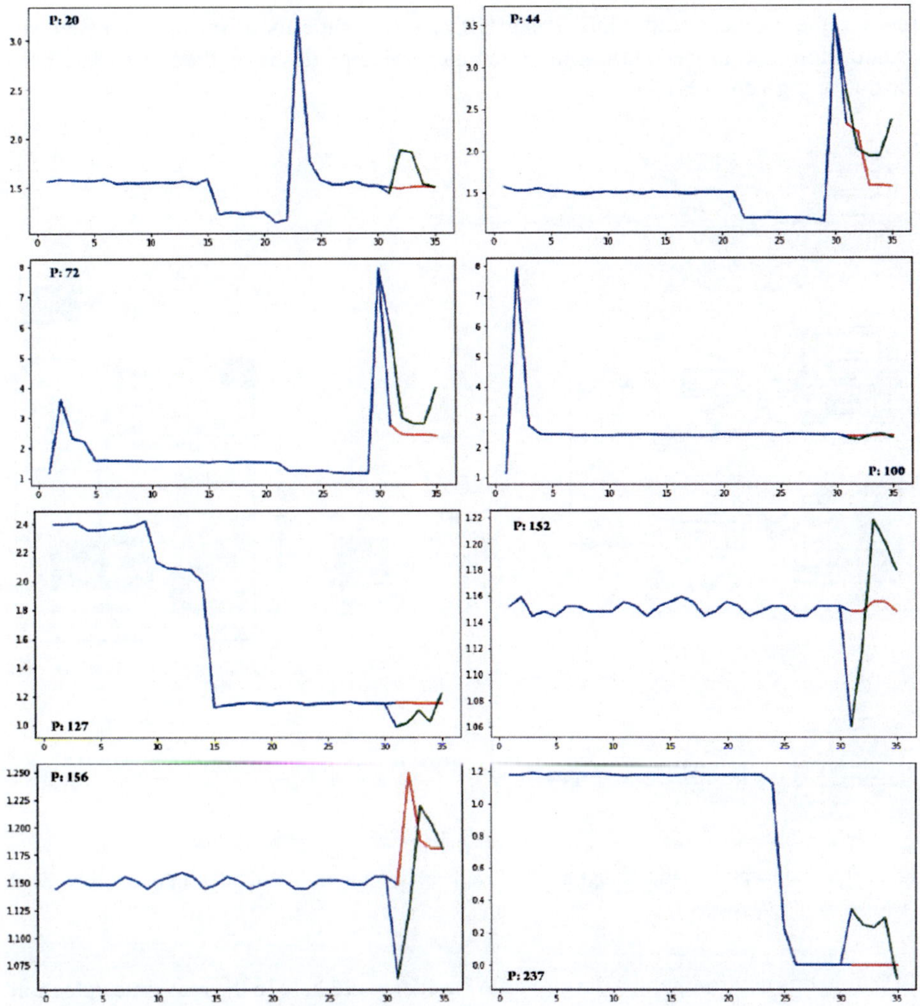

Fig. 6. Model performance comparison with real data from evaluation data set in the form of exemplary starting positions within the data set

3.1 Conceptual Deployment Approach

Each processing machine could be equipped with a proposed IoT device to acquire and publish data. The communication could be ensured via a hidden Wi-Fi network with a network Router/Switch which is connected to the client-server via Ethernet. Client-server hosts all necessary programs and codes to ensure data access via WEB App/Service. The communication could be established using the energy-efficient MQTT protocol suitable for simple and lightweight messaging, designed for constrained devices, low bandwidth, and unreliable networks. The publishing could be performed by the Eclipse Mosquitto service deployed on the server. Therefore, all messages as time-series data could be stored in the database (InfluxDB, PostgreSQL, etc.), while for advanced data analysis, visualization, and representation the Grafana service could be applied. This has been illustratively given in Fig. 7.

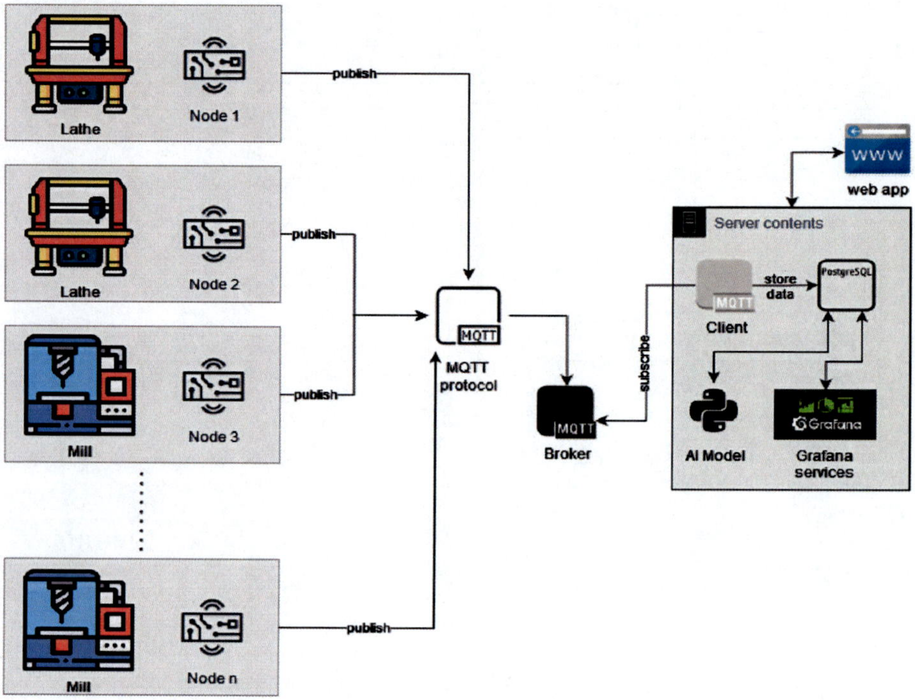

Fig. 7. The concept for the integration of IoT and referent AI model into the observed manufacturing system

Within this environment, a developed AI model could help in the real-time detection of anomalies in terms of the increase of power draw for each machine as well as point out the increasing frequency of peak loads occurrence for each specific case upon which rule-based control model could be developed to prevent situations which cause power outages within the system as a consequence of machine malfunctioning operation.

4 Conclusions

The proposed solution represents a real industrial case that combines the integration of IoT and AI intending to improve overall production efficiency by eliminating downtimes due to power outages caused by inadequate maintenance procedures. The downtime of industrial equipment accounts for heavy losses in revenue that can be reduced by making accurate failure predictions using the real-time sensor data and referent AI model. Furthermore, real-time monitoring unlocks the possibility of effectively tracking suspicious events and preventing further problem escalations. Having this in mind, an

LSTM-based model was developed, analyzed, and evaluated to predict the future states of the CNC machine tools, five steps ahead in terms of their power draw. The preferred model could be used as a data-driven prognostics approach that enables the identification of trends/patterns of a developing fault, which, in a defined timeframe, could indicate when a predetermined threshold would be reached by using information from historically treated data (trained data). However, this is the subject of further research and development.

References

1. Henning, S., et al.: Goals and measures for analyzing power consumption data in manufacturing enterprises. J. Data, Inf. Manag. **3**(1), 65–82 (2021)
2. Jacobus, H., Herman, H., Mathews, M.J., Vosloo, J.C.: Using big data for insights into sustainable energy consumption in industrial and mining sectors. J. Clean. Prod. **197**, 1352–1364 (2018)
3. Liu, X., Nielsen, P.S.: Scalable prediction-based online anomaly detection for smart meter data. Inf. Syst. **77**, 34–47 (2018)
4. Vijayaraghavan, A., Dornfeld, D.: Automated energy monitoring of machine tools. CIRP Ann. **59**(1), 21–24 (2010)
5. M. Medojević et al. "Energy management in industry 4.0 ecosystem: a review on possibilities and concerns. In: B. Katalinić (eds.) Proceedings of the 29th DAAAM International Symposium, Vienna, Austria, DAAAM International, pp. 674–680 (2018)
6. Wahiba, Y., Krishnamurthy, K., Entchev, E., Longo, M.: Recent advances in internet of things (IoT) infrastructures for building energy systems: a review. Sensors **21**(6), 2152 (2021)
7. Chen, J., et al.: Distributed collaborative control for industrial automation with wireless sensor and actuator networks. IEEE Trans. Industr. Electron. **57**(12), 4219–4230 (2010)
8. Hamidreza, S., Chin, K.W., Naghdy, F.: Coordination in wireless sensor-actuator networks: a survey. J. Parallel Distrib. Comput. **72**(7), 856–867 (2012)
9. Alvaro, L., Terrasson, G., Curea, O., Jiménez, J.: Application of wireless sensor and actuator networks to achieve intelligent microgrids: a promising approach towards a global smart grid deployment. Appl. Sci. **6**(3), 61 (2016)
10. Brkić, M., et al.: Quality assessment of system for automated multi-node environmental water parameter monitoring. In: 2019 42nd International Convention on Information and Communication Technology, Electronics and Microelectronics, MIPRO 2019 – Proceedings, pp. 163–167 (2019)
11. Vasiljević-Toskić, M., et al.: Wireless Weather Station with No Moving Parts. In: EUROCON 2019 - 18th International Conference on Smart Technologies, Institute of Electrical and Electronics Engineers Inc. (2019)
12. Andaloussi, Y., et al.: Access control in IoT environments: feasible scenarios. Procedia Comput. Sci. **130**, 1031–1036 (2018)
13. Alkhalil, A., Ramadan, R.A.: IoT data provenance implementation challenges. Procedia Comput. Sci. **109**, 1134–1139 (2017)
14. Dasgupta, A., Gill, A.Q., Farookh, H., Privacy of IoT-enabled smart home systems. In: Internet of Things (IoT) for Automated and Smart Applications, IntechOpen (2019)
15. Casola, V., et al.: A security monitoring system for internet of things. Internet Things **7**, 100080 (2019)
16. Haddad Pajouh, H., et al.: A survey on internet of things security: requirements, challenges, and solutions. Internet Things **14**, 100129 (2021)

17. Medojević, M., Tejić, B., Medojević, M., Kljajić, M.: Design and development of IIoT-based system for behavior profiling of nonlinear dynamic production systems based on energy flow theory. Therm. Sci. **26**(3A), 2147–2161 (2021)
18. Prabadevi, P.B., et al.: Deep learning for intelligent demand response and smart grids: a comprehensive survey. arXiv:2101.08013 [cs.LG]. (2021). https://arxiv.org/abs/2101.08013v1
19. Zhu, J., et al.: A novel LSTM based deep learning approach for multi-time scale electric vehicles charging load prediction. In: 2019 IEEE PES Innovative Smart Grid Technologies Asia, ISGT 2019, Institute of Electrical and Electronics Engineers Inc., pp. 3531–3536 (2019)
20. Tang, L., Yulin, Y., Yuexing, P.: An ensemble deep learning model for short-term load forecasting based on ARIMA and LSTM. In: 2019 IEEE International Conference on Communications, Control, and Computing Technologies for Smart Grids, SmartGridComm 2019, Institute of Electrical and Electronics Engineers Inc. (2019)
21. Wang, S., Xuan, W., Shaomin, W., Dan, W.: Bi-directional long short-term memory method based on attention mechanism and rolling update for short-term load forecasting. Int. J. Electr. Power Energy Syst. **109**, 470–479 (2019)
22. Ouyang, T., et al.: Modeling and forecasting short-term power load with copula model and deep belief network. IEEE Trans. Emerg. Top. Comput. Intell. **3**(2), 127–136 (2019)
23. Hu, Y., et al.: Short term electric load forecasting model and its verification for process industrial enterprises based on hybrid GA-PSO-BPNN algorithm—a case study of papermaking process. Energy **170**, 1215–1227 (2019)
24. Sideratos, G., Andreas, I., Nikos, D.H.: A novel fuzzy-based ensemble model for load forecasting using hybrid deep neural networks. Electric Power Syst. Res. **178**, 106025 (2020)
25. Ming, D., Grumbach, L.: A hybrid distribution feeder long-term load forecasting method based on sequence prediction. IEEE Trans. Smart Grid **11**(1), 470–482 (2020)
26. Gutowski, T., Dahmus, J., Thiriez, A.: Electrical energy requirements for manufacturing processes. In: 13th CIRP International Conference of Life Cycle Engineering, Leuven (2006)
27. Xing, J.T.: Energy Flow Theory of Nonlinear Dynamical Systems with Applications. Springer International Publishing, Cham (2015)
28. Medojević, M., An energy-based one step ahead of state prediction with LSTM model. In: Information Society of Serbia - ISOS, pp.128–132 (2022)
29. Olah, C.: Understanding LSTM Networks -- Colah's Blog 2015. https://colah.github.io/posts/2015-08-Understanding-LSTMs/. Accessed 21 June 2022
30. Tara, S.N., Vinyals, O., Senior, A., Sak, H.: Convolutional, long short-term memory, fully connected deep neural networks. In: ICASSP, IEEE International Conference on Acoustics, Speech and Signal Processing - Proceedings, Institute of Electrical and Electronics Engineers Inc., pp. 4580–4584 (2015)
31. Guo, Y., et al.: Attentive long short-term preference modeling for personalized product search. ACM Trans. Inf. Syst. **37**(2) (2018). https://arxiv.org/abs/1811.10155v1
32. Hochreiter, S., Schmidhuber, J.: Long short-term memory. Neural Comput. **9**(8), 1735–1780 (1997)
33. Wang, B., et al.: Parallel LSTM-based regional integrated energy system multienergy source-load information interactive energy prediction. Complexity **2019**, 1–13 (2019)
34. Gers, F.A., Schmidhuber, J., Cummins, F.: Learning to forget: continual prediction with LSTM. Neural Comput. **12**(10), 2451–2471 (2000). https://direct.mit.edu/neco/article/12/10/2451/6415/Learning-to-Forget-Continual-Prediction-with-LSTM. Accessed 21 Aug 2022

Reproductive Autonomy Conformity Assessment of Purposed AI System

Dragan Dakić(✉)

Faculty of Law, University of Kragujevac, 1 Jovana Cvijića Street, 34000 Kragujevac, Serbia
ddakic@jura.kg.ac.rs

Abstract. Using fully automated deep learning models to predict the probability of pregnancy i.e. effective and standardized embryo selection could be a promising improvement in the safety and efficacy of reproductive services. However, this aspect of artificial intelligence (AI) application in clinical decision-making tasks might be the most disputable since it interferes within complex ethical and legal issues related to reproductive choice. The increasing role of AI systems in reproductive decisioning could challenge the very paradigm of reproductive autonomy. Bearing this in mind, we tend to address reproduction-specific issues as a form of specialized conformity assessment. The results of this inquiry might be useful to programmers and developers. It could provide them with insight into normativised social values from the ambit of reproductive autonomy requiring consideration before placing purposed AI at the EU market.

Keywords: Deep learning models · Embryo selection · Reproductive services · Normative social values · EU market

1 Introduction

In Europe, the human rights paradigm of reproductive autonomy predominantly rests upon the bodily concept of autonomy. Consequently, decisioning powers in the reproductive sphere are decided by: (1a) gestational interconnection (it conferred woman with greater control over reproduction as compared to man, [1] as she is the person 'primarily concerned by the pregnancy and its continuation or termination', [2] and 'made it impossible to isolate life of unborn from that of mother' [3]; (1b) viability-in-future (in Europe abortion is temporally limited with the child's viability mostly regardless of indications) [4]; (1c) protection of the interests in life and health of the mother (The institutions of the European Convention had indicated that in the event of the conflict between mother's interests from the ambit of Article 2 and proportional interests of the unborn in initial stages of pregnancy the precedence shall be given to 'protecting the life and health of the woman').

Emerging gestating technologies such as bio-bag [5], could be capable to terminate even essentials of reproductive autonomy enabling (2a) temporal and physical separation between procreation and gestation, gestation and progenitor(s); (2b) instantaneous emergence of viability at the beginning of gestation; and (2c) preclusion of the conflict in

N. Filipovic (Ed.): AAI 2022, LNNS 659, pp. 45–57, 2023.
https://doi.org/10.1007/978-3-031-29717-5_3

life and physical health. Due to the instant viability commencement and the exclusion of the competitive rights conflict (parents vs. embryos/fetuses), it could be claimed that the artificially gestated human entity requires autonomous recognition [6] under the guarantees of the European Convention on Human Rights and Fundamental Freedoms (the European Convention) [7]. Furthermore, the exercise of negative reproductive choice might fall out of the scope of reproductive autonomy [6].

There has been an interesting debate between Romanis [8] and Colgrove [9] about conceptual differences between technologies with the same clinical objective both of which were essentially designed to substitute natural gestation in a specific way, be it for the purpose of *ex utero* gestation i.e. reproduction, be it for the purpose of incubation of prematurely born neonates i.e. neonatal intensive care. Romanis claims that those technologies should be understood as distinct which consequently merits different ethico-legal approach. She invokes three sets of arguments to support this claim: the first set refers to innate differences that are decided by technical features decisive for the scope of their application; the second set refers to different subjects of each technology decided by their behavior and social interactivity which is status decisive; and the third set refers to presumed purposes for which those technologies are going to be used in the future [8]. Colgroves is challenging the second set of arguments, claiming that subjects of both technologies should be treated equally to newborns [9].

Since we are also investigating how the application of technology (purposed AI) in reproduction might affect reproductive autonomy, the broader relation between technology and the legal status of developing human beings, as we saw, is certainly significant. However, for the purpose of our exploration, the stage of reproduction might outweigh the importance of the technology deployed. In Europe, the legal status of the developing human being as well as the scope of the reproductive choice [10], are primarily decided by the gestational age of the pregnancy [11]. Technology that was eventually used to support the process received no legal recognition as a status determinative neither it *per se* impose any restrictions to reproductive autonomy. Therefore, if we ought to analyze the relation between purposed AI and reproductive autonomy, the discussion should distinguish preimplantation from postimplantation stages. Also, the current state-of-the-art is limiting our inquiry to the former stage.

In order to conduct a comprehensive investigation, first we are going to address the state-of-the-art of AI application in reproduction together with current legal frameworks relevant to AI's assessment. Within this part of the research, we are going to briefly present the procedural group of sources that consists of basic legal rules governing the assessment process of the medical devices including software and indicate the second substantive group of sources mostly from the field of human rights. Since these two groups of sources were unequivocally interconnected via soft law sources methodology of our following inquiry coincide to that of the Assessment List for Trustworthy Artificial Intelligence. In the next part of the paper, we are assessing purposed AI system in respect to reproductive autonomy requirements. First, we analyze the implantation data and how they could be decided by legal constraints of reproductive autonomy. Also, within the same subtitle we demonstrate how dose obstacles were overcome at a regional level. In the context of reproductive autonomy and AI decision-making, perhaps the most challenging requirement is the meaningful explanation as mandated by informed consent.

Within this part of the paper, we are assessing how AI faculties correlate to the required level of explanation and offer elements of its content. In the last part of the paper we are discussing the relation between reproductive autonomy and human intervention.

2 A State-of-Art and Legal Landscape for AI Application

It is expected that AI governs the process of artificial gestation once it becomes available for humans to use. Currently, software such as IVY operates in preimplantation stage of the reproductive process enabling effective and standardized embryo selection [12] and as such, it provides information almost decisive for reproductive choice. In a technological sense, this particular software is using a fully automated system in order to predict the probability of fetal heart pregnancy based on raw time-lapse videos directly obtained [12] which distinguishes it from other compromised software and technologies based on supervised learning [13]. So, the information IVY provides is presumed to be increasingly reliable and as a result it might gain even higher prominence in reproductive decisioning. All of this firmly confirms the necessity for it's a human rights conformity assessment.

The necessity for AI alignment to human rights standards was clearly recognized at EU level. Recently the European Commission published a proposal for the so-called Artificial Intelligence Act [14], which is among other purposes, designed to safeguard fundamental rights against AI's adverse effects [15]. In this regard, a number of fundamental rights enshrined in the EU Charter of Fundamental Rights [16] (the Charter) were recognized as AI-affected including the right to human dignity (Article 1), respect for private life and protection of personal data (Articles 7 and 8), non-discrimination (Article 21) and equality between women and men (Article 23). In its Title II and Title III, the Artificial Intelligence Act labeled violation of fundamental rights as prohibited AI practices. Furthermore, Title III is addressing high-risk AI systems and identifies its two main categories one of which is stand-alone AI systems with mainly fundamental rights implications. Among AI's listed in Annex III that contains a limited but not a final number of AI systems whose risks have already materialised or are likely to materialise in the near future, Artificial Intelligence Act enumerates 'Access to and enjoyment of essential private services and public services and benefits covering AI systems intended to be used for medical aid.

As to the current legal frameworks relevant for AI's assessment/approval it is possible to allocate two general groups of sources in the European Union. The first procedural group consists of basic legal rules governing the assessment process of medical devices [17], including software [18]. They were introduced through the Directive 98/79/EC of the European Parliament and of the Council [19], i.e. Regulation (EU) 2017/745 of the European Parliament and of the Council [20], and Regulation (EU) 2017/746 of the European Parliament and of the Council [21]. Those harmonized rules are establishing criteria and govern the process of evaluation of various aspects of a new device relevant for its approval and introduction into clinical practice. The process of a conformity assessment partly depends on the type of device, but it generally encompasses a review of the manufacturer's quality system and technical documentation on the safety and performance of the device [22].

Indispensable aspects of a conformity assessment are the ethical and legal effects of the technology which connects the whole process to the second substantive group of sources mostly from the field of human rights. These two groups of sources were unequivocally interconnected via soft law sources such as Ethics Guidelines for Trustworthy AI [23], a White Paper on Artificial Intelligence [24], and the Assessment List for Trustworthy Artificial Intelligence (ALTAI) [25]. For this reason, further methodology of our inquiry will follow that of ALTAI which in its introductory part poses three core fundamental rights questions relevant for the self-evaluation of AI. Following its fundamental rights framework we tend to address reproduction-specific issues as a form of specialized conformity assessment.

3 A Reproductive Autonomy Conformity Assessment

Generally, a human rights conformity assessment is centered around the objectives of the technology and its capabilities. Herein we are going to assume that embryo selection in order to 'improve patient outcomes, increase the efficiency of healthcare diagnosis and treatment, and lower the cost of care' [26] is legally acceptable, and that AI meets the required level of analytical sensitivity, diagnostic sensitivity, analytical specificity, and diagnostic specificity as mandated by IVDMD Regulation [27] in order to focus assessment on AI's capabilities.

3.1 Implantation Data and Reproductive Autonomy

Even though a wide spectrum of AI's capabilities such as the ability to exhibit signs of rational thinking, the capability to adapt to detected changes, and the ability to engage in autonomous actions [28] could be relevant for reproductive autonomy, still those are mere outputs decisively determined by the input data. So, the list of targeted data and AI's ability to process them is the crux of the issue herein. In that sense, the most important technical feature could be the ability of the AI system to process any kind of implantation data, including genetic, which makes a list of embryo selective parameters to be non-exhaustive [12]. Bearing in mind that parameters are being used by AI in the procedure which suffers from the lack of explainability [29], but still can significantly impact reproductive autonomy [30], the application of this technology might raise some legal [31] and ethical considerations [32]. From the reproductive autonomy perspective, those considerations are almost exclusively reduced to the rights and interests of parents. Beyond the dignity realm, rights and interests of *in vitro* embryos received no recognition at a regional level [33]. Similar power distribution between parents and *in vitro* embryos is in the United States of America[34]at least for now.

But even so, reproductive autonomy at the preimplantation stage is not absolute in Europe and its limits could have constraining effects on AI. Sex for instance, as a frequent non-medical parameter for embryo selection [35] which is associated with long-reaching social and economic impacts [36], has been long ago forbidden [37] and condemned by the Council of Europe as rooted in a culture of gender inequality [38]. A Similar view was taken by the UN Committee on the Elimination of Discrimination against Women [39]. Other medically more justified filters such as malformation or hereditary disease might not be less disputable [40]. However, our assessment herein is not going to cover the broader societal perspective of AI application labeled by Smuha as 'societal harm' [15]. Rather, our inquiry could be qualified as 'individual harm' [15] assessment which we find more appropriate for the reproductive autonomy considerations.

Reproductive autonomy is a complex socio-legal construct encompassing the broad spectrum of protected values which should at least remain unaffected by AI application. To that end, the modelling process of AI should take them into consideration. Aggravating circumstances in the quest for those 'class labels' [31] are divergent approaches across EU jurisdictions as well as in their dispersion throughout hard or soft law measures at a regional level. Additional complications arise from the diversified classification of potential embryo-selecting parameters. It is not precluded that one and the same disease is listed among severe and incurable in one jurisdiction which might not be the case in the other [41]. As a result, the legal qualification of the targeted parameter for embryo selection might vary considerably from permissible/exceptionally permissible to forbidden/exceptionally forbidden. Also, the targeted parameter might not be permissible for embryo selection if obtainable via impermissible diagnostic methods under national law such as prenatal genetic tests (PGD/PGS or NIPT) [42].

On the contrary, a reproductive autonomy conformity assessment scheme looks far simpler from the regional perspective. Namely, according to the reasoning of the European Court of Human Rights in *Costa & Pavan v. Italy*, [43] any parameter that constitutes ground for abortion under the national statutory can be legitimately used in the process of embryo selection. Otherwise, national legislation would be qualified as inconsistent and any constraints in this regard would constitute a disproportionate interference with the patients' right to respect for their private and family life [44] as safeguarded through Article 8 of the European Convention on Human Rights and Fundamental Freedoms (the European Convention) [7]. Unavoidable contextualization of AI modeling norms to the safeguards of reproductive autonomy from the ambit of Article 8 of the European Convention [45] leads to further clarifications with respect to specific performance requirements.

First, unless differently specified by the manufacturer, embryo selective AI system should be able to detect all embryo health conditions (hereditary conditions, genetics, mental or physical impairment) that are relevant for reproductive choice under the national abortion legislation [46]. The troubling aspect of this requirement is the generalized wording of abortion defenses in national statutory so relevant health conditions are mostly decided by clinical practice. Second, detection must occur in a (timely) [47] manner, sufficient to ensure the protection of parents' interests [48] which should not be a problem since embryo profiling occurs prior to implantation. For all reasons above, the assessment process should cover each parameter separately, and AI should be customized to enable their application on a case-by-case basis.

3.2 AI and Informed Consent

Still, even if AI is safe and meets performance requirements, its technical proprieties still might not be compliant with guarantees from the domain of Article 8 of the European Convention. Namely, the results obtained from testing might not be followed by a meaningful explanation. Reasons for this are in the lack of the answer regarding whether and how some of the results predict the health of a future child [34],as well as in the lack of AI explainability about 'the characteristics and features these results are based upon, and the respective underlying assumptions' [49]. While the former reason is substantive and as such fully attributable to the incompleteness of medical knowledge, the latter one even if inseparable from the former, is technical and should be discussed herein.

In general, AI application in medicine mandates that 'the user should be informed how the artificial intelligence will react in critical situations, as well as be made accurately aware of all drawbacks, possible errors, misdiagnosis, and things that can go wrong when relying on it' [44] which is not always possible [50]. Ferretti, Schneider, and Blasimme[51] are classifying reasons for AI's opacity into three categories. First refers to the lack of disclosure which is not the case herein. The second is epistemic opacity and it is related to the question of how an AI system provides a specific outcome [51]. The third is explanatory opacity related to the question of why an AI system provides a specific outcome [51]. A possible way to overcome the gap between technical characteristics and legal requirements could be found by Amann, J., Blasimme, A., Vayena, E. *et al.* who are considering that physicians should be able to provide explanations on: '(1) the agent view of AI, i.e. what it takes as input; what it does with the environment; and what it produces as output, and (2) explaining the training of the mapping which produces the output by letting it learn from examples—which encompasses unsupervised, supervised, and reinforcement learning' [49].

This appears to be a safe way to meet very basic requirements of informed consent in medicine. In the reproductive sphere, however, the required degree of explainability might be higher due to the severity of 'the consequences of erroneous or otherwise inaccurate output to human life' [49]. As we could properly understand this it is useful to refer to the current situation where approximately 40 percent of healthy embryos are unnecessarily discarded due to inadequate interpretation of results obtained from the genetic test [34]. As a consequence, the hopes of many people to achieve biological parenthood have been squandered, [34] for, it might not be enough that physicians provide explanations on implantation data and their output (as reliable as medical knowledge allows it), and the training/learning process of AI. In order to maintain consent validity, a meaningful explanation should additionally provide at least information on the basis for AI decision including the factors, the logic, and the techniques that produced the outcome [52]. Also, any shortcomings of the AI system's capabilities and limitations should be communicated to parents.

This claim could be firmly supported by the guarantees steaming from a new right from the group of so-called digital rights [53]. Namely, in parallel to the development of new technologies occurs the process of new rights formation [54], and the development of AI is no exception. One of those rights that might be relevant herein is the right to algorithmic transparency. Typically [55], the process of its formation commenced in soft law sources [52] such as: the Universal Guidance for AI (2018) that sets a Right to Transparency conferring the individuals concerned with the right to know the basis for an AI decision including factors, logic and techniques behind it [52]; the OECD Recommendation on Artificial Intelligence (2019) [56] which enumerates 'Transparency and Explainability' within the list of its recommendations and mandates human intervention by non-experts (ability to challenge AI outcome by those who are affected), and sets the easy-to-understand standard in respect to the quality of information on the factors and logic that produced recommendation/decision; and the UNESCO Recommendation on the ethics of Artificial Intelligence (2021) [57] which puts transparency in the broader societal context of peace, justice, democracy and inclusion. Further provisions of this right could be found in the Convention for the protection of individuals with regard to the processing of personal data by the Council of Europe [58] as well as in the Regulation (EU) 2016/679 (General Data Protection Regulation, GDPR) [59].

According to Rotenberg, 'a right to algorithmic transparency is now firmly established as a fundamental right and cornerstone for the regulation of Artificial Intelligence' [52]. Its guarantees are intended to explain the rules applied by AI in decision making i.e. to disclose epistemic opacity as well as to enlighten the link between implantation data and the decision (explanatory opacity disclosure) [51]. But it should be noted here that even when comprehensive opacity disclosure is not achievable, the application of AI in decision-making could be acceptable both from the GDPR aspect as well as from the aspect of medical standards [51]. Still, for the purpose of reproductive autonomy protection, the provisions of the General Data Protection Regulation might pose some additional AI modelling requirements beyond the opacity disclosure. It is accepted that GDPR applies when medical devices process personal data [60]. What could be challenging in this regard is the determination of the notion of'data subject' since it is not clear if GDPR covers embryos [61]. Also, subsuming health information on embryos under the personal data of parents might look like the too extensive interpretation of the concept of data subjects even though the European Court of Justice embraces a broad understanding of that notion [62]. As to the *in vitro* embryos – European nonpersons, any change of their status would be an isolated exception which needs to be explicitly stated in the text. Since the wording of the GDPR consists of no suchlike *expressis verbis*, it is sensible to conclude that general rule applies herein and that *in vitro's* have no 'natural person' status for the purpose of the GDPR. Consequently, they cannot be considered for data subjects.

As to the parents – undisputable natural persons, health information on embryos could fall within data concerning their health from the ambit of Article 4 (15) of the GDPR. Namely, in the view of the European Court, a child's condition could endanger their mental health [63] which is in the context of reproductive choice safeguarded as a positive right from the ambit of Article 8 of the European Convention [64]. Tangent to the right to mental health [65] is the right to mental integrity from the scope of Article 3(1) of the Charter which by the analogy, could be also threatened by the child's condition. Accordingly, GDPR could be applicable based on Article 1(2) that safeguards the fundamental rights and freedoms of natural persons. If this is correct, along to the requirements arising from the principle 'protection by design' introduced in Article 35, Article 22 (2) of the GDPR further requires embryo selective AI to be calibrated in a manner to enable human intervention, parents should be able to express their point of view on results obtained; and to consent to the proposed choice [66].

3.3 Reproductive Autonomy and Human Intervention

As we can see the second requirement – the ability to challenge AI reasoning/decision, and the third requirement - explicit consent, are indisputably modeling AI systems in reproductive autonomy favorable manner. What might not be so clear is how human intervention requirements could affect it. Namely, this particular requirement directly correlates with decision making and if reduced to physicians-AI interaction it could neglect the parental point of view and potentially passivate them. In the GDPR realm, human intervention is a right protecting data subjects from being left to a fully automated decision. But beyond the GDPR and opacity disclosure, the scope and the content of this requirement are legally undecided [67] so the answer to the questions of what human intervention exactly means and when should it occur in medical decisioning could be debatable. Solaiman and Bloom [68] are offering their interpretation of the term and according to them human intervention could mean 'humans replacing automated decisions without algorithmic help; a human decision taking into account the algorithmic assessment, or humans monitoring the input data based on a person's objections and a new decision made solely by the network' [68]. An answer to the question when should humans get involved, they have located within risk-based approach reducing it to procedural rather than substantive validation [68].

Therefore, when combined with other two parents-empowering requirements (ability to challenge AI reasoning/decision and explicit consent), human intervention as described above fits within frameworks of physician-patient relationship enshrined within one of the shared decision-making models: the informative model of care, the interpretative model of care or the deliberative model of care [69]. Each of those patient-centered models is respectful toward the established hierarchy of decision-making in the reproductive sphere and leaves no space for direct or indirect situating the AI as another instance in the chain. Accordingly, human intervention has no adversarial effects on the scope of reproductive autonomy. Furthermore, human intervention could be one of its main strongholds in the future. The troubling shortcoming of the Artificial Intelligence Act from the human rights perspective is its omission to clearly ensure the application of necessity and proportionality tests as well as to 'consistently allocate legal responsibility for the wrongs and harms of AI' [70]. So, without human intervention necessity and proportionality tests - fundamentals in human rights reasoning essential for reproductive choice; as well as rules on legal responsibility would remain unenforceable.

4 Conclusion

A reproductive autonomy conformity assessment of embryo selective AI system was conducted with respect to implantation data it could use, informed consent and human intervention. As to the implantation data, it was demonstrated how they receive different legal qualification across EU jurisdictions. Also, it was concluded that any parameter that constitutes ground for abortion under the national statutory can be legitimately used in the process of embryo selection. In order to meet requirements from Article 8 of the Convention, the AI system should be able to detect all embryo health conditions that are relevant to reproductive choice under the national abortion legislation in a timely manner. Due to the diverse and evolving legal landscape with respect to implantation data, it was suggested that the assessment process cover each intended parameter separately, and to customize the AI system in a way enabling the application of different parameters on a case-by-case basis. As to the informed consent, it was concluded that meaningful explanation requires disclosure of epistemic as well as explanation opacity i.e. provision of information on the basis for AI decision including the factors, the logic, the techniques that produced the outcome, and to communicate any shortcoming of the AI system's capabilities and limitations. The analysis of GDPR requirements beyond opacity disclosure disclosed that AI should be calibrated in a manner to enable human intervention, parents should be able to express their point of view on results obtained; and to consent to the choice proposed. This brought us to the discussion about human intervention and reproductive autonomy which resulted in conclusions that it has no adversarial effects on the scope of reproductive autonomy and that without of it necessity and proportionality tests as well as rules on legal responsibility could remain unenforceable in the future.

5 Summary

It was explored in this chapter if the application of AI for the purpose of standardized embryo section infringers the scope or content of reproductive autonomy as conceptualized under the European Convention. More precisely, this research presents human rights conformity assessment of the purposed AI from the perspective of potentially the most affected interests. It was presumed that embryo selection itself is legally acceptable, and that the proposed AI meets safety and diagnostic requirements. The analysis further focused on assessing AI's technical faculties: the ability to process any kind of implantation data as well as the lack of explainability. The former trait was investigated with respect to the legal status accorded to certain implantation data such as genetic information, while the latter feature was assessed with respect to informed consent supplemented with a right to algorithmic transparency and GDPR requirements.

References

1. Sheldon, S.: Gender equality and reproductive decision-making. Feminist Legal Stud. **12**, 303–316, 312 (2004)
2. Boso v Italy, no. 50490/99. https://hudoc.echr.coe.int/fre?i=001-23338. Accessed 15 Sept 2022
3. Paton v The United Kingdom, Application No. 8416/78, Decision of the Commission 1980, 19
4. Report of the Library of Congress Abortion Legislation in Europe. http://www.loc.gov/law/help/abortion-legislation/europe.php
5. Partridge, E.A., et al.: An extra-uterine system to physiologically support the extreme premature lamb. Nat. Commun. (2017) https://doi.org/10.1038/ncomms15112
6. Dakić, D.: The scope of reproductive choice and ectogenesis: a comparison of European regional frameworks and Canadian constitutional standards. ELTE Law J. 127–145 (2017). ISSN 2064 4965. https://eltelawjournal.hu/the-scope-of-reproductive-choice-and-ectogenesis-a-comparison-of-european-regional-frameworks-and-canadian-constitutional-standards/
7. European Convention on Human Rights and Fundamental Freedoms. https://www.echr.coe.int/Pages/home.aspx?p=basictexts&c
8. Romanis, E.C.: Artificial womb technology and the frontiers of human reproduction: conceptual differences and potential implications. J. Med. Ethics **44**, 751–755 (2018)
9. Colgrove, N.: Subjects of ectogenesis: are 'gestatelings' fetuses, newborns or neither? J. Med. Ethics **45**, 723–726 (2019)

10. Dakić, D.: Introduction into principles of unborn life protection under the European Convention on Human Rights. UDC 341 231 14:347.158 at: KOLARIĆ, Dragana (ed.), et al. Archibald Reiss Days: International Scientific Conference, Belgrade, 3–4 March 2014: Thematic Conference Proceedings of International Significance. Vol. 3. Belgrade: Academy of Criminalistic and Police Studies; Bonn: German Foundation for International Legal Cooperation (IRZ), vol. III, pp. 343–352 (2014)
11. Dakić, D.: Temporal dimension of reproductive choice and human rights issues. UDC: 342.7. J. Crim. Justice Secur. 152–166 (2015). ISSN 1580-0253
12. Berntsen, J., Rimestad, J., Lassen, J.T., Tran, D., Kragh, M.F.: Robust and generalizable embryo selection based on artificial intelligence and time-lapse image sequences. PLoS ONE **17**(2), e0262661 (2022). https://doi.org/10.1371/journal.pone.0262661
13. Lapuschki, S., Wäldchen, S., Binder, A., Montavon, G., Samek, W., Müller, K.-R.: Unmasking Clever Hans predictors and assessing what machines really learn. Nat. Commun. **10**(1), 1096 (2015). https://doi.org/10.1038/s41467-019-08987-4
14. EUR-Lex - 52021PC0206 - EN - EUR-Lex (europa.eu)
15. Smuha, N.A.: Beyond the individual: governing AI's societal harm. Internet Policy Rev. **10**(3) (2021). https://doi.org/10.14763/2021.3.1574
16. EU Charter of Fundamental Rights. https://eur-lex.europa.eu/legal-content/EN/TXT/?uri=CELEX:12012P/TXT
17. Forthcoming changes listed at Factsheet for Authorities in non-EU/EEA States on Medical Devices and in vitro Diagnostic Medical Devices. https://ec.europa.eu/docsroom/documents/33863
18. Guidance on Classification Rules for in-vitro Diagnostic Medical Devices for Regulation (EU) 2017/746, Medical Device Coordination Group Document (2022). https://ec.europa.eu/health/system/files/2022-01/md_mdcg_2020_guidance_classification_ivd-md_en.pdf
19. Directive 98/79/EC of the European Parliament and of the Council. https://eur-lex.europa.eu/legal-content/EN/TXT/?uri=celex%3A01998L0079-20120111
20. Regulation (EU) 2017/745 on medical devices (MDR)
21. Regulation (EU) 2017/746 of the European Parliament and of the Council. https://eur-lex.europa.eu/eli/reg/2017/746/oj
22. Medical devices. www.ema.europa.eu/en/human-regulatory/overview/medical-devices
23. European Commission: High-Level Expert Group on Artificial Intelligence 18, 2019. https://digital-strategy.ec.europa.eu/en/library/ethics-guidelines-trustworthy-ai. Accessed 22 Aug 2022
24. European Commission, White Paper on Artificial Intelligence – A European Approach to Excellence and Trust (2020). https://ec.europa.eu/info/sites/default/files/commission-white-paper-artificial-intelligence-feb2020_en.pdf. Accessed 24 Aug 2022
25. European Commission, The Assessment List for Trustworthy Artificial Intelligence (ALTAI) for Self Assessment (2020). https://ec.europa.eu/digital-single-market/en/news/assessment-list-trustworthyartificial-intelligence-altai-self-assessment
26. Lehmann, L.S.: Ethical challenges of integrating AI into healthcare. In: Lidströmer, N., Ashrafian, H. (eds.) Artificial Intelligence in Medicine, pp. 139–144. Springer, Cham (2022). https://doi.org/10.1007/978-3-030-64573-1_337
27. Regulation (EU) 2017/746 of the European Parliament and of the Council of 5 April 2017 on in vitro diagnostic medical devices and repealing Directive 98/79/EC and Commission Decision 2010/227/EU. Devices for screening for congenital disorders in the embryo or foetus are subject to class C requirements under Rule 3 (l). https://single-market-economy.ec.europa.eu/single-market/european-standards/harmonised-standards/iv-diagnostic-medical-devices_en
28. Varkonyi, G.G.: Operability of the GDPR's consent rule in intelligent systems: evaluating the transparency rule and the right to be forgotten. J. Ambient Intell. Smart Environ. 206–215 (2019). https://doi.org/10.3233/AISE190044

29. Amann, J., Blasimme, A., Vayena, E., et al.: Explainability for artificial intelligence in healthcare: a multidisciplinary perspective. BMC Med. Inform. Decis. Mak. **20** (2020). https://doi.org/10.1186/s12911-020-01332-6
30. Siermann, M., et al.: A systematic review of the views of healthcare professionals on the scope of preimplantation genetic testing. J. Community Genet. **13**(1), 1–11 (2022). https://doi.org/10.1007/s12687-021-00573-w
31. European Union Agency for Fundamental Rights, Getting the future right: artificial intelligence and fundamental rights: report, Publications Office of the European Union (2020). https://data.europa.eu/doi/10.2811/774118
32. Chapman, C.R., Mehta, K.S., Parent, B., Caplan, A.L.: Genetic discrimination: emerging ethical challenges in the context of advancing technology. J. Law Biosci. **7**(1) (2020). https://doi.org/10.1093/jlb/lsz016
33. Dakić, D.: Kopaonik school of natural law perception of dignity and legal discourse in Europe. Belgrade Law Rev. **64**(3), 287–312 (2016). https://doi.org/10.5937/AnaliPFB1603287D. ISSN 2406-269
34. Kraschel, K.: Regulating devices that create life. In: Cohen, I., Minssen, T., Price II, W., Robertson, C., Shachar, C. (eds.) The Future of Medical Device Regulation: Innovation and Protection, pp. 203–214. Cambridge University Press, Cambridge (2022). https://doi.org/10.1017/9781108975452.016
35. Bayefsky, M.: AMA J. Ethics **20**(12), E1160–1167 (2018). https://doi.org/10.1001/amajethics.2018.1160
36. Chao, F., Gerland, P., Cook, A.R., et al.: Projecting sex imbalances at birth at global, regional and national levels from 2021 to 2100: scenario-based Bayesian probabilistic projections of the sex ratio at birth and missing female births based on 3.26 billion birth records. BMJ Global Health **6**, e005516 (2021). https://doi.org/10.1136/bmjgh-2021-005516
37. Article 14 of the Convention on Human Rights and Biomedicine (the 'Oviedo Convention'). https://rm.coe.int/168007cf98
38. Resolution 1829 by the Parliamentary Assembly (2011). https://uniteforreprorights.org/wp-content/uploads/2018/01/Prenatal-Sex-selection.pdf
39. UN, Concluding comments of the Committee on the Elimination of Discrimination against Women: China, 17, 21, U.N. Doc. CEDAW/C/CHN/CO/6 (2006); United Nations, Concluding comments of the Committee on the Elimination of Discrimination against Women: India 38, U.N. Doc. CEDAW/C/IND/CO/3 (2007)
40. Ouellette, A.: Selection against disability: abortion, ART, and access. J. Law Med. Ethics **43**, 211 (2015)
41. Löwy, I.: ART with PGD: risky heredity and stratified reproduction. Reprod. Biomed. Soc. Online **5**(11), 48–55 (2020). https://doi.org/10.1016/j.rbms.2020.09.007
42. European Union Agency for Fundamental Rights, Getting the future right: artificial intelligence and fundamental rights: report, Publications Office, pp. 101–105 (2020). https://data.europa.eu/doi/10.2811/774118
43. Costa & Pavan v. Italy, Application No. 54270/10 Merits, 2012
44. Perc, M., Hojnik, J.: Social and Legal Considerations for Artificial Intelligence in Medicine. In: Lidströmer, N., Ashrafian, H. (eds.) Artificial Intelligence in Medicine, pp. 129–138. Springer, Cham (2022). https://doi.org/10.1007/978-3-030-64573-1_266
45. Codarcea v Romania, Application No. 31675/04 at para 101 and refer to Pretty v the United Kingdom, Application No. 2346/02, at para 61 and 63, ECHR 2002-III
46. R.R. v Poland, Application no. 27617/04, Merits 26 May 2011, pp. 148–162
47. P. and S. v Poland, Application No. 57375/08, Merits 30 October 2012 at para 111
48. A. K. v Latvia, Application No 33011/08 Merits 24 June 2014 para 94

49. Amann, J., Blasimme, A., Vayena, E., et al.: Explainability for artificial intelligence in healthcare: a multidisciplinary perspective. BMC Med. Inform. Decis. Mak. **20**, 310 (2020). https://doi.org/10.1186/s12911-020-01332-6
50. The Assessment List for Trustworthy Artificial Intelligence, p. 15. https://digital-strategy.ec.europa.eu/en/library/assessment-list-trustworthy-artificial-intelligence-altai-self-assessment
51. Ferretti, A., Schneider, M., Blasimme, A.: Machine learning in medicine. Eur. Data Prot. Law Rev. **4**(3), 320–332 (2018). https://doi.org/10.21552/edpl/2018/3/10
52. Rotenberg, M.: Artificial intelligence and the right to algorithmic transparency. In: Ienca, M., Pollicino, O., Liguori, L., Stefanini, E., Andorno, R. (eds.) The Cambridge Handbook of Information Technology, Life Sciences and Human Rights (Cambridge Law Handbooks), pp. 153–165. Cambridge University Press, Cambridge (2022). https://doi.org/10.1017/9781108775038.015
53. Custers, B.: New digital rights: imagining additional fundamental rights for the digital era. Comput. Law Secur. Rev. **44**, 105636 (2022). https://doi.org/10.1016/j.clsr.2021.105636
54. Von Arnauld, A., Von der Decken, K., Susi, M. (eds.): The Cambridge Handbook of New Human Rights: Recognition, Novelty, Rhetoric, pp. 7–20. Cambridge University Press, Cambridge. https://doi.org/10.1017/9781108676106.002
55. Von der Decken, K., Koch, N.: Recognition of new human rights: phases, techniques and the approach of 'differentiated traditionalism'. In: Von Arnauld, A., Von der Decken, K., Susi, M. (eds.) The Cambridge Handbook of New Human Rights: Recognition, Novelty, Rhetoric, pp. 7–20. Cambridge University Press, Cambridge (2020). https://doi.org/10.1017/9781108676106.002
56. OECD, Recommendation of the Council on Artificial Intelligence, OECD/LEGAL/0449 (2020). https://legalinstruments.oecd.org/api/print?ids=648&lang=en
57. Report of the Social and Human Sciences Commission (SHS) - UNESCO Digital Library
58. Convention for the protection of individuals with regard to the processing of personal data by the Council of Europe. https://rm.coe.int/convention-108-convention-for-the-protection-of-individuals-with-regar/16808b36f1
59. Regulation (EU) 2016/679. https://gdpr-info.eu/
60. Regulation 2017/745 Recital 47, arts. 62(4)(h), 72(3), 92(4), 110(1)–(2) (EU)
61. European Union Agency for Fundamental Rights, Getting the future right: artificial intelligence and fundamental rights: report, Publications Office, pp. 101–105 (2018). https://data.europa.eu/doi/10.2811/774118
62. European Court of Justice: Case of Patrick Breyer v Bundesrepublik Deutschland, C-582/14 (2016)
63. Boso v Italy, Application No. 50490/99, Decision from 5 September 2002
64. Tysiac v. Poland (Appl. no. 5410/03), judgment, 20 March 2007
65. Bensaid v. United Kingdom (Appl. no. 44599/98), judgment, 6 February 2001; Dolenec v. Croatia (Appl. no. 25282/06), judgment, 26 November 2009, p. 165
66. Ienca, M., Pollicino, O., Liguori, L., Stefanini, E., Andorno, R. (eds.) The Cambridge Handbook of Information Technology, Life Sciences and Human Rights (Cambridge Law Handbooks), pp. 179–180. Cambridge University Press, Cambridge (2022). https://doi.org/10.1017/9781108775038
67. European Commission, Ethics Guidelines for Trustworthy AI: High-Level Expert Group on Artificial Intelligence 9 (2019)
68. Solaiman, B., Bloom, M.: AI, explainability, and safeguarding patient safety in Europe: toward a science-focused regulatory model. In: Cohen, I., Minssen, T., Price II, W., Robertson, C., Shachar, C. (eds.) The Future of Medical Device Regulation: Innovation and Protection, pp. 91–102. Cambridge University Press, Cambridge (2022). https://doi.org/10.1017/9781108975452.008

69. Marinković, V., Rogers, H.L., Lewandowski, R.A., Stević, I.: Shared decision making. In: Kriksciuniene, D., Sakalauskas, V. (eds.) Intelligent Systems for Sustainable Person-Centered Healthcare. Intelligent Systems Reference Library, vol. 205, pp. 71–90. Springer, Cham (2022). https://doi.org/10.1007/978-3-030-79353-1_5
70. Smuha, N.A., et al.: How the EU Can Achieve Legally Trustworthy AI: A Response to the European Commission's Proposal for an Artificial Intelligence Act (2021). SSRN. https://ssrn.com/abstract=3899991. https://doi.org/10.2139/ssrn.3899991

Baselines for Automatic Medical Image Reporting

Franco Alberto Cardillo(✉)

National Research Council, Institute for Computational Linguistics, Pisa, Italy
francoalberto.cardillo@ilc.cnr.it

Abstract. Despite the high number of machine learning models presented in the last few years for automatically annotating medical images with deep learning models, clear baselines to compare methods upon are still missing. We present an initial set of experimentations of a standard encoder-decoder architecture with the Indiana University Chest X-ray dataset. The experiments include different convolutional architectures and decoding strategies for the recurrent decoder module. The results here presented could potentially benefit those tackling the same task in languages with fewer linguistic resources than those available in English.

Keywords: Computer Vision · Neural Language Generation · Image Classification · Medical Imaging

1 Introduction

The demand for image-based medical examinations has been rising for the past years and has nowadays become so high as to make it impossible for a radiology department to report on the acquired images in a timely manner[1]. Such bottleneck represents a major problem since a short turnaround of written reports from radiologists to clinicians is a key factor from several points of view: it enables early planning of correct treatment, increasing the likelihood of healthier clinical courses, it reduces the economic burden associated to the treatments necessary in later stages of the disease and, generally, it improves the patient experience[2]. For such reasons, computer-based approaches supporting radiologists in the inspection of the collected visual data represent an important topic of a large amount of past and current research. This work focuses on the deep learning models able to generate free-form texts describing an input medical image. More specifically, the task of automatic medical image reporting (AMIR) consists in the generation of a narrative text, expressed in natural language, describing the diagnostic content of one or more medical images given as input data to a computer program. It is an inherently multi-modal task involving images and texts, whose solution requires

[1] Radiology Review, A national review of radiology reporting within the NHS in England, CareQuality Commission, July 2018. https://l.cnr.it/nhseng18.

[2] American College of Radiology, Qualified Clinical Data Registry, January 2022. https://l.cnr.it/acrqdr22.

N. Filipovic (Ed.): AAI 2022, LNNS 659, pp. 58–74, 2023.
https://doi.org/10.1007/978-3-031-29717-5_4

a successful combination of computer vision (CV) and Natural Language Processing (NLP) algorithms [1, 2]. Indeed, an AMIR system can be thought of as a pipeline of two stages. In the first stage, the image is processed and its features computed; in the second step, the image features are used as the initial seed for a text generator, that outputs a verbal description of the image of an appropriate surface form, with the correct lexicon and grammar.

Here the focus is on deep learning (DL) models for text generation in the medical domain, whose performances currently represent the state of the art in the field [3]. Since the seminal paper [4], most of the DL models for AMIR have been using an encoder-decoder architecture, with a convolutional neural network (CNN) acting as the image encoder and a recurrent neural network (RNN) acting as the decoder generating text. The CNN is optionally followed by a multi-layer perceptron (MLP) trained in a supervised multi-label fashion to assign one or more classes (also called tags or labels) to the input images. During training, an image is processed by the CNN, optionally by the MLP for the classification step, then the convolutional features and, if available, the classes assigned by the MLP are combined to the word embeddings (that are vector representations of the words in the vocabulary) and passed to the RNN. More sophisticated approaches [5], based on large language models originating from the Transformer architecture [6], that have improved upon previously encoder-decoder models, are not considered in this study as the interest is on smaller neural architectures, that are better suited to elaborate datasets in languages with scarce medical textual resources, like, e.g., Italian. Indeed, despite the large number of models published, AMIR still lacks a clear assessment of baseline performances and a standard way for comparing the results obtained with different approaches. As noted also in [1], new complex models are compared to old complex ones without a reference baseline and often exploiting specific characteristics of the input data. For example, it is not clear whether the use of different CNNs leads to different results or if the presented results depend too much on the specific partitions used in training and evaluating the model. When developing a new model or experimenting with a new dataset, a clear indication of what components mostly influence the final performance might lead to faster and more sustainable activities, saving time and computational resources. The experiments discussed in the next sections try to clarify what basic components of an encoder-decoder system for text generation mostly influence the final performance in terms of the quality of the generated text. In fact, previously published papers, described in the next paragraphs, use, for example, different CNNs, different sets of image labels or different vocabularies, thus making it hard to understand if the differences in performance are due to the model architecture or to other factors.

The research on AMIR has closely followed the work done within the more general domain of image labelling and captioning, with only a slight delay due to the initial unavailability of medical datasets. The first published works presented algorithms and models for annotating medical images with word-level descriptions [7, 8], like single words or bags of words. They first apply a clustering procedure to define a set of labels and then train an CNN-based image classifier to map image on those labels. Clustering is performed using only the text data [7] or aligned images and texts in a multi-modal fashion [8]. Using clustering it is possible to implement systems that are language-agnostic, but their results tend to be difficult to interpret by radiologists and too sensitive to the set of parameters used to form clusters. In the meantime, NLP tools tailored for the English language led to the publication of the CheXpert [9] and the NegBio [10] labelers, used for building datasets enabling the training of deep learning models, such as [11], able to classify the (not yet public) MIMIC-CXR dataset (described in Sect. 2) on the classes extracted from the medical reports by receiving a frontal and a lateral view of the same subject as input.

[12] introduces an encoder-decoder architecture to AMIR and represents the first model able to generate a sentence instead of single words. An image encoder based on the GoogleNet architecture is followed by a recurrent module based on a Long Short-Term Memory (LSTM) [13] or a GRU [14] recurrent layer. They experiment with the Indiana University chest X-ray dataset (hereafter named IU-CHEST) [15] (described in Sect. 2), but used only the 13 most frequent MeSH terms[3] plus other four terms (selected from unique combinations of tags occurring at least 30 times in the dataset), for a total of 17 tags, out of the original 119, thus restricting their experiments to 40% of the available data in IU-CHEST. The generated description contains five words at the most. Results are not directly comparable with subsequent approaches due to a limited amount of training data and very short generated captions. [12] stresses and highlights the difficulties in processing the IU-CHEST dataset, characterized by an extremely imbalanced distribution of image classes. Attentional mechanisms [16] are first introduced in [17]. The authors experiment with several datasets, including a proprietary one with 1000 bladder cancer images acquired from 32 patients. Each image is annotated with five paragraphs, corresponding to five visual features indicative of the disease, and classified using four labels. The model used ResNet or GoogleNet as CNN for feature extraction, followed by a MLP for the classification of the convolutional features. Their output seeds a single-layer LSTM trained to generate all five paragraphs.

Extending the model presented in [18], [19] introduced an attention-based encoder-decoder architecture with a VGG19 [20] CNN as an image encoder extracting visual features, followed by a MLP for image classification (semantic features). The decoder is composed of a hierarchical LSTM module with two single-layer LSTM cells: the first one, called sentence LSTM, generates topic vectors from the visual and the semantic features coming from the CNN using self-attention. The second LSTM, named word LSTM, generates words starting from the topic vector generated by the sentence LSTM. The model is also able to learn how many sentences need to be generated using a MLP trained on the hidden states of the sentence LSTM. The model was tested with the IU-Chest dataset reaching an impressive performance and was later extended in [21] with dual decoders, each trained to generate sentences on normal (i.e., images with no apparent diseases) or abnormal images (i.e. images with one apparent disease at least), classified as such by a MLP. Combining the ideas from [11] and [17], [22] introduces a model based on a ResNet-152 CNN, pretrained on ImageNet, to extract convolutional features from the input images. In their experimentations with the IU-CHEST dataset, the visual features are used to generate the first sentence, that corresponds to the "Findings"

[3] MeSH is a controlled vocabulary prepared by the American National Library of Medicine. Further details are provided in Sect. 2.

section of the reports. Then, the embedding of the first sentence is passed to a Bi-LSTM network (or a 1D CNN) that exploits an attentional layer observing the embedding of the previous sentence and the visual features to generate a context vector that is used for generating further sentences until an empty sentence is output. The reported results are good, considering also that they used the full vocabulary of IU-CHEST with 2218 unique words. It should be noted that this method first generates the "Impression" section (usually composed by a single sentence), that is then used to generate the "Findings" section, even if the findings section should logically precede the conclusive "impression" text. From the previous description, it is clear that it is very hard to compare the results in order to establish what would be better suited for a new scenario under investigation.

The rest of the paper is organized as it follows. Section 2 describes the available datasets providing details on IU-CHEST, used in this work. Section 3 describes the models and the experimental settings. Finally, Sects. 4 and 5 wrap up the results and delineate further experimentations.

2 Materials

Even if there are several public dataset for image captioning [1, 3], some of which even contains medical images, only two datasets are actually well-suited and, indeed, often used for the task of AMIR: the Indiana University chest X-ray [15] and the MIMIC-CXR v2 (plus its JPG version MIMIC-CXR-JPG) datasets [23, 24]. Other available resources are the PEIR Digital library[4], that contains images of different body parts and different modalities annotated with a single sentence making it more similar to an image captioning dataset, and the ICLEF-CAPTION [25] dataset, populated with images and text automatically extracted from the medical literature on PubMed Central. Even if PEIR Digital Library has been used in previous works on AMIR (e.g. [19]), its being composed by different body parts and modalities could lead the inductive model to focus its learning more on the distinguishing the features describing the body part or modality than on expressing the visual content of a medical image. The ICLEF-CAPTION dataset is considered very noisy [1] and has been used almost exclusively by the ImageCLEF community. The IU-CHEST and the MIMIC-CXR are the only two datasets with labelled images associated to real free-text reports that respect the form and the structure commonly used in the current clinical practice of various countries. In this work all the experiments use the IU-CHEST dataset.

The Indiana University Chest X-ray dataset[5] is a collection of chest radiographies paired with their de-identified full reports collected from two large hospitals within the American Indiana Network for Patient Care. The IU-CHEST dataset contains 7470 images and 3955 reports from 3955 different patients. Each report, corresponding to an examination of a single patient, is an XML file containing (among other information):

- four narrative sections with free-form text:
  - *indication:* describing the clinical reasons for performing the examination;

[4] https://peir.path.uab.edu/library/index.php?/category/106.
[5] Freely accessible at https://openi.nlm.nih.gov/faq.

– *comparison:* describing the differences with previous examinations, if any;
– *findings:* describing what the radiology actually observed during the visual inspection of all the images belonging to the study;
– *impression:* containing the final diagnosis.

- A set of Medical Text Indexer (MTI) tags automatically extracted by the tool MTI[6] from the sections described at the previous points. The full dataset contains 523 unique MTI tags after normalization[7].
- A set of tags belonging to the Medical Subject Headings (MeSH)[8] controlled vocabulary, plus some additional terms from the Radiology Lexicon (RadLex)[9], manually assigned by radiologists. The entire dataset contains 119 unique MeSH tags.

Table 1. Statistics of the IU-CHEST dataset. "Normal" corresponds to studies without any reported diseases that are associated to the only label "Normal". "Non-normal" corresponds to studies that have one disease at least, that are associated to one or more labels other than "Normal".

	All	Normal	Non-Normal
Number of images	7244	2656	4588
Avg num. of sentences and std	5.87 ± 2.18	4.79 ± 1.64	6.5 ± 2.22
Avg. length of the 1st sentence in impression	4.32 ± 3.44	3.77 ± 1.85	4.64 ± 4.05
Avg length of the 1st sentence in findings	6.31 ± 4.22	5.92 ± 3.59	6.5 ± 4.53
Avg number of tags and std	117 (total number of tags)	1	2.55 ± 1.67

It is worth noting that the narrative texts and the two sets of tags are associated to the study/report and not to the individual images. The images are of different sizes and of different aspect ratios.

2.1 Pre-processing

The IU-CHEST images are available both in raw DICOM[10] format and in pre-processed Portable Network Graphics (PNG) formats. This work uses the PNG images without further processing their intensity values. The free-form texts in the "Findings" and "Impression" sections are pre-processed with a minimal set of filters: all the words are

[6] MTI is a tool used by the National Library of Medicine to index PubMed/MEDLINE citations.
[7] Even if automatically extracted, the MTI tags need to be normalized using a vocabulary distributed with the same dataset.
[8] https://www.ncbi.nlm.nih.gov/mesh/.
[9] https://www.rsna.org/practice-tools/data-tools-and-standards/radlex-radiology-lexicon.
[10] https://www.dicomstandard.org/.

lower-cased, punctuation symbols and parentheses are removed, patterns corresponding to numbers (either integer or decimal) and measures (e.g.1cm) are substituted with special unique tokens (e.g. _NUM_ for an integer number). The sections "Indication" and "Comparison" are ignored as in all the other previously published approaches. We decided to use only the set of MeSH tags as image labels for several reasons. The controlled MeSH vocabulary is commonly used in the reporting procedures adopted by the healthcare system of many countries. Indeed, many countries translate MeSH terms in their own language[11]. MTI labels are limited to the English-speaking countries as they are extracted from the free-form texts by automatic tools that are explicitly built and finely tailored for the English language, characterized by an abundance of medical texts. The study of MeSH labels is thus likely to provide results that are more general than those based on MTI labels and more useful for extending automatic approaches in countries that are currently developing their electronic resources for healthcare. Furthermore, MeSH terms are used as image labels in almost all of the previous works using IU-CHEST. Images associated with reports without any free-form texts and images labelled with any of the MeSH tags "technical quality of image unsatisfactory" and "no indexing" are dropped and not used in the experiments. After the pre-processing stage, the IU-CHEST dataset contains 7244 images: 2656 normal (with no reported diseases, i.e.associated to the only label "Normal") images (37%) and 4588 (63%) non-normal (with one reported disease at least, i.e. associated to one or more tags other than "Normal") images. Non-normal images are associated to 2.55 tags (MeSH terms) on average. The difficulty with this dataset is the co-occurrence of high-frequency and low-frequency tags. For example, the 20 most frequent terms co-occur with other terms in more than 60% of the cases and some very frequent tags appear with others in approximately 80% of the dataset [12]. The "normal" tag does not co-occur with others. Both the imbalance among tags and the co-occurrence of frequent and rare tags makes the image classification task very difficult, if not impossible. This aspect will be discussed in the next section.

3 Experimentations

All the experiments presented here were organized in order to guarantee the reproducibility of the results and highlight the differences in the final performance that are possibly due to a specific selection of training and test data. Specifically, the IU-CHEST dataset is partitioned into three data splits, each containing a subset of examples used for training, validating, and testing the models that are used in all the experiments. In all the data splits, the percentages of normal and non-normal images are kept equal to the original dataset (37%–63%). Furthermore, all the experiments use two random seeds for initializing the model parameters (weights) and shuffling the data. Due to the high number of experiments and in order to reduce the time spent in training the models, the three criteria for early stopping are GL, PQ, and UP [26]. Let E be the error function, $E_{tr}(t)$, $E_{va}(t)$ the average errors on the training and the validation sets at epoch t, respectively. The generalization error at epoch t, GL(t), is defined as the relative increase of the validation error at epoch t over the optimum so far:

[11] For example, the Italian National Institute of Health (ISS) offers an official translation of the MeSH dictionary: https://w3.iss.it/site/mesh/

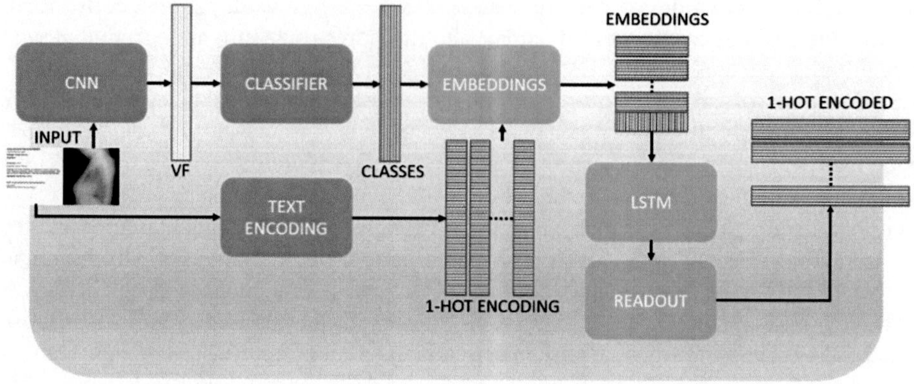

Fig. 1. A schematic view of the experimented model, rectangles with vertical stripes correspond to the visual pathway, rectangles with horizontal stripes correspond to the text pathway. Given an input composed by an image and the related text, the image is processed by a CNN, that outputs visual features (VF in the figure), VF is then passed to a classifier (linear or MLP) mapping it on the selected set of MeSH terms. The text is first encoded with 1-hot vectors with a dimension equal to the size of the vocabulary (including the special tokens), then both the visual embeddings and the word embeddings are passed to a LSTM, whose output is mapped on a space with a dimension equal to the vocabulary size by a readout layer (linear or MLP).

$$GL(t) = 100 \cdot \left(\frac{E_{va}(t)}{E_{opt}(t)} - 1 \right) \tag{1}$$

where $E_{opt}(t) = min_{t' \leq t} E_{va}\left(t'\right)$ is the lowest validation error up to epoch t. The early stopping criterion GL activates (and the training is stopped) as soon as $GL(t)$ exceeds a user-specified threshold α_{GL}. The criteria PQ and GL are defined in terms of training strips of length k, defined as k consecutive training epochs $n + 1, \ldots, n + k$, with n divisible by k. The training progress $P_k(t)$, measured after a training strip of length k at epoch t, measures, basically, how much larger the average E_{tr} during the strip is with respect to its minimum during the training strip, and is defined as:

$$P_k(t) = 1000 \cdot \left(\frac{\sum_{t'=t-k+1}^{t} E_{tr}\left(t'\right)}{k \cdot \min_{t'=t-k+1}^{t} E_{tr}(t')} - 1 \right) \tag{2}$$

The early stopping criterion PQ stops the training as soon as the ratio:

$$PQ(t) = \frac{GL(t)}{P_k(t)} \tag{3}$$

exceeds a user-specified threshold α_{PQ}. The criterion PQ does not activate when the training is still unstable but stops the training when the progress $P_k(t)$ is very low, so that additional training epochs might likely not be very effective in learning the target

Table 2. Experimental settings. BOS and EOS are two special tokens standing for, respectively, begin and end of sentence.

CNN Models:	ConvNext-Base, DenseNet-121, ResNet-152, ResNet-18, VGG-11, VGG-19	
RNN Models:	LSTM	Learning rate: CNN: 1e-4, 1e-5; RNN: 1e-4
Loss functions:	CNN, binary cross-entropy. weighed cross-entropy RNN, cross-entropy loss Optimizer, Adam for both the CNN and the RNN	
Early Stopping:	CNN and RNN, GL, PQ, UP, training strips of length 5	
Image labels:	41 including "Normal"	Vocabulary: 1000 words + special tokens Sentence length: 10,15 + bos/eos
Training augmentations:	images resized to 300x300, then random rotation of ± 5 deg, finally 224x224 random crop	
Test preprocessing:	resize to 300x300, 224x224 center crop	
Image classification:	linear layer or Multilayer Perceptron	
LSTM readout layer:	linear or multilayer perceptron	

function and not be worth the extra time (and computational resources). The last criterion *UP* may be considered a form of smoothed patience and is defined recursively:

$$UP_1(t): \text{stop at the end of the first training strip if } E_{va}(t) > E_{va}(t-k);$$

$$UP_s(t): \text{stop at epoch } t \text{ if and only if } UP_{s-1} \text{ stops after epoch } t-k \text{ and } E_{va}(t) > E_{va}(t-k).$$

The *UP* criterion stops the training when the validation error always increases at the end of the last s strips, meaning that the learner is likely overfitting the training data. The strip length for the three criteria is set to 5, as in [26] with the thresholds set as follows $\alpha_{gl} = 3$, $\alpha_{PQ} = 0.5$. The criteria are then evaluated (also GL) every five epochs starting from epochs 30 and 20 for, respectively, the CNN and the RNN modules. The batch size was fixed at 128 in order not to add variations in performance due to such hyperparameter. Smaller batches would have increased the training time, while larger ones could not be used with the larger models due to the size of the available memory on the Graphics Processing Unit (GPU). However, additional experiments (not reported here) did not show changes in the results with different batch sizes (in the range 32–256), only the training time was affected.

In the following we will describe the experimental settings aimed at evaluating the single components of previously published approaches. The basic architecture used in all the experiments is a classical encoder-encoder architecture. Such architecture is trained to generate the first sentence of the section "Findings" or the single sentence in "Impression". Given an input image I to the model, the image is first encoded by a CNN-based encoder. The visual features $VF = Conv(I)$ correspond to the output of the convolutional layers *Conv* of the CNN, basically to the output of the last layer performing a pooling operation, linearized if needed (it depends on the specific CNN). The encoder may also include a classifier *C* that maps the *VF* onto a set of classes, in this case a set of MeSH terms, providing, for each class *c*, the likelihood the input image *I* belongs to *c*. The encoding is then passed to a decoder module, based on a single LSTM recurrent neural network, that also receives the embeddings of the words *[w1, ..., wN]* associated to the input image *I*. The recurrent module is trained with teacher forcing, i.e. during training the word given as input to the decoder at time $t + 1$ is the target word at time t and not the last output of the decoder. We used a vocabulary with the 1000 most-frequent words, as in [19], that cover approximately 98% of the input text. Words not in the vocabulary are substituted with a special out-of-vocabulary (OOV) token. The maximum number of tokens in the sentences used in training is fixed at 10 and 15, that correspond to the length of the first sentence in, respectively, the "Findings" and the "Impression" section for the 0.8 percentile. Sentences longer than the maximum number of tokens are truncated. The encoding of each sentence is then extended with two special tokens, begin-of-sentence (BOS) and end-of-sentence (EOS), placed, respectively, before the first token and after the last one. After the extension with the two special tokens, sentences shorter than the maximum number of tokens + 2 are padded with null values.

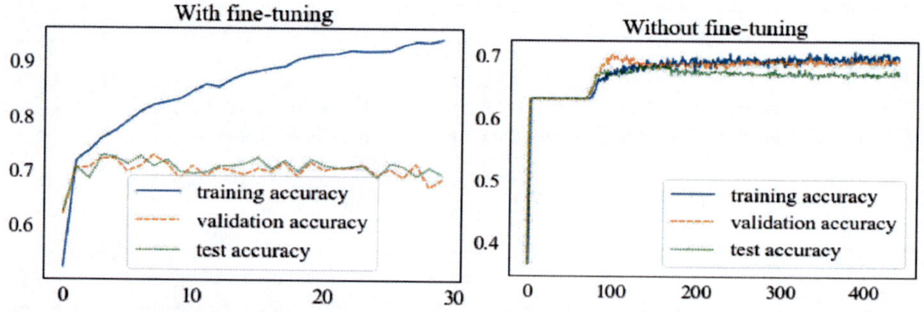

Fig. 2. Accuracy of a single ResNet-18 trained with the IU-Chest dataset, with and without fine-tuning the convolutional layers, on the binary classification task of distinguishing normal images from the rest. The x-axis corresponds to the training epochs.

3.1 Normal vs Non-normal

As described in [15], the IU-CHEST collection has been prepared in two passes. In the first pass, the studies were simply classified as normal (without diseases) or non- normal (at least one disease). Being able to differentiate between the two classes would be helpful also in an automatic approach since the text associated to normal and non-normal images tends to be different in size (Table 1) and in the vocabulary used. Indeed, [21] introduced a model based on dual decoders, with each decoder trained to annotate only normal or non-normal images, reaching very good results. We used two CNNs pretrained on ImageNet, a ResNet-18 [27] and a VGG-19, and trained them on the binary classification task of distinguishing normal images from non-normal, but results were poor. The models were trained with and without fine-tuning the convolutional layers. The fine-tuned models reached a very high accuracy (close to 95%) on the training set but presented an unstable accuracy on the validation and test sets, with a maximum accuracy only slightly higher than the baseline (most frequent value in the set). The performance of the models trained without fine tuning was even lower, with a maximum accuracy on the three partitions of about 70%. Early breaking criteria interrupted the training after about 30 epochs when fine-tuning models and about 350 epochs when not fine-tuning the models. Since IU-CHEST is a small dataset, whose dimension can easily lead CNNs to overfit the data, we tried to train the same models on images from the MIMIC-CXR dataset, that contains over 300 thousand images, but the results were similar in dynamics (number of epochs before early breaking and stability of the accuracy metric) and performance. Due to the poor performance of the two CNNs with the two datasets, the dual decoder approach has not been investigated any further. Two plots of accuracy over the training epochs corresponding to a single training run of a ResNet-18 CNN on the IU-CHEST dataset are shown in Fig. 2.

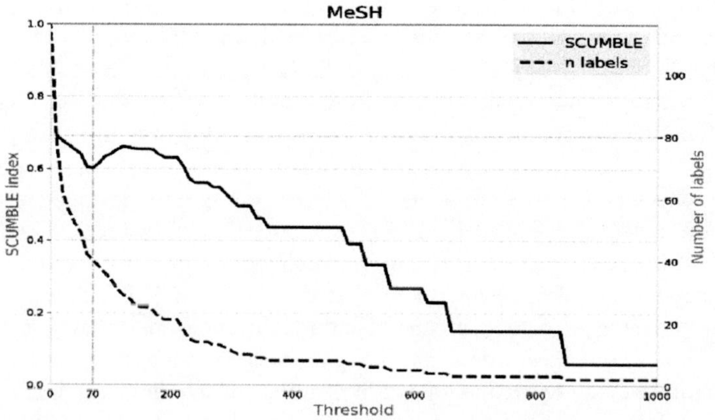

Fig. 3. SCUMBLE index of the IU-CHEST dataset for different thresholds on the minimum number of occurrences. The left y-axis corresponds to the SCUMBLE index, the secondary y-axis on the right corresponds to the number of labels remaining after applying the thresholds on the x-axis. The solid line plots the value of the SCUMBLE index, while the dashed line plots the number of labels.

3.2 Fine-Tuning CNNs on IU-CHEST

In training the decoder, previous approaches used either visual features extracted by a CNN trained on the ImageNet dataset, without any fine-tuning on the IU-CHEST images, or visual features extracted by a CNN first trained on the ImageNet dataset and then fine-tuned on IU-CHEST. In order to fine-tune a CNN, we need to train a new classifier C for the visual features using the specific classes of IU-CHEST. The new classifier can then be used to provide semantic features to the decoder, as explained previously. IU-CHEST is an extremely imbalanced dataset. In non-normal images, only one label ("lung") occurs more than 1000 times, while 84 out of 116 labels occur less than 100 times. Furthermore, rare labels co-occur with common ones, making the classification task even harder. The number of labels, however, need to be reduced, as done also in other works such as [12]. Instead of choosing the minimum frequency for a class to be kept, we used the SCUMBLE index [28] of the IU-CHEST dataset. Such an index, in the range [0,1], provides a quantitative evaluation of how frequently rare and common labels co-occur on the same images. A plot of such an index at varying thresholds on the label frequency is shown in Fig. 2. We selected a threshold equal to 70, that leaves 41 labels. Removed labels are joined in a new "miscellanea" label. It should be noted that the SCUMBLE value for a threshold set to 70 stays quite high, being equal to 0.6. However, in order to reduce the SCUMBLE index to a low value, such as 0.2, the number of labels should be reduced to 20, populating new category "miscellanea" with instances of too many different classes.

We trained four different CNN models: ConvNext-Base, DenseNet-121, ResNet-152, ResNet-18, VGG-11, and VGG-19 substituting their final classifier with an equivalent one using the sigmoid activation on the output layer. All the CNNs were trained with two different learning rates (1e-4, 1e-5) and two different loss functions: the binary cross-entropy loss and the weighed cross-entropy loss as defined in [29], that should help the learner when the number of negative samples is far greater than the positive ones. For each model, we report the results of fine-tuning and no fine-tuning the convolutional layers. Each model was trained on the same three data splits in training, validation, and test examples. During training, the images are first resized to 300x300, randomly rotated $\pm 5°$, then a random crop 224×224 in size is given as input to the encoder. During the test stage, the images are first resized to 300×300, then a 224×224 center crop is given as input to the encoder. Images are normalized to zero mean and unit variance using the mean and standard deviation computed on the IU-CHEST dataset.

The macro-average accuracy, precision and recall scores are reported in Table 2. As it is clear, there are not huge differences among the tested models, that have, in particular, a low precision in every experimental setting. In the subsequent experiments, the fine-tuned ResNet-18 is used as the encoder since its memory footprint is much smaller than the other models, thus allowing to train more models concurrently on a single GPU.

3.3 Decoders

In the last experimentations, we train a decoder using the output of different encoders. Specifically, we aim at evaluating the performance in text generation when using:

- the visual features of an encoder trained on ImageNet without any fine-tuning on the IU-Chest dataset;
- the visual features of an encoder trained on ImageNet and fine-tuned on the IU-CHEST dataset (Table 3);

the visual features and the semantic features of an encoder trained on ImageNet and fine-tuned on the IU-CHEST dataset. As previously said, the encoder is based on a LSTM cell, whose hidden layer at time t is mapped to an index of a word in the 1000-word vocabulary by a readout layer (with output dimension equal to the size of the vocabulary). Usually, such mapping is performed with an affine transformation via a dense layer. We experimented also a readout layer providing a non-linear transformation

Table 3. Macro average of Accuracy, Precision and Recall for the experimented CNNs.

	Accuracy		Precision		Recall	
	Fine-tune	No FT	Fine-tune	No FT	Fine-tune	No FT
ConvNext-Base	0.87 0.85	0.85 0.83	0.07 0.05	0.05 0.05	0.35 0.35	0.33 0.33
DenseNet-121	0.86 0.87	0.83 0.83	0.06 0.04	0.04 0.04	0.32 0.29	0.29 0.31
ResNet-152	0.84 0.85	0.81 0.84	0.05 0.05	0.04 0.05	0.34 0.34	0.29 0.34
ResNet-18	0.84 0.84	0.79 0.80	0.05 0.05	0.03 0.03	0.34 0.32	0.31 0.29
VGG-11	0.85 0.85	0.85 0.86	0.05 0.06	0.05 0.05	0.34 0.35	0.32 0.33
VGG-19	0.83 0.84	0.86 0.86	0.05 0.05	0.06 0.05	0.34 0.35	0.32 0.32

via a trainable MLP. We adopt the definition of the learning task provided in [22], where the goal is to compute a set of parameters θ maximizing the probability of generating the correct sequence of words in the first sentence of the "Impression" and "Findings" section:

$$P(w_1, w_2, \ldots, w_n|I; \theta) = P(w_1|I; \theta)\prod_{t=2}^{N} P(w_t|I, \theta, w_1, \ldots, w_{t-1}) \tag{4}$$

where I is the input image and the w_i stand for the words in the sentence. The definition relies on the Markov assumption for sentence generation with a 2-g model, i.e. the word generated at time t depends only on the previous word observed at time *t-1*. The decoder is trained with a cross-entropy loss applied to the softmaxed output of its readout layer. A dropout of 0.3 is applied on the hidden state of the LSTM. At the beginning of the training, the hidden and cell states are zeroed. The vector of visual features, projected onto a space with the same dimension of the word embeddings by a fully connected layer, is used as initial input, followed by the embedding of the BOS token. Word embeddings are initialized randomly and are trainable.

In the test stage, text generation uses a greedy strategy: at time step 0, the hidden and cell state of the LSTM cell are first zeroed, then the same cell receives the embedding of the visual features, followed at time 1 by the embedding of the BOS token. At the subsequent time steps t ($t > 1$), the decoder receives the token generated at time step *t-1*, that correspond to the index of the maximum value in the softmaxed output at time *t-1*. The generation continues as long as the decoder does not generate the EOS token or it is interrupted when the maximum number of tokens has been generated.

When the decoder receives not only the visual features *VF*, but also the semantic features *SF*, the two vectors *VF* and *SF* are first concatenated, then projected on a space with the same dimension of the word embeddings by a fully connected layer.

Due to the time needed to train sequential models (even for a single sentence the training can last more than one day), each decoder was trained on a single split of the dataset, chosen randomly. The learning rate was fixed at 1e-4. The dimension of the word embeddings was set to 256. We experimented with two different sizes of the hidden state of the LSTM: 256, 512. As previously said, the readout layer is either a linear fully connected layer or a non-linear MLP. When required, an encoder fine-tuned on the same split is used. We recall here that early breaking criteria apply also to the training of the decoders. As previously said, we used a fine-tuned ResNet-18 in all the experiments plus additional CNNs to test whether the results would change.

The quality of the generated text is evaluated with the BLEU [30], METEOR [31] and the ROUGE [32] precision and recall. The computation of the BLEU score uses the smoothing function defined by the NIST[12]. Despite the fact that previous works report very high BLEU scores on generation tasks more complex than the one studied in this work, for example in the generation of multiple sentences, we have found that the all the decoders generally perform quite poorly. Generated sentences are generally syntactically well-formed, but quite different from the target ones, thus leading to poor values of the evaluation metrics. In some cases, the generated sentence is semantically equivalent to the target one, but in most of the cases the sentence is simply different. Considering the number of encoder-decoder combinations and the search of the hyperparameter space we do not provide a huge table with the full listing of the results because they are basically all equivalent but for the training set, where some combinations reach high values of the evaluation metrics in the same number of epochs of other models.

3.3.1 Text in *"Findings"*

Training the decoders to generate the first sentence in the "Findings" section using as encoder a CNN pretrained on ImageNet, not finetuned on IU-CHEST, the scores at the end of the training for all the LSTM tested, the BLEU-2 score on the validation and test sets is approximately 0.18, the METEOR score is approximately 0.23, and the ROUGE precision and recall are, respectively, about 0.31 and 0.32. Values do not radically change using a linear or non-linear readout layer, with the only differences being on the training set, where non-linear readouts generally reach much higher values. We might think that non-linear readout layers overfit the training data, but values on the test or validation sets are not different from those obtained with linear readouts. The values reported above were obtained with various CNN: ResNet-18, VGG-19, DenseNet121.

When using a CNN pretrained on ImageNet, fine-tuned on the IU-CHEST dataset, but using only visual features, the evaluation metrics are higher. At the end of the training, the BLEU-2 score on the validation and test sets is approximately 0.21, the METEOR score is approximately 0.28, and the ROUGE precision and recall are, respectively, about 0.42 and 0.43. If we use both the visual and the semantic features extracted by a finetuned ResNet-18, the three scores do not change as if the semantic features did not add any information that the decoder can exploit to generate text.

[12] American National Institute of Standards and Technology, https://www.nist.gov/.

3.3.2 Text in *"Impression"*

Despite being trained with longer sentences, results are much higher when the decoder is trained to generate the first sentence in the *Impression* section. In this case we trained the decoders using only encoders fine-tuned on the IU-CHEST dataset. When using only the visual features, the decoder reaches a BLEU-2 score on the validation and test set of, respectively, about 0.30 and 0.28, a METEOR of approximately 0.29, and a ROUGE precision and recall of, respectively, about 0.36 and 0.35. When we add semantic features, metrics do increase. The BLEU-2 score on the validation and test set is approximately 0.37, the METEOR score is about 0.42 and the ROUGE precision and recall are, respectively, about 0.53 and 0.52.

4 Discussion

The results presented in the previous section seem to suggest that the performance of an encoder-decoder architecture depends more on the data distribution than on the specific base models, layers or decoding strategies used in the full model. First, in none of the experiments the CNN has been able to discriminate between "normal" and "non-normal" images using either IU-CHEST or MIMIC-CXR-JPG. This is quite surprising considering that one of the best published methods [21], using two different LSTMs each dedicated to generate normal or non-normal texts, relies on the CNN discriminating between those two classes. However, it should be noted that in [21] the entire model is trained end-to-end, i.e., the encoder and the decoder are trained simultaneously, using a loss function that combines the errors made on the visual and textual features. Quite likely the contribution to the loss coming from the LSTM was the reason of the improved performance of their model. The investigation of this aspect will be the object of future experiments. The experiments aimed at evaluating how important is the fine-tuning of convolutional layers proved that the IU-CHEST is particularly hard to classify due to the extreme imbalance of the label distribution. Indeed, the final precision and recall are very low despite an aggressive reduction of the number of MeSH categories via the application of the SCUMBLE index. However, if we consider only the accuracy on the test set, our method improves upon it [12] (that reports only training and validation accuracies and that is the only paper that measured the classification performance of the CNN), even if they use a smaller subset of images and labels selected to reduce the class imbalance. The LSTM-based decoder that generates the final text, the sentence in the "Impression" section or the first sentence in the "Findings" section, has been trained with visual features computed by fine-tuned and not fine-tuned encoders, with and without the semantic features, that correspond to MeSH terms assigned by the CNN. Results are in line with previously published approaches, even if experimental settings are not the same due to, for example, different vocabulary sizes. Our simple encoder-decoder architecture reaches a BLEU-2 score of 0.21 and 0.28 on the test set for the "Findings" and "Impression" sections respectively, that are lower than results obtained in [19, 21] on the generation of multiple sentences but using much more complex architectures. However, our results are better than those obtained with architectures of a similar complexity as our model, like [12]. The results confirm that the text varies less in the "Impression" sentence than in the "Findings". The final improvement in performance obtained including the semantic features is quite surprising considering how low the precision and the recall of all the fine-tuned CNNs are.

The results suggest to better investigate the statistical properties of the text to be generated and a deeper experimentation to fully understand the contribution of the text in training the convolutional neural network. It should be noted that results vary significantly across different data splits and, when using different vocabularies and also when using different implementation of the metrics (e.g. BLEU). In order to build a common ground to compare new proposals upon, the construction of resources with a clear and unambiguous data splits, vocabulary, labels and metrics should be considered a priority.

5 Conclusions

This paper presented an extensive set of experimentation of neural language generator models, based on an encoder-decoder architecture, and trained on the IU-CHEST dataset. The final performance is far from the state-of-art models based on more complex encoder-decoder architecture but it is in line with or better than models of similar complexity. However, the goal of this paper was to prepare an initial set of experiments aimed at evaluating how important the different data and neural modules are in an AMIR task. We think that the results might help researchers working on new datasets, possibly in languages with limited textual medical resources, in avoiding useless training steps and in limiting the search space of their experiments. It should be noted that the experiments here discussed required months of GPU time to be accomplished. Indeed, results suggest that a greater attention should be paid to two central aspects when working in this field: obtaining a balanced dataset that can lead to effective image classifiers and studying the word distribution in the texts that are supposed to be learnt and generated. Indeed, an important topic, planned for future studies, is an analysis of the distribution of the sentences in the training and test splits.

Acknowledgment. This work has been partially supported by the EU Horizon 2020 DeepHealth Project (GA No. 825111).

We wish to thank Centro Servizi CNR of the ICT-SAC Department of the National Research Council for the precious computing services and resources they made available. We wish to address a special thanks to Ing. Giorgio Bartoccioni (ICT-SAC) for his technical support.

References

1. Kougia, V., Pavlopoulos, J., Androutsopoulos, I.: A survey on biomedical image captioning. In: Proceedings of the Second Workshop on Shortcomings in Vision and Language, Minneapolis, Minnesota, pp. 26–36, June 2019. https://doi.org/10.18653/v1/W19-1803
2. Bernardi, R., et al.: Automatic description generation from images: a survey of models, datasets, and evaluation measures (extended abstract). In: Proceedings of the Twenty-Sixth International Joint Conference on Artificial Intelligence, Melbourne, Australia, pp. 4970–4974, August 2017. https://doi.org/10.24963/ijcai.2017/704
3. Monshi, M.M.A., Poon, J., Chung, V.: Deep learning in generating radiology reports: a survey. Artif. Intell. Med. **106**, 101878 (2020). https://doi.org/10.1016/j.artmed.2020.101878

4. Vinyals, O., Toshev, A., Bengio, S., Erhan, D.: Show and Tell: A Neural Image Caption Generator. arXiv, 20 April 2015. https://arxiv.org/abs/1411.4555. Accessed 26 July 2022
5. Alfarghaly, O., Khaled, R., Elkorany, A., Helal, M., Fahmy, A.: Automated radiology report generation using conditioned transformers. Inform. Med. Unlocked **24**, 100557 (2021). https://doi.org/10.1016/j.imu.2021.100557
6. Vaswani, A., et al.: Attention is all you need. In: Advances in Neural Information Processing Systems, vol. 30 (2017). https://proceedings.neurips.cc/paper/2017/hash/3f5ee243547dee91fbd053c1c4a845aa-Abstract.html. Accessed 21 July 2022
7. Shin, H.C., Lu, L., Kim, L., Seff, A., Yao, J., Summers, R.M.: Interleaved Text/Image Deep Mining on a Large-Scale Radiology Database for Automated Image Interpretation. arXiv, 04 May 2015. https://arxiv.org/abs/1505.00670. Accessed 24 Oct 2022
8. Wang, X., et al.: Unsupervised Category Discovery via Looped Deep Pseudo-Task Optimization Using a Large Scale Radiology Image Database. arXiv, 25 March 2016. https://arxiv.org/abs/1603.07965. Accessed 24 Oct 2022
9. Irvin, J., et al.: CheXpert: A Large Chest Radiograph Dataset with Uncertainty Labels and Expert Comparison. arXiv, 21 January 2019. https://arxiv.org/abs/1901.07031. Accessed 07 July 2022
10. Peng, Y., Wang, X., Lu, L., Bagheri, M., Summers, R., Lu, Z.: NegBio: a high-performance tool for negation and uncertainty detection in radiology reports. AMIA Summits Transl. Sci. Proc. **2018**, 188–196 (2018)
11. Rubin, J., Sanghavi, D., Zhao, C., Lee, K., Qadir, A., Xu-Wilson, M.: Large Scale Automated Reading of Frontal and Lateral Chest X-Rays using Dual Convolutional Neural Networks. arXiv, 24 April 2018. https://arxiv.org/abs/1804.07839. Accessed 24 Oct 2022
12. Shin, H.-C., Roberts, K., Lu, L., Demner-Fushman, D., Yao, J., Summers, R.M.: Learning to read chest X-rays: recurrent neural cascade model for automated image annotation. In: 2016 IEEE Conference on Computer Vision and Pattern Recognition (CVPR), Las Vegas, NV, USA, pp. 2497–2506, June 2016. https://doi.org/10.1109/CVPR.2016.274
13. Hochreiter, S., Schmidhuber, J.: Long short-term memory. Neural Comput. **9**(8), 1735–1780 (1997). https://doi.org/10.1162/neco.1997.9.8.1735
14. Cho, K., van Merriënboer, B., Bahdanau, D., Bengio, Y.: On the properties of neural machine translation: encoder–decoder approaches. In: Proceedings of SSST-8, Eighth Workshop on Syntax, Semantics and Structure in Statistical Translation, Doha, Qatar, pp. 103–111, October 2014. https://doi.org/10.3115/v1/W14-4012
15. Demner-Fushman, D., et al.: Preparing a collection of radiology examinations for distribution and retrieval. J. Am. Med. Inform. Assoc. **23**(2), 304–310 (2016). https://doi.org/10.1093/jamia/ocv080
16. Luong, T., Pham, H., Manning, C.D.: Effective approaches to attention-based neural machine translation. In: Proceedings of the 2015 Conference on Empirical Methods in Natural Language Processing, Lisbon, Portugal, pp. 1412–1421, September 2015. https://doi.org/10.18653/v1/D15-1166
17. Zhang, Z., Xie, Y., Xing, F., McGough, M., Yang, L.: MDNet: A Semantically and Visually Interpretable Medical Image Diagnosis Network. arXiv, 08 July 2017. https://arxiv.org/abs/1707.02485. Accessed 07 July 2022
18. Krause, J., Johnson, J., Krishna, R., Fei-Fei, L.: A Hierarchical Approach for Generating Descriptive Image Paragraphs. arXiv, 10 April 2017. http://arxiv.org/abs/1611.06607. Accessed 07 July 2022
19. Jing, B., Xie, P., Xing, E.: On the automatic generation of medical imaging reports. In: Proceedings of the 56th Annual Meeting of the Association for Computational Linguistics (Volume 1: Long Papers), Melbourne, Australia, pp. 2577–2586 (2018). https://doi.org/10.18653/v1/p18-1240

20. Simonyan, K., Zisserman, A.: Very Deep Convolutional Networks for Large-Scale Image Recognition. arXiv, 10 April 2015. https://doi.org/10.48550/arXiv.1409.1556
21. Harzig, P., Chen, Y.-Y., Chen, F., Lienhart, R.: Addressing Data Bias Problems for Chest X-ray Image Report Generation. arXiv, 06 August 2019. https://arxiv.org/abs/1908.02123. Accessed 07 July 2022
22. Xue, Y., et al.: Multimodal recurrent model with attention for automated radiology report generation. In: Frangi, A.F., Schnabel, J.A., Davatzikos, C., Alberola-López, C., Fichtinger, G. (eds.) MICCAI 2018. LNCS, vol. 11070, pp. 457–466. Springer, Cham (2018). https://doi.org/10.1007/978-3-030-00928-1_52
23. Johnson, A.E.W., et al.: MIMIC-CXR, a de-identified publicly available database of chest radiographs with free-text reports. Sci. Data **6**(1), 317 (2019). https://doi.org/10.1038/s41597-019-0322-0
24. Johnson, A.E.W., et al.: MIMIC-CXR-JPG, a large publicly available database of labeled chest radiographs. arXiv, 14 November 2019. https://arxiv.org/abs/1901.07042. Accessed 25 July 2022
25. Eickhoff, C., Schwall, I., Muller, H.: Overview of ImageCLEFcaption 2017 – Image Caption Prediction and Concept Detection for Biomedical Images, p. 10 (2017)
26. Prechelt, L.: Early Stopping | but when? p. 15
27. He, K., Zhang, X., Ren, S., Sun, J.: Deep residual learning for image recognition. Presented at the Proceedings of the IEEE Conference on Computer Vision and Pattern Recognition, pp. 770–778 (2016). https://openaccess.thecvf.com/content_cvpr_2016/html/He_Deep_Residual_Learning_CVPR_2016_paper.html. Accessed 20 Aug 2022
28. Charte, F., Rivera, A.J., del Jesus, M.J., Herrera, F.: Dealing with difficult minority labels in imbalanced mutilabel data sets. Neurocomputing **326–327**, 39–53 (2019). https://doi.org/10.1016/j.neucom.2016.08.158
29. Rajpurkar, P., et al.: CheXNet: Radiologist-Level Pneumonia Detection on Chest X-Rays with Deep Learning. arXiv, 25 December 2017. https://arxiv.org/abs/1711.05225. Accessed 23 July 2022
30. Papineni, K., Roukos, S., Ward, T., Zhu, W.-J.: Bleu: a method for automatic evaluation of machine translation. In: Proceedings of the 40th Annual Meeting of the Association for Computational Linguistics, Philadelphia, Pennsylvania, USA, pp. 311–318, July 2002. https://doi.org/10.3115/1073083.1073135
31. The Meteor metric for automatic evaluation of machine translation | SpringerLink. https://link.springer.com/article/https://doi.org/10.1007/s10590-009-9059-4. Accessed 20 Aug 2022
32. Lin, C.-Y.: ROUGE: a package for automatic evaluation of summaries. In: Text Summarization Branches Out, Barcelona, Spain, pp. 74–81, July 2004. https://aclanthology.org/W04-1013. Accessed 08 Aug 2022

HR Analytics: Serbian Perspective

Dragan V. Vukmirović[1](✉), Željko Z. Bolbotinović[2], Tijana Z. Comić[2], and Nebojsa D. Stanojević[1]

[1] Faculty of Organizational Science, 154 Jove Ilića Street, 11000 Belgrade, Serbia
dragan.vukmirovic@fon.bg.ac.rs, ns20205003@student.fon.bg.ac.rs
[2] Tekijanka d.o.o, 50 Dunavska Street, 19320 Kladovo, Serbia
{zeljko,tijana}@tekijanka.com

Abstract. The explosion of new business models is not only the consequence of the accelerated development of information and communication technologies but also of a global crisis. Disruptions such as the COVID-19 pandemic have made managers realize faster that change is necessary and has already taken place. Traditional Human Resources (HR) models were not ready to respond to the challenge of digital transformation. The "new normal" implies a new set of necessary skills, capabilities, and a different working environment. Companies face the challenge to innovate HR functions by developing new career paths and creating more flexible models and relationships with different stakeholders. Most business leaders have become aware that it is impossible to carry out a business transformation without quantifying HR functions. HR analytics is on the business agenda, and it is the fastest-growing area of HR Management today, driven significantly by global crises, and it will be more crucial for organizations' decision-making on how to proceed to the "next normal". The paper defines HR 5.0 as the future of HR, in the framework of Industry 5.0. It attempts to study the state-of-the-art HR in order to explore the current status and perspectives of HR analytics in Serbia and points out the approach that radically transforms the use of human data by policymakers; moreover, it defines one set of conceptual recommendations for successful HR implementation.

Keywords: HR analytics · Industry 5.0 · HR 5.0 · Digital transformation · Artificial intelligence

1 Introduction

Digital transformation is no longer a buzzword. The explosion of new business models that emerged during the COVID-19 pandemic is not an exclusive consequence of global crises but also of disruptions such as COVID-19 and the war in Ukraine, which have made managers realize faster that change is necessary and that it has already taken place. "*New normal*" is mentioned for the first time in the reports of the consulting company McKinsey, as a condition in which we found ourselves globally due to the consequences of the COVID-19 pandemic. Later, this term was replaced by "next normal", precisely to underline the importance of the speed of emergence and obsolescence of changes in the socio-economic environment [1].

N. Filipovic (Ed.): AAI 2022, LNNS 659, pp. 75–89, 2023.
https://doi.org/10.1007/978-3-031-29717-5_5

Agility has become the mantra of digital transformation. Agility starts with knowing the company's talent ecosystem. Back in 2014, Joseph Solis, originally said digital transformation is powered by the human side of the business. "Digital transformation is less about digital and more about people, culture, and new leadership to create business value in a post-industrial market." [2].

However, traditional HR models are not ready to respond to the challenge of digital transformation. This confirms the finding that only one-third of the research companies collected information on the individual skills of their employees in 2021 (a global study of 10,910 participants, of which: 930 were C-Suite executives, 1,736 h Leaders, 8,244 Employees, 16 Geographies and 13 Industries) [3].

The good news is that most respondents have become aware that it is impossible to carry out a transformation without quantifying the skills. Working with data has a long history. For years it has been the subject of discussions in science and professionals like mathematicians, statisticians, data scientists and others (managers, philosophers, sociologists, librarians, e.g.). [4].

The "next normal" means business models shift with a new set of necessary skills and capabilities. Companies face the challenge of making innovations in finding new talent and keeping what they already have by developing new career paths and creating more flexible models and relationships with different stakeholders [5]. That includes forming partnerships, joint ventures, strategic vendor relationships and alliances, and even collaboration with competitors.

HR analytics is on the business agenda, and it is the fastest-growing area of Human Resources Management today, driven significantly by the COVID-19 pandemic, and it will be more crucial when organizations have to decide how to proceed in the "next normal" [6].

The paper discusses HR analytics, the situation and perspectives, with an emphasis on implementation in Serbia, primarily in the field of small and medium enterprises (SMEs).

The initial hypotheses were set from the macro and micro aspects of HR analytics:

H1: Advanced HR analytics (HR 5.0) should be considered within Industry 5.0, as HR 5.0.

H2: HR analytics should be considered in the function of enterprise transformation.

This paper tries to study the macro aspect - the state of HR analytics in the world in order to explore the perspectives of HR analytics in Serbia and defines a set of conceptual recommendations for successful HR analytics implementation.

In the research, we combined a literature review and empirical research about the usage of information and communication technologies in the Republic of Serbia, conducted according to the standardized methodology of EUROSTAT, which enables representativeness at the level of Serbia and comparability between EU countries [7].

2 HR Analytics

In the "new normal" conditions, which the world introduced in the "next normal" where "black swan" [8] is the rule rather than the exception, there is a clear transition to new business models working from home and using platforms and digital skills that most did not even know existed. Also, changes in lifestyle models caused by erasing the boundaries between private and business, home and office, are evident. This has caused a lot of anxiety because of the rapid transition to new skills and changing lives and shopping habits. The evaluation of the HR function indicates that a revolution or at least an evolution of the HR profession is needed. In the context of the future of work, HR must focus on creating an agile workforce. Therefore, it is not surprising that this time practice was faster than theory, looking to solve the problems for which science does not have an established methodology.

Big data, a concept that has long been defined as a practice without theory, announced changes, not only as a new business but also a life philosophy, barely breaking through as a scientific discipline [9]. It seems that a parallel can be drawn with HR analytics.

Findings from a 2017 study by van den Heuvel and Bondarouk suggest that, by 2025, HR analytics will have become an established discipline [10]. Regardless of whether this will be realized within the specified deadline, the impact of HR on business outcomes, as well as its strong influence on strategic and operational decision-making, will most likely be proven before that. The development of HR analytics is determined by integration with data and information communication technology (ICT) infrastructure. Moreover, the HR analytics function may be subsumed under a central analytics function - transcending individual disciplines such as marketing and finance, which is not the case now. At this point, it is certain that analytics in the field of HR does not sufficiently monitor the development of sales analytics and easy marketing [11]. This paper shows this phenomenon in the example of the Republic of Serbia.

Many definitions related to HR analytics can be found in white papers and scientific journals. Different scholars have defined HR analytics in numerous ways. For the purposes of this paper, we will use one of the simplest:

"HR analytics is the application of analytical logic for the HRM function" [12].

2.1 HR X.0

Starting from the cited definition, that HR analytics is the application of analytical logic for the HRM function, and the set hypotheses H2, we defined the taxonomy of HR x.0 based on the Gartner analytical maturity model [13] as a function of analytics.

Descriptive analytics tools may include: descriptive statistics, including graphs and plots, benchmarking tools, KPIs-based methods (scorecards), business intelligence (BI) dashboards, and advanced survey analytics.

Predictive analytics tools may include regression and correlation time series analysis, classification methods (decision trees, logistic regression, discriminant analysis), clustering (K-nearest neighbors K-means), anomaly detection, profiling, association rules, link analysis, causality modelling (Bayesian networks), text analysis and NLP and Attrition modelling. Predictive analytics are based on lagging metrics as outputs of events and

Table 1. Development/taxonomy of HR analytics

HR x.0	Type of analytics	Description	Ind. x.0	Data sources
HR 0.0	No analytics	Past tense	Ind. 1.0	No data
HR 1.0	Descriptive Analytics	Focus on the past: Analysis of raw numbers to figure out WHAT happened	Ind. 2.0	Primary data sources Secondary data sources
HR 2.0	Diagnostic Analytics	Focus on the past: Advanced analysis to figure out WHY this happened	Ind. 3.0	Consulting firms' data, Macro-industry data
HR 3.0	Predictive Analytics	Focus on the present in relation to the future: What WILL happen	Ind. 4.0	Big data Open data
HR 4.0	Prescriptive Analytics	Focus on the future: HOW we can make it happen	Ind.5.0 evolution	Micro (personal) data
HR 5.0	Cognitive Analytics	Future tense: follow Industry 5.0. WHEN it will happen	Ind. 5.0	All data

focusing on what has already happened. They are useful for understanding if intended results have yet been achieved but are often of limited use for predicting future trends [14].

Prescriptive analytics uses almost the same tools like predictive analytics, but its focus has shifted to real-time solutions: HOW we can make it happen. Prescriptive analytics is based on leading metrics as a measure of the input that has a direct influence on an outcome. These metrics show the status of things now and can be course-corrected in real-time to meet business objectives. They include things like employee engagement and satisfaction. Employers could monitor employees' emotional and cognitive states by living nudges, mandating breaks, or even making promotion and termination decisions based on collected data [15].

Cognitive Analytics is based on artificial intelligence (AI) tools such as neural networks, deep learning, blockchain, etc. In the near future, AI will be one of the most influential human resource technology trends. Already now AI supports routine and repetitive activities and monitoring operations, leaving managers with more time to plan, strategize, learn, and work on personal development. AI is also gaining ground in the actual screening process where companies can potentially gain a massive efficiency boost by using AI recruiting tools able to reduce a pool of several hundred applicants down to a shortlist of 5–6 with the highest potential [16].

Beyond hiring (talent acquisition), AI is increasingly used to track and assess workers in their jobs in the following areas: Onboarding, Learning and Training, Cognitive-Supporting Decision-Making, Leadership Coaching, and Automating Administrative Tasks [15].

Table 1 is named Development/taxonomy of HR analytics. This refers to the inheritance of previous concepts. In particular, HR 2.0 Diagnostic Analytics did not exclude the previous HR 1.0 Descriptive Analytics, but improved it. The same applies to the following stages in the development of HR analytics (HR 3.0 vs HR 2.0; HR 4.0 vs HR 3.0; HR 4.0 vs HR 5.0). The same is true for data sources. The use of data from the previous HR x-1 phase is inherited, specifically, in HR 2.0 except in accordance with the resulting firm data and macro industry data sources, the analysis is based on primary and secondary data sources.

HR 5.0 is the HR analytics umbrella and brings together all HR analytics tools from HR 1.0, (Descriptive Analytics) to HR 4.0 (Prescriptive Analytics). Accordingly, since HR 5.0 relies on Industry 5.0 concepts, it will be more closely defined through this concept.

3 HR Analytics Framework

Building an HR analytics framework is the first step to applying and using HR analytics. A prerequisite is to know that useful HR analytics use a process that helps the organization define and align goals and then collect meaningful data that can be acted upon. Many businesses achieve this through the well-known LAMP framework: Logic (articulate the connections between talent and strategic success, and the conditions that predict individual and organizational behavior), Analytics (how data can provide answers), Measures (the numbers and indices calculated from data systems) and Process (communication and knowledge transfer mechanisms through which the information becomes accepted and acted upon by key organization decision-makers [17].

HR 4.0 is a framework to bring about changes in people's strategies of organizations changing the way work is experienced [18] and shaping people's strategies in Industry 4.0 as an initial response to the changing role of organizations in the context of this challenge [19]:

- HR 4.0 explores why Industry 4.0 creates the impetus for transformation in people's strategies and HR practices;
- HR 4.0 outlines what business leaders—including Chief People Officers, Chief Human Resource Officers, CEOs, and other C-suite leaders—can do to respond; and
- HR 4.0 describes how organizations are already responding to the need for change, with examples of emerging roles, technologies, and critical skills for the future of HR.

Industry 4.0 is based on data [20]: how it is collected, analyzed, synthesized, interpreted, and applied to make the right decisions, predict outcomes, and improve performance, has become a competitive factor [21]. HR 4.0 approach integrates HR and data science perspectives. The holy grail of HR 4.0 is prescriptive analytics in the Industry 4.0 framework. "In simple terms, it is the e-action plan based on the data." [22].

Generally, Industry 4.0 has brought a new future of work (new normal) disrupted by technological advancements. Successful organizations, or those who want to be, are ready for a change: 53 percent of companies are introducing next-generation ERP systems, intelligent ERP, which includes artificial intelligence, machine learning, blockchain technology, and Big Data [9] with a focus on user experience, cloud computing and intelligent Internet [23]. At the same time, 33 percent of companies are already in the process, while only 4 percent have implemented such a solution [24].

4 Step Forward to Industry 5.0 and HR 5.0

By analogy, if HR 4.0 follows Industry 4.0, the HR 5.0 follows Industry 5.0. Numerous technologies and applications are expected to help Industry 5.0, such as supercomputing, edge computing, digital twins, collaborative robots, Internet of Things, blockchain, augmented reality, 6G, etc. Industry 5.0 is the next industrial evolution (not revolution), and its objective is to leverage the creativity of human experts in collaboration with efficient, intelligent, and accurate machines, in order to obtain resource-efficient and user-preferred manufacturing solutions compared to Industry 4.0 [25]. These three categories (experts' knowledge, user experience, and customization) are added value compared to Industry 4.0. Consequently, Industry 5.0. Brings people back into focus - human experts, which proves the need for a redefined HR analytics function within HR 5.0.

Survey data from 2021, conducted by software company Sage, with 500 h and business leaders, across the UK, US, Canada, and Australia (respondents were from midsize global companies, in traditionally high-growth, high-skill sectors such as technology, business services, and not-for-profit) found that more than 80% of the C-suite leaders said they would not have been able to operate effectively during the pandemic without HR technology. The same percentage of HR leaders said HR technology enabled them to be more flexible and responsive to changing priorities while helping their businesses become more resilient. Also, they had to scale HR technology to manage and operate effectively during the pandemic as remote working became pervasive across organizations [26].

HR analytics is different from the analysis of sales figures or logistics efficiency [27]. It is about people who work together in complex organizations and even more complex living and working conditions. The development of HR 5.0 is characterized by integration, with data and ICT infrastructure integrated across disciplines and even across organizational boundaries. HR 5.0 is expected to be very well incorporated in a central analytics function - transcending individual disciplines such as marketing, finance, and HRM [10].

On the other hand, Deloitte's research (completed by more than 3,600 executives in 96 countries, the report included responses from more than 1,200 C-suite executives and board members) shows that executives are gradually abandoning the idea of optimization solely through automation, as well as focusing more on integrating people and technology to ensure their complementarity and organizational advancement. This is in line with Industry 5.0. We believe that the reason for returning people to the centre of decision-making lies in this result: Almost three-quarters (72%) of executives identified "the ability of their people to adapt, reskill and assume new roles" as a priority for navigating future disruptions [28].

Wellness management is one of the trends in HR 5.0 function. Its main goal is the management of employees' mental and physical wellness to improve the health and well-being of employees, setting realistic expectations in terms of performance, and to increase the chances of achieving business success. Special attention must be paid to people with disabilities [29]. Long working hours led to 745,000 deaths from stroke and ischemic heart disease in 2016, a 29 percent increase since 2000, according to the latest estimates by the World Health Organization and the International Labor Organization [30].

The growing influence of data processing technologies, especially for their storage (mostly cloud technologies) has required the need to protect sensitive and personal data. Recently adopted laws supporting privacy and data security (such as the EU's GDPR) have also contributed to this becoming one of the most important HR technology trends. It causes the development of HR technology (including data and their analytics) to focus not only on increased security as an additional software function but also to force companies to adopt new procedures. If a company wants to implement HR analytic systems, it must develop solutions that will securely manage data. Blockchain solutions enable data integrity and workplace transparency.

5 CS: Serbia

Considering that at this moment, based on the author's secondary research, there are no representative data on the use of HR for Serbia, we have used these proxy variables from existing research. The first source is the official statistical data from the publication entitled: "Usage of information and communication technologies in the Republic of Serbia, 2021 Households / Individuals Enterprises", published and printed by the Statistical Office of the Republic of Serbia [7].

Target population according to the Classification of Activities has been in use according to the Regulation on the Classification of Activities (this classification is harmonized with NACE rev.2):

- Enterprises with 10 and more employees
- Section C: Manufacturing
- Sections D and E: Electricity, gas, steam and air conditioning supply; Water supply and sewerage
- Section F: Construction
- Section G: Wholesale and retail trade; repair of motor vehicles and motorcycles
- Section H: Transportation and storage
- Section I: Accommodation and food service activities
- Section J: Information and communications
- Sections L and M: Real estate activities; Professional, scientific and technical activities
- Sections N and division 95: Administrative and support service activities; Repair of computers
- Banks and insurance companies

The survey was carried out in March 2021. Phone interviews with 1.573 enterprises (representative stratified sample size) were used. The response rate was 82.6%.

The key findings:

- 100% of enterprises have a fixed broadband Internet connection in Serbia.

  - 96.6% of large enterprises have a website;
  - 94.1% of medium enterprises have a website;
  - 81.6% of small enterprises have a website.

The use of the company's website in the function of HR is not statistically significant. The website of the enterprise misdescribes description services and pricelists (85.3%), links or references to the enterprise's social media profiles (41.3%) and online ordering or reservation or booking of goods/services (19.2%). To assess the level of use of a web business analytics tool, we used the proxy variable: Tracking or status of orders placed (10.1%).

- 22.3% of enterprises use ERP software in Serbia. Small (10–49 employees): 17.6%; Medium (10–49 employees): 35.2% and Large (more than 250 employees).

In 2017, Delloit researchers, as part of a global study involving 10,000 business and HR leaders from 140 countries, found that 72% of companies in Serbia said that Digital HR and employee analytics are very important for their business. At the same time, 17% stated that they largely use it to measure, manage, and improve the strategic role of the HR department, and 8% stated that they have helpful data that they use in HR evaluating the strategy of attracting talent in the direction of cognitive recruitment considering the options of social networks (32%), the use of analytics for forecasting (11%) and the use of games and simulations to attract and evaluate potential candidates (14%) [31].

Serbia 2021: According to the RSO study, the answer to the question: "Does your enterprise use any of the following Artificial Intelligence?" is shown in Table 2. The following usage options are offered: Technologies performing analysis of written language (text mining), Technologies converting spoken language into machine-readable format (speech recognition), Technologies generating written or spoken language (natural language generation), Technologies identifying objects or persons based on images (image recognition, image processing), Machine learning for data analysis, Technologies automating different workflows or assisting in decision-making (AI-based software) and Technologies enabling physical movement of machines via autonomous decisions based on observation or surroundings (autonomous robots, self-driving vehicles, autonomous drones).

Table 2. Used AI technologies in companies in Serbia [7]

Technologies	Small (10–49 employees)	Medium (10–49 employees)	Large (more than 250 employees)	% of enterprises in Serbia
Text mining	-	0.7	1.9	0.2
Speech recognition	0.0	0.6	2.6	0.2
Natural language generation	0.1	0.4	2.3	0.2
Image recognition, image processing	0.1	0.7	0.4	0.2
Machine learning (deep learning)	0.1	1.5	1.4	0.4
AI-based software	0.1	1.9	1.0	0.5
Autonomous robots, self-driving vehicles, autonomous drones	0.0	0.7	0.6	0.2

For human resources management or recruiting, AI technologies (e.g., candidates, pre-selection screening, automation of recruiting based on machine learning, employee profiling or performance analysis based on machine learning, chatbots based on natural language processing for recruiting or supporting human resources management, etc.) are used by 8.0% of enterprises in Serbia (Fig. 1). Note: Only for enterprises that answered "Yes" to the previous question "Does your enterprise use any of the following Artificial Intelligence?".

Enterprises		Size class			Region				Total
		Small (10–49)	Medium (50–249)	Large (250+)	Beogradski	Vojvidne	Šumadije i Zapadne Srbije	Južne i Istočne Srbije	Total
For human resources managment or recruiting									
Activity	Manufacturing	-	-	-	-	-	-	-	-
	Electricity, gas, steam and air conditioning supply; Water supply and sewerage	-	-	-	-	-	-	-	-
	Construction	-	-	-	-	-	-	-	-
	Wholesale and retail trade	-	-	-	-	-	-	-	-
	Transportation and storage	-	-	59.4	50.0	-	100.0	-	27.5
	Accommodation and food service activities	-	100.0	-	100.0	-	-	-	100.0
	Information and communications	-	5.9	-	-	-	-	20.0	2.2
	Real estate activities; Professional, scientific and technical activities	-	-	-	-	-	-	-	-
	Administrative and support service activities; Repair of computers	100.0	-	-	55.8	-	-	-	55.8
Total		14.2	3.8	15.9	14.2	-	4.1	9.8	8.0

Fig. 1. Does your enterprise use any of the following Artificial Intelligence?" for human resources management or recruiting (only for enterprises that answered "Yes" to the question "Does your enterprise use any of the following Artificial Intelligence?") [7]

In the introductory part, we stated that the use of HR analytics lags behind other business functions, especially marketing and sales. The results that confirm this in the Republic of Serbia are presented in Fig. 2.

Furthermore, the results of the research indicate that the citizens of Serbia are not too worried about Privacy and the protection of personal data. Table 3 shows the results of the research based on the question: "Have you carried out any of the following to manage access to your personal data?".

	Enterprises	Size class			Region				Total
		Small (10–49)	Medium (50–249)	Large (250+)	Beogradski	Vojvodine	Šumadije i Zapadne Srbije	Južne i Istočne Srbije	
For marketing or sales									
Activity	Manufacturing	-	9.7	37.5	-	-	38.7	-	11.9
	Electricity, gas, steam and air conditioning supply; Water supply and sewerage	-	-	-	-	-	-	-	-
	Construction	-	-	-	-	-	-	-	-
	Wholesale and retail trade	-	-	-	-	-	-	-	-
	Transportation and storage	-	-	18.7	-	-	100.0	-	8.7
	Accommodation and food service activities	-	100.0	-	100.0	-	-	-	100.0
	Information and communications	26.6	-	-	11.0	-	33.3	-	12.6
	Real estate activities; Professional, scientific and technical activities	-	100.0	-	-	-	100.0	-	100.0
	Administrative and support service activities; Repair of computers	100.0	-	-	55.8	-	-	-	55.8
Total		37.0	10.2	12.5	16.7	-	42.8	-	16.3

Fig. 2. Does your enterprise use Artificial Intelligence software or systems for marketing or sales? (only for enterprises that answered "Yes" to the question "Does your enterprise use any of the following Artificial Intelligence?") [7]

Thus, 35.8% of the online population in Serbia has restricted or refused access to their geographical location in 2021, which is more compared to the previous year, when this percentage was 30.9%.

Other activities on Privacy and protection of personal data are even less represented [7].

Generally, the obtained results indicate that the use of AI technologies in companies in Serbia is at the level of statistical error, and HR analytics is even below this level.

Table 3. Activity did manage access to personal data – Serbia 2021 [22]

Activity to manage access to personal data	%
Read privacy policy statements before providing personal data	32.0
Restricted or refused access to your geographical location	35.8
Restricting access to profile or content on social network sites or shared online storage	32.4
Checked that the website where you provided personal data was secure (e.g *HTTPS* website, safety logo or certificate)	7.0
Asked the administrator of the website or search engine to access data they hold about you to update or delete it	0.6

6 Conclusions and Recommendations

Based on the presented analysis, we can conclude that the next generation of HR, which exists in the Industry 5.0 environment, is actually HR 5.0. At the moment, HR 5.0 is in its infancy in terms of the ways it is being used by HR. It is indicative that the HR function transcends the boundaries of an organization. Megatrends indicate that Industry 5.0 is characterized by the human factor, and this implies the need for serious consideration and socio-political support. H1: "Advanced HR analytics should be viewed in Industry 5.0 as HR 5.0" should be considered correct.

Proving the hypothesis H2: "HR analytics should be considered in the function of the overall digitalization of the company" remains to be proven through future works and, above all, future practice. At this point, HR development is certainly determined by data. Using data science to build business analytics models on IT infrastructure contributes to improving business results and supporting operational and strategic decision-making. It is indicative that the HR function should be integrated into other business functions, with the prospect of taking a central place in the function of business analytics. From the point of view of science and profession, HR goes beyond the framework of an individual discipline, such as HR, economics, marketing, finance, e-business, etc. in the first place, which makes it a serious candidate for multidisciplinary discipline.

6.1 Recommendations for Action

It is unlikely that independent HR analytics (that is if it exists) will function well in a vacuum. That is, the development of HR in companies with a low degree of digitalization, which characterizes the SME sector in Serbia, cannot be expected. Large companies generally have implemented HR functions.

According to the findings of the presented research, HR 5.0 is still at the level of statistical error in Serbia. At this moment, it is more certain that SMEs will skip the HR

3.0 link and make preparations for HR 4.0, to reduce the backlog. We found support for this recommendation (among others) in the conclusions of the 2020 survey that Human Resource Analytics was successfully implemented by only a few organizations due to poor data management, a dearth of analytical skills, and a lack of organizational adaptability [18].

For the development of HR in SMEs in Serbia, several support measures are needed.

Government level:

- Establishment of – legislative-strategic frameworks by the government
- Further support for the digitization process: training, software, the establishment of start-ups
- Creating public-private-academic partnerships
- Open data initiative
- Modernization of curricula

SME level:

- Digital transformation of the company (in progress)
- Implementation and integration of HR 4.0

Serbia has high-quality ICT specialists with competitive wages that are attractive to foreign companies looking to outsource [32]. Serbia's tech sector is expected to continue to grow by more than 20 percent a year. Still, an expansion is hampered by a lack of skilled people—with foreign firms hiring as quickly as the educational system can produce them. Universities are churning out engineers, but it is estimated that the country needs at least 15,000 more to meet the rising demand [33].

Finally, it is necessary to add the need to raise the general Digital literacy of the Serbian population. In the list of European countries for 2020, Serbia ranks 19th with a DESI index value of 46.1 (the EU average is 50.6) [34].

References

1. Bolbotinović, Ž., Radojičić, S., Dragović, N., Vukmirović, D.: Challenges of digital transformation of small and medium enterprises in the conditions of "new normality, YU info 2022, 28th Conference and exhibition, accepted for publication (2022)
2. Solis, B.: Digital Transformation Is More About Humans Than Digital. Brian Solis blog, 2 June 2022. https://www.briansolis.com/2022/06/digital-transformation-is-more-about-humans-than-digital/. Accessed 10 Aug 2022
3. Mercer, Global talent trends. Marsh & McLennan Companies (2021). https://www.marshmclennan.com/insights/publications/2021/october/global-talent-trends-survey-a-look-at-the-tech-industry.html. Accessed 10 Aug 2022
4. Stanojević, N.: Analysis of statistical census data in Serbia. Master's thesis, University of Belgrade, Faculty of Organizational Science, Belgrade (2019)
5. McKinsey, The path to the next normal. Leading with resolve through the coronavirus pandemic. McKinsey & Company (2020). https://www.mckinsey.com/~/media/McKinsey/Featured%20Insights/Navigating%20the%20coronavirus%20crisis%20collected%20works/Path-to-the-next-normal-collection.pdf. Accessed 10 Aug 2022

6. Torre, T., Sarti, D., Antonelli, G.: People analytics and the future of competitiveness: which capabilities HR departments need to succeed in the "next normal." In: Mondal, S.R., Di Virgilio, F., Das, S. (eds.) HR Analytics and Digital HR Practices, pp. 1–24. Springer, Singapore (2022). https://doi.org/10.1007/978-981-16-7099-2_1
7. Republic Statistical Office, Usage of information and communication technologies in the Republic of Serbia, Households/Individuals Enterprises. Statistical Office of the Republic of Serbia (2021). ISSN 1820-9084
8. Taleb, N.N.: The blackswan: Second Edition: the impact of the highly improbable. Penguin Books, Random House U.K (2010)
9. Comić, T.: Improvement of official statistics by applying the concept of Big Data. Doctoral dissertation. The University of Belgrade, Faculty of Organizational Science, Belgrade (2019)
10. Van den Heuvel, S., Bondarouk, T.: The rise (and fall?) of HR analytics: a study into the future application, value, structure, and system support. J. Organ. Eff. People Perform. **4**(2), 127–148 (2017). https://doi.org/10.1108/JOEPP-03-2017-0022
11. Bolbotinović, Ž.: Big data application in the retailing industry. Master's thesis, University of Belgrade, Faculty of Organizational Science, Belgrade (2018)
12. Bhattacharyya, D.K., Analytics, H.R.: Understanding Theories and Applications. SAGE Publications, New Delhi (2017)
13. Gartner, What Is Data and Analytics? https://www.gartner.com/en/topics/data-and-analytics
14. Sage, Research report HR at the moment: Impact through insights. How HR can create more value and increase business impact. The Sage Group (2021). https://www.sage.com/en-us/sage-business-cloud/people/resources/research-analyst-reports/hr-in-the-moment-impact-through-insights/
15. Eisenstadt, L.F.: #MeTooBots and the AI WorkplaceUniversity of Pennsylvania Journal of Business Law, Forthcoming (2021). https://ssrn.com/abstract=3921186
16. Muzyka, B.: Top 15 Human Resource Technology (HR Tech) Trends in 2022, TechMagic (2021). https://www.techmagic.co/blog/top-10-human-resource-technology-trends. Accessed 10 Aug 2022
17. Lawler, E., Boudreau, J.W.: Chapter 9. HR Analytics and Metrics Effectiveness". Achieving Excellence in Human Resources Management: An Assessment of Human Resource Functions, pp. 75–82. Stanford University Press, Redwood City (2009). https://doi.org/10.1515/9780804771238-012
18. Gaur, B.: HR4.0: an analytics framework to redefine employee engagement in the fourth industrial revolution. In: 2020 11th International Conference on Computing, Communication and Networking Technologies (ICCCNT), pp. 1–6 (2020). https://doi.org/10.1109/ICCCNT49239.2020.9225456
19. World Economic Forum, HR4.0: Shaping People Strategies in the Fourth Industrial Revolution. White Paper. World Economic Forum (2019). https://www.weforum.org/reports/hr4-0-shaping-people-strategies-in-the-fourth-industrial-revolution
20. Vukmirović, D.: Introductory Data Science for Managers and Business Leaders –Workshop. Conference: SYMORG (2020). https://doi.org/10.13140/RG.2.2.27339.41764
21. Nagy, J., Oláh, J., Erdei, E., Máté, D., Popp, J.: The role and impact of industry 4.0 and the internet of things on the business strategy of the value chain—the case of Hungary. Sustainability **10**, 3491 (2018). https://doi.org/10.3390/su10103491
22. Roper, J.: The possibilities for prescriptive analytics in HR. October 2019 HR Technology Supplement. HR Magazine (2019). https://www.hrmagazine.co.uk/content/other/the-possibilities-for-prescriptive-analytics-in-hr#:~:text=%E2%80%9CThe%20final%20stage%2C%20and%20arguably,HR%20analytics%20awareness%20is%20low. Accessed 12 Aug 2022
23. Vukmirović, D., Comić, T., Bolbotinović, Ž., Dabetić, D.J., Jovanović Milenković, M.: MIP - Prototype of the model of intelligent enterprise. In: SYM-OP-IS 2019, XLVI Symposium on Operational Research, Kladovo, Serbia (2019)

24. Financesonline, 11 Top ERP Software Trends for 2022/2023: New Predictions & What Lies Beyond? Finance online, Review for business (2021). https://financesonline.com/erp-software-trends/. Accessed 10 Aug 2022
25. Maddikunta, P.K.R., et al.: Industry 5.0: a survey on enabling technologies and potential applications. J. Ind. Inf. Integration **26** (2022). https://doi.org/10.1016/j.jii.2021.100257. ISSN 2452-414X
26. Sage, Research report HR in the moment: Impact through insights. How HR can create more value and increase business impact. The Sage Group (2021). https://www.sage.com/en-us/sage-business-cloud/people/resources/research-analyst-reports/hr-in-the-moment-impact-through-insights/. Accessed 12 Aug 2022
27. De Dood, M.: HR 4.0: The sense and nonsense of HR analytics. Centric (2021). https://insights.centric.eu/en/themes/hr/the-sense-and-nonsense-of-hr-analytics/. Accessed 15 Aug 2022
28. Deloitte's 2021 Global and Middle East Human Capital Trends report, "The social enterprise in a world disrupted" (2021). https://www2.deloitte.com/xe/en/pages/about-deloitte/articles/61-executives-double-pre-covid19-levels-focused-transforming-work.html. Accessed 10 Aug 2022
29. Comić, T.: Challenges and experiences in reporting on SDGs on people with disabilities - Serbian Perspective, International Seminar on Data for Sustainable Development Goals: Data Disaggregation, United Nations Statistics Division (UNSD), Statistics Korea (KOSTAT), Seoul, Republic of Korea, 3–4 November 2016 (2016). ESA/STAT/AC.324/7
30. Pega, F., et al.: Global, regional, and national burdens of ischemic heart disease and stroke attributable to exposure to long working hours for 194 countries, 2000–2016: A systematic analysis from the WHO/ILO Joint Estimates of the Work-related Burden of Disease and Injury. Environ. Int. **154** (2021). https://doi.org/10.1016/j.envint.2021.106595. ISSN 0160-4120
31. Deloitte istraživanje globalnih trendova u oblasti ljudskih resursa. Pisanje novih pravila za digitalnu eru. Delloite University Press (2017). https://www2.deloitte.com/content/dam/Deloitte/rs/Documents/human-capital/global-hc-trends-deloitte-serbia-2017.pdf. Accessed 9 Aug 2022
32. Vukmirović, D.: Data Science Education - Serbian perspective. Data Science Conference/4.0, 18–19 September 2018, Belgrade, Serbia (2018)
33. Serbia - Country Commercial Guide. Information and Communications Technology Market. International Trade Administration (2022). https://www.trade.gov/country-commercial-guides/serbia-information-and-communications-technology-market. Accessed 10 Aug 2022
34. Radojičić, S., Bolbotinović, Ž., Dragović, N., Stanojević, N., Vukmirović, D.: Digitally literate population as the basis of social and economic development - the example of the Republic of Serbia. YU info 2022, 28th Conference and exhibition, accepted for publication, 2022 (2022)

Ontology-Based Analysis of Job Offers for Medical Practitioners in Poland

Paweł Lula[1](✉) and Marcela Zembura[2]

[1] Cracow University of Economics, 27 Rakowicka Street, 32-510 Kraków, Poland
pawel.lula@uek.krakow.pl

[2] Medical University of Silesia, 15 Poniatowskiego Street, 40-055 Katowice, Poland

Abstract. In the paper, the results of the analysis of the labor market for medical specialists in Poland were presented. The research was focused on the demand side of the market. The data describing required competencies were retrieved from job offers published on the Internet. Detailed information about requirements were identified with the use of ontology-based tools. Next, graph models were built to show the importance of individual competencies and sets of them. Finally, the analysis of relationships between required competencies and other attributes of the labor market was conducted. All software tools used during the research were prepared in the *R* language.

Keywords: Exploratory analysis of text documents · Ontology-based analysis of text documents · Labor market models · Network analysis · Bipartite graph models

1 Introduction

According to the newest Eurostat data from 2020, which presents an overview of European Union (EU) statistics on physicians, in 2020, 1.75 million physicians were practicing in the EU [1]. In Poland, the number of practicing physicians per 100 000 inhabitants was in fact the lowest in the EU (237.8 in 2017).

Regarding the EU population's ageing, the demand for healthcare is going to extend substantially, given the fact that over a third of all doctors in the EU are aged 55 years and above [1]. The covid-19 pandemic was the greatest challenge faced by the healthcare system, it highlighted how important healthcare workers are to the system [2]. Migration is also an important factor, as it may lead to the rectification of labor market imbalances between countries as well as to exacerbation of imbalances [1]. Nowadays, a substantial inflow of healthcare practitioners from Ukraine and Belarus is observed in Poland. Therefore, it is essential to assess the demand for medical doctors with respect to specialization, regions, health care institutions, and labor market dynamics.

The Global Strategy on Human Resources for Health—Workforce 2030 was introduced by World Health Organization in 2016 [3]. It aims to "improve health, social and economic development outcomes by ensuring universal availability, accessibility,

N. Filipovic (Ed.): AAI 2022, LNNS 659, pp. 90–102, 2023.
https://doi.org/10.1007/978-3-031-29717-5_6

acceptability, coverage, and quality of the health workforce through adequate investments to strengthen health systems, and the implementation of effective policies at national, regional and global levels" [3]. The framework by Sousa et al. [4], which is a key part of WHO Strategy, provides a basis for Health labor market analysis (HLMA). The framework consists of two parts, one is related to the education sector, and one regards the health labor market [2]. It also contains policies for addressing issues in the healthcare labor market. HLMA was used in many countries in order to assess the dynamics of the health labor market and investigate policies implemented to address issues in the healthcare system [5, 6].

Few studies assessed job offers for health practitioners. In the study by Bagat et al. physician labor market in Croatia with respect to internship and the implementation of the State Program for Intern Employment Stimulation was analyzed [7]. In the study by Gaidarov et al., vacancies and job offers for doctors from state medical organizations of the Irkutsk region were assessed [8].

Two studies aimed to analyze health care labor market in the UK, according to Rimmer, nearly half of advertised consultant vacancies across the UK were unfilled in 2020, and action should be taken in order to avoid medical staff shortages in the NHS [9]. On the other hand, the study by Dosani et al. examined non-standard grade posts in the NHS [10].

To our knowledge this is the first study analyzing job offers for medical practitioners in Poland.

The aim of this study was to build a system for automatically retrieving and analyzing of job offers for medical practitioners, to analyze the demand for medical doctors in Poland with respect to specializations, regions, and health care institutions, and to analyze the specificity of different aspects of a demand side of the Polish labor market.

2 Research Methodology

The research process was composed of the following steps:

- Job offers retrieving,
- Semantic annotations,
- Building a bipartite graph model,
- Analysis of nodes with respect to their strength and specificity,
- Identification of communities

2.1 Job Offers Retrieving

Job offers published on the Internet were used as the main data source. They were retrieved by web scraping technique using *Docker*[1] container and *RSelenium*[2] package and saved in plain text format.

---

[1] https://www.docker.com/.

[2] https://CRAN.R-project.org/package=RSelenium.

2.2 Semantic Annotation

The ontology-based approach was used for the semantic annotation process. For the analysis process, three ontologies were prepared:

- *Specializations* – containing a list of medical specializations,
- *Locations* – with a list of towns and cities in Poland,
- *Institutions* – including names of public and private medical health centers and institutions in Poland.

All ontologies were built from scratch and had a form of lists of concept definitions defined as:

```
Concept name:
```

- pattern_1
- ...
- pattern_n

where patterns were defined as regular expressions defining a set of strings corresponding to a given concept. Ontologies were stored in *yaml* format and the *quanteda*[3] package was used to identify concepts included in the corpus of job offers. The concept identification process allowed to build three document-feature matrices informing about medical specializations, locations and medical institutions mentioned in offers.

2.3 Bipartite Graph Models

A bipartite graph, also called a bigraph, is a set of graph vertices decomposed into two disjoint sets such that no two graph vertices within the same set are adjacent [11].

A bipartite graph model was used for presenting relationships existing between concepts defined in two selected ontologies used for analysis of a given corpus of documents. Bipartite graphs are widely used in social network analysis [12] and ecological models [13, 14]. A survey on bipartite models in biology and medicine is delivered in [15].

Let us make an assumption that:

$$U = \{u_1, u_2, \ldots, u_n\} \tag{1}$$

is a set of concepts belonging to the first ontology, and:

$$V = \{v_1, v_2, \ldots, v_m\} \tag{2}$$

defines concepts from the second ontology.

The graph representing relationships between concepts defined in U and V is presented in Fig. 1.

[3] https://CRAN.R-project.org/package=quanteda.

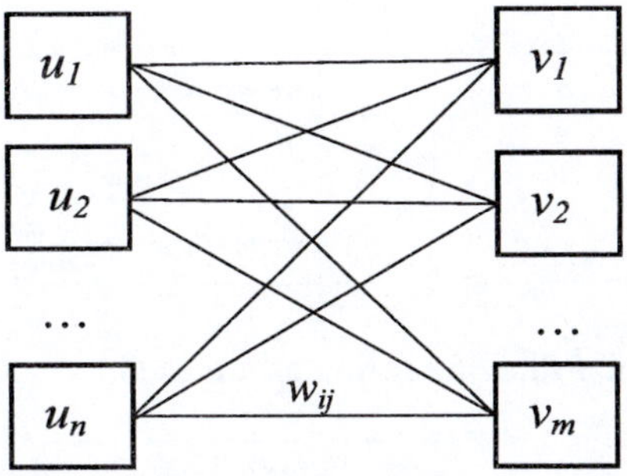

Fig. 1. Bipartite graph showing relationships between two sets of concepts (source: own elaboration)

Weights w_{ij} inform how many times concepts u_i and v_j occurred in the same document.

During further steps, two bipartite models were built. The first is for showing relationships between medical specializations and workplace location. And the second – represents links between medical specializations and health care institutions which wanted to hire medical practitioners.

The analysis of bipartite models included calculation and interpretation of statistics describing the relationship between set of concepts and identification of strongly connected components (communities).

2.3.1 Bipartite Graph Statistics

Let's assume that $U = \{u_1, u_2, \ldots, u_N\}$ and $V = \{v_1, v_2, \ldots, v_M\}$ are two sets of nodes occurring in a bipartite graph defined by an interaction matrix:

$$W = \begin{bmatrix} w_{11} & \ldots & w_{1M} \\ \ldots & \ldots & \ldots \\ w_{N1} & \ldots & w_{NM} \end{bmatrix} \tag{3}$$

where element w_{ij} informs about the number of interactions between nodes u_i and v_j.

The role of every node in a bipartite graph is characterized by some simple statistics:

- connectance – informs about the density of the network and is calculated as the number of links divided by the number of possible links (product of N and M value),
- node degree – defines the number of elements from the other set connected with a given node,
- node strength – the normalized number of nodes from the opposite set which are connected to a given node [16] or the sum of interactions with elements belonging to the opposite set [17]. In the case of Bascompte's method, the following calculations should be performed:

for v_j nodes	for u_j nodes
$d_{ij}^{H} = \frac{w_{ij}}{\sum_{k=1}^{M} w_{ik}}$	$d_{ij}^{L} = \frac{w_{ij}}{\sum_{k=1}^{N} w_{kj}}$
$strength(v_j) = \sum_{p=1}^{N} d_{pj}^{H}$	$strength(u_i) = \sum_{p=1}^{M} d_{ip}^{L}$

for Barrat's method the following formulas are used:

for v_j nodes	for u_j nodes
$strength(v_j) = \frac{\sum_{i=1}^{N} w_{ij}}{\sum_{k=1}^{N}\sum_{p=1}^{M} w_{kp}}$	$strength(u_i) = \frac{\sum_{j=1}^{M} w_{ij}}{\sum_{k=1}^{N}\sum_{p=1}^{M} w_{kp}}$

- node specificity – variations of interactions represented by w_{ij} normalized to the [0; 1] range, where 0 indicated lack of variability (lack of specificity) and values close to 1 inform about very high variability (and high specificity),
- niche overlap index is calculated for two nodes belonging to the same set and informs about the similarity of sets of nodes from the opposite set which are connected to each of the considered nodes [13]. Value 0 means that two opposite sets have no common nodes whereas value 1 indicates that the two given nodes have the same partners from the opposite set,
- specialization index H_2' is a measure which informs about the variability of all w_{ij} values of a given bipartite graph. The H_2' index can be treated as an entropy value normalized to the [0; 1] range, where 0 represents low and 1 informs about the high specificity of a network [13].

2.3.2 Community Detection

Community in a graph can be defined as a set of vertices which are densely connected to each other and simultaneously are very loosely connected to vertices in the other communities. Many techniques used by community detection algorithms are based on modularity measures which compare the number of edges in the existing group and in the null model which forms a graph with the same set of vertices connected by edges distributed at random. The idea of modularity for bipartite graph was introduced in [18] and can be measured by the Q index defined as [19]:

$$Q = \frac{1}{L}\sum_{i=1}^{N}\sum_{j=1}^{M}\left(A_{ij} - P_{ij}\right)\delta\left(u_i, v_j\right) \tag{4}$$

where L is a number of edges in the graph, A_{ij} is an element of the adjacency matrix, P_{ij} is a probability that an edge exists between elements u_i and v_j, and $\delta\left(u_i, v_j\right)$ is the Kronecker delta function which is equal to one if nodes u_i and v_j belong to the same module or zero otherwise.

For weighted bipartite graph, the weighted bipartite modularity can be defined [19]:

$$Q_W = \frac{1}{L}\sum_{i=1}^{N}\sum_{j=1}^{M}\left(w_{ij} - E_{ij}\right)\delta\left(u_i, v_j\right) \tag{5}$$

where w_{ij} is a weight assigned to an edge between elements u_i and v_j, and E_{ij} is an expected value corresponding to the w_{ij} in the null model (null model is a graph with the same number of nodes as an original one in which nodes have the same degree but edges were rewired at random).

In [19] the most popular algorithms for community identification by maximizing modularity measure were proposed. These two methods are implemented in the *bipartite* package for R language.

3 Analysis of Job Offers for Medical Practitioners in Poland

3.1 General Description of the Research Process

During the research process about 1676 job offers retrieved at the end of 2021 from the most popular web portals publishing offers for medical specialists: *pracuj.pl*, *mp.pl* and *konsylium24.pl* were used. All of them were saved in a plain text format.

After preparing a corpus with offers, three ontologies were defined: *Specializations*, *Locations* and *Institutions* (Fig. 2).

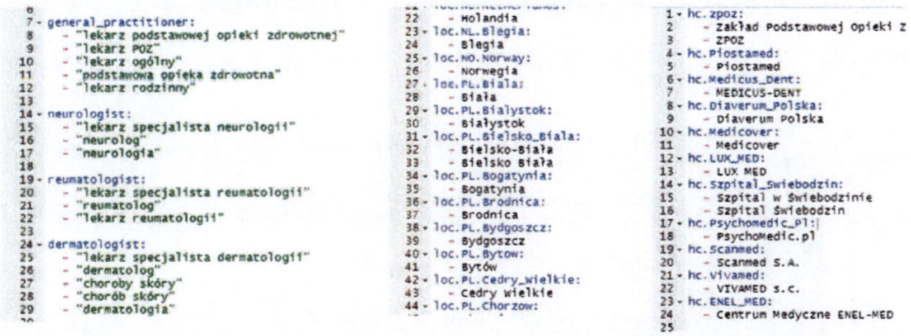

Fig. 2. Ontologies in yaml format created for job offers analysis (source: own elaboration)

In the initial phase of the study, an analysis of the need for medical professionals by specialty was conducted (Fig. 3).

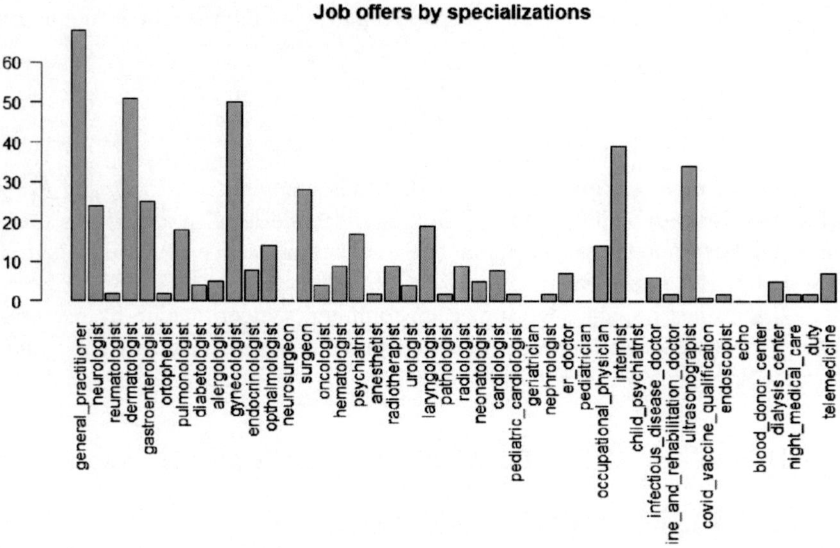

Fig. 3. The distribution of job offers over medical specializations (source: own elaboration)

Results show that the highest demand is observed for general practitioners, internists, dermatologists, gynecologists, ultra-sonographers and surgeons.

3.2 Analysis of the Specialization-Locations Bipartite Model

As expected, the largest number of job offers were in large metropolitan areas (Fig. 4).

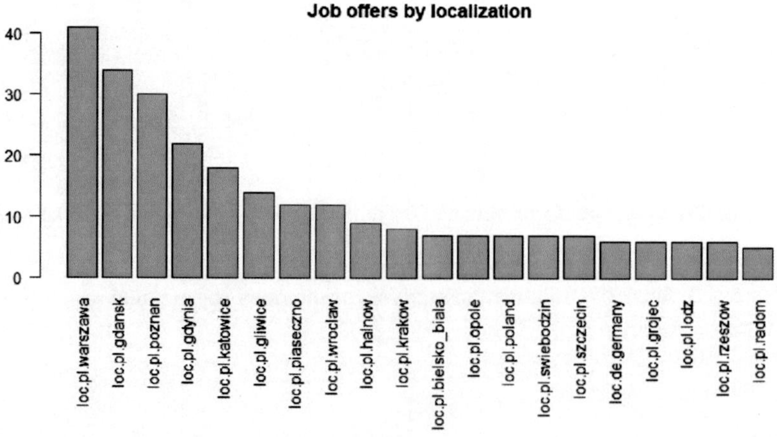

Fig. 4. Distribution of job offers over locations for 20 the most common locations (source: own elaboration)

To study links between required specializations of medical doctors and location a bipartite graph model was build (Fig. 5).

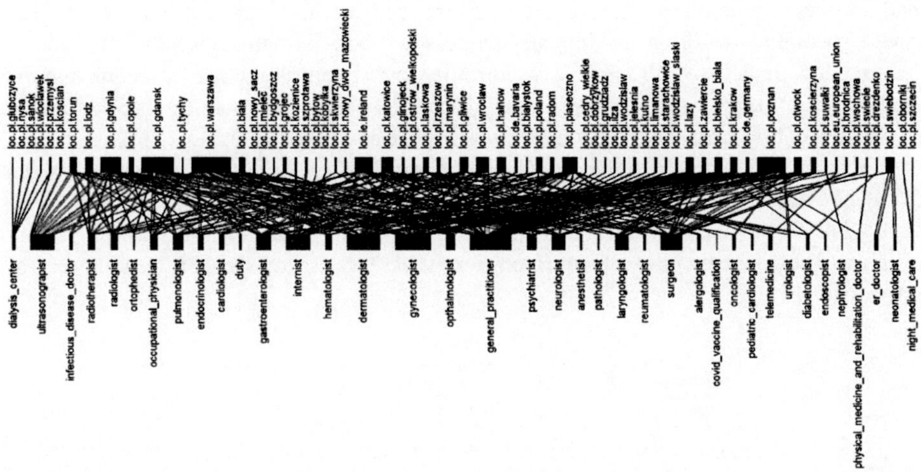

Fig. 5. Bipartite graph model showing relationships between required medical specializations and locations (source: own elaborations)

The main statistics characterizing the constructed model are shown in Table 1.

Table 1. Main network statistics for specialization-location bipartite graph model

Statistics	Value
Number of nodes (locations)	66.0000
Number of nodes (specializations)	38.0000
Connectance	0.1023
Mean number of links (locations)	9.3282
Mean number of links (specializations)	14.9195
Number of compartments	2.0000
Mean number of shared partners (locations)	0.7832
Mean number of shared partners (specializations)	1.2831
Niche overlap (locations)	0.1600
Niche overlap (specializations)	0.1471
$H_2^{'}$ index	0.2948

Source: own elaboration

The density of the network is relatively small (connectance index is equal to 0.1023). The value of the $H_2^{'}$ index suggest low specificity level of concepts. Mean number of shared partners and niche overlap index show that similarity between partners' sets is negligible.

The highest values of node strength statistics (calculated according to Bascompte's method) were obtained for general practitioners (14.92), internists (6.89), gynecologists (5.62), specialists working for dialysis centers (4.5) and dermatologists (3.75).

All specialists listed above possessed rather low specificity index (it means that the need for their employment exists in many locations). Only for practitioners demanded by dialysis centers the specificity index was higher and equal to 0.43.

Localizations with the highest value of strength index were the following: Ireland (3.64), Warsaw (3.42), Gdańsk (3.05), Poznań (2.73), Świebodzin (2.18) and Piaseczno (2.13). For all the above localizations the specificity index was relatively low.

Next, the identification of communities, with the use of Beckett's method, was performed (Fig. 6).

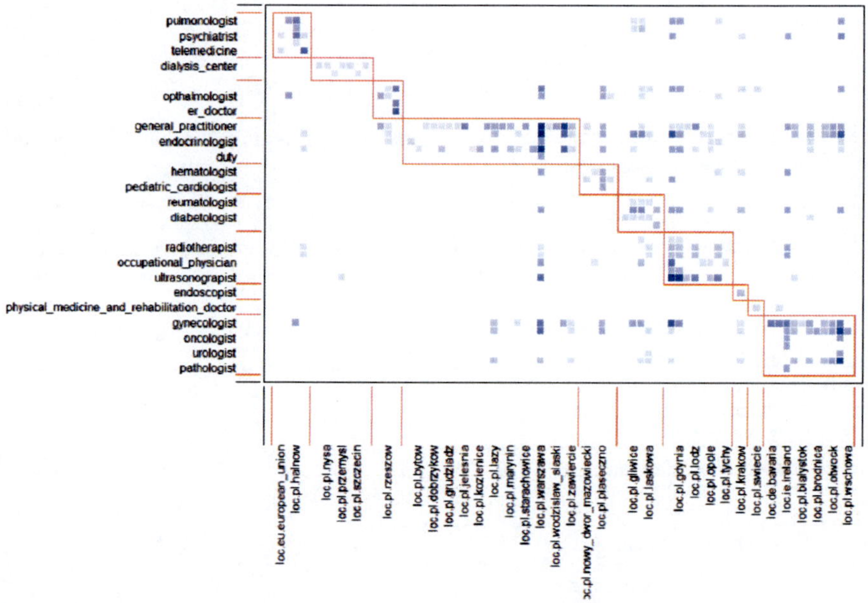

Fig. 6. Communities identified in the specialization-location bipartite model (Source: own elaboration)

3.3 Analysis of the Specialization-Institutions Bipartite Model

The second bipartite model was built to show relationships between medical specializations of doctors and institutions that wanted to hire them (Fig. 7). It is worth noting that only in a relatively small group of offers, the name of the company wishing to hire doctors was given.

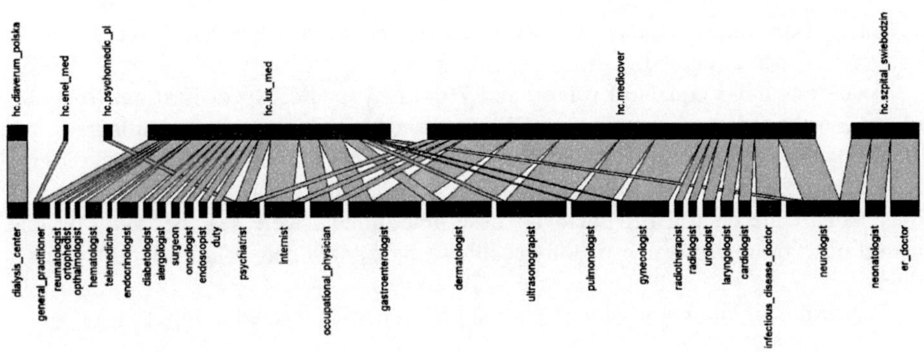

Fig. 7. Bipartite graph model showing relationships between required medical specializations and medical institutions (source: own elaborations)

Table 2 shows some main statistics describing the constructed model.

Table 2. Main network statistics for specialization-institution bipartite graph model

Statistics	Value
Number of nodes (institutions)	6.0000
Number of nodes (specializations)	31.0000
Connectance	0.2312
Mean number of links (institutions)	15.6066
Mean number of links (specializations)	1.8087
Number of compartments	2.0000
Mean number of shared partners (institutions)	0.9333
Mean number of shared partners (specializations)	0.7290
Niche overlap (institutions)	0.0661
Niche overlap (specializations)	0.4336
H_2' index	0.5930

Source: own elaboration

Measures presented in Table 2, especially H_2' index equal to 0.593, indicate a higher level of specificity than in the specialization-location model. The niche overlap index for institutions suggests a high degree of their diversity.

The strength of nodes representing health care institutions was as follows: *Lux-Med*: 15.55, *Medicover*: 11.67, *Hospital in Świebodzin*: 2.28, *Diaverum-Polska*: 1.00, *Psychomedic.pl*: 0.25 and *Enel-Med*: 0.25.

For three health care institutions (*Psychomedic.pl*, *Enel-Med* and *Diaverum-Polska*) the specificity index was equal to one. It means that these institutions intended to hire medical practitioners of only one specialization. For two huge private medical health care institutions the specificity index was low (for *Medicover* was equal to 0.2596 and for *Lux-Med* was equal to 0.1787).

Specificity index equal to 1 was obtained for many medical specializations (rheumatologist, orthopedist, diabetologist, allergist, endocrinologist, ophthalmologist, surgeon, oncologist, hematologist, radiotherapist, urologist, laryngologist, hematologist, neonatologist, cardiologist, infection disease doctor, endoscopist, dialysis center doctor, telemedicine doctor, ER doctor). These practitioners were searched only by one institution. The lowest value of the specificity index was calculated for psychiatrists (0.46).

The results of community identification process are presented in Fig. 8.

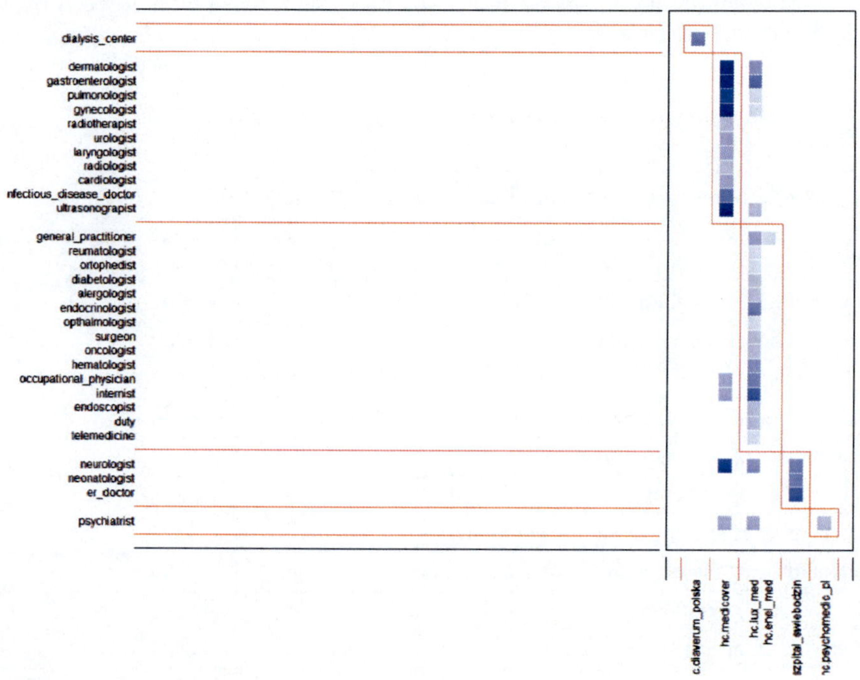

Fig. 8. Communities identified in the specialization-institution bipartite model (Source: own elaboration)

The results allow to identify clusters composed of specializations and institutions and confirm previous findings: two institutions (*Psychomedic.pl* and *Diaverum-Polska*) have a homogenous profile and the profile of two of them (*Medicover* and *Lux-Med*) is heterogenous but vastly different.

4 Conclusions

It seems worth to define general concluding remarks on three issues:

1. *the form of offers for employment of medical practitioners*
   Job offers for medical practitioners are very often very short and condensed (sometimes only with specialization name), with no information about required competencies and proposed salary. Therefore, the analysis of expected competencies is not possible.
2. *tools used for analysis*
   all tools (*Docker* container and *RSelenium* package used for offer retrieving) and *R* language and *quanteda* package for semantic annotation and for performing all analysis) deserve a positive evaluation
3. *models for analysis*
   ontology-based exploratory models for analysis job offer content allowed to identify crucial pieces of information (related to specialization, location, and health care institutions) and bipartite graph models gave insights into main labor market attributes and relationships existing between them.

The research process allowed to:

- identify the most demanded medical specializations in Poland,
- evaluate the needs of specific locations due to the doctors they are looking for,
- access the level of specificity of medical specializations, locations, and health care institutions,
- find communities specialization-location and specialization-institution networks.

Acknowledgements. This project has been financed by the Minister of Education and Science within the "Regional Initiative of Excellence" Programme for 2019–2022. Project no.: 021/RID/2018/19. Total financing: 11 897 131,40 PLN.

References

1. Healthcare personnel statistics - physicians (2022). https://ec.europa.eu/eurostat/statistics-explained/index.php?title=Healthcare_personnel_statistics_-_physicians&oldid=551980. Accessed 15 Aug 2022
2. World Health Organization, Global strategy on human resources for health: Workforce 2030 (2021). https://apps.who.int/iris/bitstream/handle/10665/250368/9789241511131-eng.pdf. Accessed 15 Aug 2022
3. Sousa, A., Scheffler, R.M., Nyoni, J., Boerma, T.: A comprehensive health labour market framework for universal health coverage. Bull. World Health Organ. **91**(11), 892–894 (2013). https://doi.org/10.2471/BLT.13.118927
4. Sousa, A., et al.: Health labour market policies in support of universal health coverage: a comprehensive analysis in four African countries. Hum. Resour. Health **12**(1), 55 (2014). https://doi.org/10.1186/1478-4491-12-55

5. World Health Organization; Country Office for Sri Lanka, Health labour market analysis: Sri Lanka. World Health Organization. Regional Office for South-East Asia (2018). https://apps.who.int/iris/handle/10665/324911. Accessed 17 Aug 2022
6. Bagat, K., Sekelj Kauzlarić, K.: Physician labor market in Croatia. Croatian Med. J. **47**(3), 376–384 (2006)
7. Gaidarov, G.M., Makarov, S.V., Yu Alekseeva, N., Maevskaya, I.V.: Analysis of vacancies and job offers for doctors in state medical organizations of the Irkutsk Region. Acta Biomedica Scientifica (East Siberian Biomed. J.) **3**(4), 101–108 (2018). https://doi.org/10.29413/ABS.2018-3.4.14
8. Rimmer, A.: Physician vacancies are at highest level in almost a decade, colleges find. BMJ n2810 (2021). https://doi.org/10.1136/bmj.n2810
9. Dosani, S., Schroter, S., MacDonald, R., Connor, J.: Recruitment of doctors to non-standard grades in the NHS: Analysis of job advertisements and survey of advertisers. Br. Med. J. **327**(7421) (2003). https://doi.org/10.1136/bmj.327.7421.961
10. Weisstein, E.W.: Bipartite Graph. From MathWorld–A Wolfram Web Resource (2022). https://mathworld.wolfram.com/BipartiteGraph.html. Accessed 16 Aug 2022
11. Lattanzi, S., Sivakumar, D.: Affiliation networks. In: Proceedings of the Annual ACM Symposium on Theory of Computing (2009). https://doi.org/10.1145/1536414.1536474
12. Dormann, C.F., Fründ, J., Blüthgen, N., Gruber, B.: Indices, graphs and null models: analyzing bipartite ecological networks. Open Ecol. J. **2**, 7–24 (2009). https://doi.org/10.2174/1874213000902010007
13. Saavedra, S., Reed-Tsochas, F., Uzzi, B.: A simple model of bipartite cooperation for ecological and organizational networks. Nature **457**(7228) (2009). https://doi.org/10.1038/nature07532
14. Pavlopoulos, G.A., Kontou, P.I., Pavlopoulou, A., Bouyioukos, C., Markou, E., Bagos, P.G.: Bipartite graphs in systems biology and medicine: a survey of methods and applications. GigaScience **7**(4) (2018). https://doi.org/10.1093/gigascience/giy014
15. Bascompte, J., Jordano, P., Olesen, J.M.: Asymmetric coevolutionary networks facilitate biodiversity maintenance. Science **312**(5772) (2006). https://doi.org/10.1126/science.1123412
16. Barrat, A., Barthélemy, M., Pastor-Satorras, R., Vespignani, A.: The architecture of complex weighted networks. Proc. Natl. Acad. Sci. USA **101**(11) (2004). https://doi.org/10.1073/pnas.0400087101
17. Barber, M.J.: Modularity and community detection in bipartite networks. Phys. Rev. E Stat. Nonlinear Soft Matter Phys. **76**(6) (2007). https://doi.org/10.1103/PhysRevE.76.066102
18. Beckett, S.J.: Improved community detection in weighted bipartite networks. R. Soc. Open Sci. **3**(1) (2016). https://doi.org/10.1098/rsos.140536
19. Dormann, C.F., Strauss, R.: A method for detecting modules in quantitative bipartite networks. Methods Ecol. Evol. **5**(1) (2014). https://doi.org/10.1111/2041-210X.12139

Synergizing Four Different Computing Paradigms for Machine Learning and Big Data Analytics

Veljko Milutinović(✉) and Jakob Salom

IPSI Ltd, Belgrade, Serbia
vm@etf.rs

Abstract. This article presents and analyses four computing paradigms that are present in today's IT programming world - Control Flow, Data Flow, Diffusion Flow, and Energy Flow. It compares their main properties, points out what purposes each has, and describes what are their advantages and disadvantages. In the third part of this article, the Authors speculate on the possible architecture of a supercomputer on a chip and in the fourth part, they suggest the optimal distribution of resources for a specified set of Civil engineering applications.

Keywords: Data Flow · Control Flow · Diffusion Flow · Energy Flow · Maxeler DFE · WSN · BioMolecular computing · QuantumMechanical computing · Computing paradigms

1 Introduction

The computing scene nowadays includes four different computing paradigms and related programming models. Some of the paradigms/models are on the rise, and others are on stable grounds. These 4 paradigms are Control Flow (MultiCores like with Intel and ManyCores like with NVidia), Data Flow (Fixed ASIC-based like with Google TPU and flexible FPA-based like initially with Maxeler DFE and lately with many others), Diffusion Flow (like with IoT, Internet of Things, and WSNs, Wireless Sensor Networks), and Energy Flow (like with BioMolecular and QuantumMechanical computing). For more details, see the references [1–9].

Each one of the paradigms has different characteristics, as far as (a) Speed, (b) Power, (c) Size, (d) Potential for high precision, and Ease of Programming. Each one of the paradigms is best suited for a given set of problems. Some paradigms are better suited to serve as hosts, others as accelerators. However, they all are best used through a proper type of synergy.

This article first presents the pros and cons of each one and then discusses possible ways for them to synergize.

N. Filipovic (Ed.): AAI 2022, LNNS 659, pp. 103–108, 2023.
https://doi.org/10.1007/978-3-031-29717-5_7

2 Comparison of Four Computing Paradigms

The Control Flow paradigm is based on the research of von Neumann. It is well suited for transactional computing and could be effectively used as a host in hybrid machines that combine all the paradigms mentioned above. In the case when a Control Flow MultiCore machine is used as a host, the transactional code is best run on the Control Flow host, while the other types of problems are best crunched on accelerators based on other types of paradigms. In the case when the code works on data organized in 2D, 3D, or nD structures, a good level of acceleration could be achieved by a Control Flow ManyCore accelerator. The programming model is relatively easy to comprehend. Speed, Power, Size, and potential for high precision of Control Flow machines are well understood.

The Data Flow paradigm was inspired by the research of Richard Feynman and others and insists on the fact that computing is most effective if data are being transferred, during the computational process, over infinitesimal distances, as in the case of execution-graph-based computing. Compared with Control Flow, this approach brings speedups, power savings, smaller size of machinery, and larger potentials for higher precision, but it utilizes a more complex programming model, which could be lifted on the higher levels of abstraction, in which case a part of the claimed advantages could disappear.

The Diffusion Flow paradigm is based on research in massive parallelism (IoT), possibly enhanced with sensors (WSNs). One intrinsic characteristic of this approach is a large area or geographical coverage, which means that it is theoretically impossible to move data over small distances, during the computing process. Yet, some level of processing is necessary, maybe for data reduction purposes, or for some kind of pre-processing, during the “diffusion” of the collected data towards the host, for the final processing of the Big Data type. If the energy is scavenged, the power efficiency is high, while the size is negligible, as well as the potential for the highest precisions. On the other hand, the programming model has evolved since the initial MIT PROTO approach and has to be mastered properly, which could be a challenge.

The Energy Flow paradigm is meant only for the acceleration of the algorithms that are best suited for one of the existing sub-paradigms. No matter if the BioMolecular or QuantumMechanical approach is used, the processing is based on the energy transformations, and the corresponding programming model must respect the intrinsic essence if the best possible performance is needed. For the doable algorithms, the speedup is enormous, the needed power is minimal, the size is acceptable, and the potential for precision is unthinkably big. The programming models are on the rise.

The synergy of AI and the presented computing paradigms has two dimensions: (a) These paradigms could serve as accelerators for AI applications, and (b) AI can help decrease the number of iterations in simulation experiments and other applications that need the acceleration a lot more intensive than offered by the utilised paradigms.

3 Possible Architecture of a Supercomputer on a Chip

At the current state of the technology, with over 100 billion transistors (BTr) on a chip, or a trillion transistors (TTr) on a wafer, it is possible to place (on a single chip) both the above-mentioned Control Flow engines and both above-mentioned Data Flow engines.

However, possible enhancers (in the form of IoT or WSN) and possible accelerators (in the form of BioMolecular and/or QuantumMechanical) must be off-chip, but easily accessible via proper interfaces.

Of course, memory and classical I/O must be partially on the chip and partially off the chip, again connectable with proper interfaces.

Therefore, no matter if 100BTr or 1TTr structures are involved, the internal architecture, on the highest level of abstraction, should be as in Fig. 1. However, the distribution of resources could be drastically different from one such chip to another, due to different application demands (transactions-oriented or crunching-oriented), and due to different data requirements (memory intensive for massive Big Data of the static type, or streaming oriented for massive Big Data of dynamic type - coming and going via the Internet or other protocols).

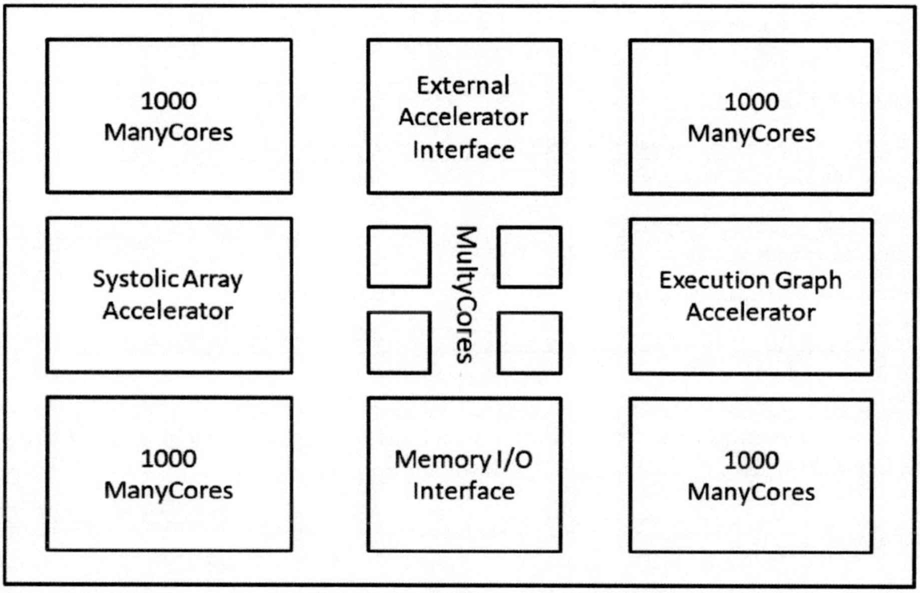

Fig. 1. Generic structure of a future Supercomputer-on-a-Chip with 100 billion Transistors [10]

Examples that follow cover simulations of Big Data problems needing Machine Learning and are related to complex problems in Civil Engineering, or related fields, namely:

1. NBCE - Nature-Based Construction Engineering
2. GNBE - Genomics Supporting NBCE
3. EQIS - Earth Quake Information Systems for Prediction and Alarm
4. NCEM - Creation of new civil engineering materials, for CO2 and EQ (Earth Quakes)

The algorithm used in the above areas could be:

1. Statistical and stochastic processes mimicking nature
2. NW or NE or similar
3. PDEs of the type FE or FD or hybrid (FE = Finite Element, FD = Finite Difference)
4. Tensor calculus and mathematical logic or hybrid

For such a set of applications, we presume that the optimal distribution of resources would be as in Table 1.

Table 1. Chip hardware type and estimated transistor count [10]

Chip Hardware Type	Estimated Transistor Count
One ManyCore with Memory	3.29 million
One Systolic Array	<1 billion
4000 ManyCores with Memory	11 800 million
One Reprogrammable Ultimate Dataflow	<69 billion
One MultiCore with Memory	1 billion
Interface to I/O with external Memory	<100 million
4 MultiCore with Memory	4 billion
Interface to External Accelerators	<100 million
TOTAL	<100 billion

It is important to underline that for applications of interest, data come either from the internal memory system or from an IP stream.

4 Elaboration

In NBCE, it is better to use biological structures that grow fast, and are populated with insects that generate nano-materials, than to build concrete walls that emit CO2 and are EQ-sensitive. Also, it is better to use fish and plankton than metal nets, to protect underwater structures. Before each investment of this type, a feasibility study has to be performed, based on simulation. However, such simulations could be very time-consuming and could last for years. The solution is in switching from Control Flow to a proper combination of the other three computing paradigms.

In GNBE, the genetics of species, and related processes may take years to generate the desired effects, as far as civil engineering goals. However, computer simulations on Control Flow engines, based on enough details, could take even more time. Again, the solution is in proper synergies of the four paradigms.

In EQIS, models do exist of cities, based on bricks and cement, but simulations of earthquakes with these models as inputs may take a century on the fastest Control

Flow machine today. The simulation process could be drastically accelerated only if a proper Data Flow accelerator is used. They are suited for PDEs of the type FE (Final Element) needed for predictions and for PDEs of the type FD (Final Difference) needed for alarming in emergencies.

In NCEM, new materials with desired properties are best found if ML algorithms are combined with classical algorithms used in materials research. Such hybrid algorithms are computing-intensive, so again, the solution is in the synergy of several paradigms.

Rather than adding a section on selected applications of Machine Learning and Big Data analytics, here we direct the interested readers to the former publications of the authors of this article [11, 12]; these references are especially focussed on on-chip implementations.

5 Conclusion

This article sheds light on the potentials coming from the synergistic interactions of four different computing algorithms.This approach is discussed in the context of civil engineering but could be easily ported to other different contexts.This article could be used for educational or research purposes, in academia and industry.This approach (the approach advocated in this article) is best implemented on a chip that includes some of the paradigms that are used more frequently and effectively interfaces to the other paradigms used less frequently!

References

1. Henning, J.L.: SPEC CPU2000: measuring CPU performance in the New Millennium. Computer **33**(7), 28–35 (2000). https://doi.org/10.1109/2.869367.,July
2. Mittal, S., Vetter, J.S.: A survey of CPU-GPU heterogeneous computing techniques. ACM Comput. Surv. **47**(4), 1–35 (2015). https://doi.org/10.1145/2788396. Article No. 69
3. Kumar, S., et al.: Scale MLPerf-0.6 models on Google TPU-v3 Pods. Computer Science, Cornel University. https://doi.org/10.48550/arXiv.1909.09756
4. Wang, Y.E., Wei, G., Brooksar, D.:"Benchmarking TPU, GPU, and CPU Platforms for Deep Learning. Computer Science, Cornel University (2019). https://doi.org/10.48550/arXiv.1907.10701
5. Trifunovic, N., Milutinovic, V., Salom, J., Kos, A.: Paradigm shift in big data SuperComputing: DataFlow vs. ControlFlow. J. Big Data **2**(1), 1–9 (2015). https://doi.org/10.1186/s40537-014-0010-z
6. Srivastava, A., Mishra, P.K.: A survey on WSN issues with its heuristics and meta-heuristics solutions. Wirel. Pers. Commun. **121**, 745–814 (2021). https://doi.org/10.1007/s11277-021-08659-x
7. Centenaro, M., Costa, C.E., Granelli, F., Sacchi, C., Vangelista, L.: A survey on technologies, standards and open challenges in satellite IoT. IEEE Commun. Surv. Tutor. **23**(3), 1693–1720 (2021). https://doi.org/10.1109/COMST.2021.3078433
8. Schlick, T., Portillo-Ledesma, S., Myers, C.G., et al.: Biomolecular modeling and simulation: a prospering multidisciplinary field. Ann. Rev. Biophys. **50**, 267–301 (2021). https://doi.org/10.1146/annurev-biophys-091720-102019

9. Baiardi, A., Grimmel, S.A., Steiner, M., et al.: Expansive quantum mechanical exploration of chemical reaction paths. Laboratory of Physical Chemistry, ETH Zurich, Acc. Chem. Res. (2022). https://doi.org/10.1021/acs.accounts.1c00472
10. Milutinović, V., Azer, E.S., Yoshimoto, K., et al.: The ultimate DataFlow for ultimate SuperComputers-on-a-chip, for scientific computing, geo physics, complex mathematics, and information processing. In: 10th Mediterranean Conference on Embedded Computing (MECO), pp. 1–6 (2021). https://doi.org/10.1109/MECO52532.2021.9459725
11. Milutinović, V., Kotlar, M.: Handbook of research on Methodologies and Application of Supercomputing. IGI Global (2021)
12. Milutinović, V., Salom, J., Trifunović, N., Giorgi, R.: Guide to DataFlow Supercomputing. Springer, Cham (2015). https://doi.org/10.1007/978-3-319-16229-4

Pose Estimation and Joint Angle Detection Using Mediapipe Machine Learning Solution

Katarina Mitrović(✉) and Danijela Milošević

Faculty of Technical Sciences, Department of Information Technologies, University of Kragujevac, 65 Svetog Save Street, 32000 Čačak, Serbia
{katarina.mitrovic,danijela.milosevic}@ftn.kg.ac.rs

Abstract. Health is one of the central aspects of life and innovative ways for its improvement are constantly being studied. Artificial intelligence has an extensive application and its contribution to health and medicine is widely recognized. In this paper, the application of machine learning algorithms in the field of health care is presented. A model for physical activity injury prevention based on the MediaPipe solution for body pose tracking has been developed. The solutions for pose estimation and detection of joint angles and angles relative to the horizontal are integrated into a comprehensive system that detects all key body landmarks and angles during the movement of the observed person. In addition, one of the goals of this research is to develop a flexible system with the ability to process a variety of inputs in terms of video content and format. The system is trained and tested on video inputs and can process front, left, and right perspectives. In the processing phase, a graph of posture and angle estimation is generated. The graph represents detected joints and the corresponding angles that vary depending on the observed perspective. The input is integrated with the graph and thus provides valuable information about body posture and alignment. The results provide support to professionals in physical activity monitoring and injury prevention.

Keywords: Artificial Intelligence · Machine Learning · MediaPipe · Pose Estimation · Angle Detection

1 Introduction

Throughout history and in all parts of the world, health has always been considered a central aspect of life. Good health is imperative for progress and a basis for a high-quality life. It directly affects all other aspects of life such as productivity, economic and career advancement, as well as social and family life. A number of factors can have a positive impact on health, therefore many people practice them daily to develop good habits and improve their lives. Quality nutrition, consistent hygiene, sufficient sleep, stress management, as well as regular physical activity are some of the crucial factors that promote human wellbeing.

The health potential of physical activity is substantial due to a great number of health conditions that are affected by physical activity [1]. It brings benefits to the cardiovascular system, skeletal muscles, tendons and connective tissues, the skeleton, metabolic

N. Filipovic (Ed.): AAI 2022, LNNS 659, pp. 109–120, 2023.
https://doi.org/10.1007/978-3-031-29717-5_8

functions and psychological functions [2]. Physical activity can be categorized into sports activity (fitness, athletics, football, yoga and similar), habitual activity (everyday physical activities such as walking, gardening and cleaning) and work-related activity (physical activities that are a part of the job such as fireman, repairman, paramedic) [3]. The adverse effects of physical activity on health are shown to be small and mostly preventable [1]. One of the most common downsides of physical activities is the risk of potential injury. Guided practicing of physical activities is one of the preferred methods to organize a routine for injury prevention [4]. The main obstacles in implementing the coaching program are lack of time in the schedule of the person conducting the supervision as well as the high costs of individual supervision.

Various fields of artificial intelligence (AI), especially machine learning (ML), have both an important role and a big potential in the field of medicine. Numerous ML models that can diagnose whether a person is suffering from a certain health issue have been developed based on different inputs such as images, biomarkers and other types of data. A classifier based on convolutional neural networks (CNNs) that successfully recognizes infected cells can add great value to traditional medical methods of disease identification and treatment. [5] Furthermore, computer-aided detection and diagnosis in medical imaging offers a beneficial second opinion to the doctors and assists them in the screening process [6]. Technological advancement, especially in the domain of information technologies and AI, created the necessary conditions for transitioning from traditional physical activity injury prevention to computer-aided and automated prevention systems.

Deep CNNs have found many applications in building fine-tuned models for implementation in vision-related tasks [7]. Papers [7, 8] and [9] propose intelligent fitness trainer systems based on human pose estimation without the help of a trainer that provide instant feedback about posture. A lightweight 2D human pose estimation for a fitness coaching system is described in [10]. The results in [11] indicate that human pose estimation is maturing and produce viable results for the detection of specific technique-related issues which are highly associated with risk of injury during common exercises.

One of the important additions to pose estimation is detecting joint angles and identifying irregularities when preforming a specific exercise. In the research [12] a wavelet neural network that uses vertical ground reaction forces is proposed for ankle, knee, and hip joint angle estimation. Paper [13] compares the performance of multilayer perceptron, long short-term memory, and CNNs for the prediction of joint kinematics and kinetics based on the inputs from inertial measurement unit (IMU) sensors. In the research [14] a recurrent neural network is used for joint position estimation from surface electromyography (sEMG) data measured at the forearm. Most research conducted on the topic of joint angle assessment is based on individual joint angle identification and uses different inputs such as the IMU sensor or sEMG data.

The main goal of this paper is to create a solution for exercise monitoring and to deliver a comprehensive but legible and user-friendly system. The main contribution of this paper is integrating various techniques such as pose estimation and angle detection as well as extending the current pose estimation model with multiple solutions. Additionally, the software provides identification of critical landmarks and joint angles depending on the perspective of the person performing the exercise, which addresses some of the problems faced by researchers on this topic. Determination of these elements is defined in cooperation with domain experts. In addition to the basic joint angles, the angles relative to the horizontal are also identified in order to complete the image of the body pose and perform a proper estimation of the posture. Another contribution of this paper is defining equations for scaling the graph to input data.

This paper is structured as follows. In the next chapter, the method used in this paper is presented. In the third chapter, the results of the proposed model are given. In the final chapter, main conclusions of the paper are highlighted.

2 Methods

In this chapter, the main elements of the proposed methodology are presented. Primarily, input data is described. Further, the implementation of the MediaPipe solution for pose estimation is presented. In the third subchapter, the detection of joint angles is explained. In the fourth subchapter, the image scaling of the input videos is described. Finally, an overview of software and hardware requirements for the implementation of the proposed system is given.

2.1 Input Data

In recent years, the focus of research in the field of neural networks and ML has shifted to achieving greater flexibility of the model performance regarding inputs. The goal is to create models that can achieve high accuracy and speed, preferably in real time, in conditions where segmentation of inputs and forwarding of controlled uniform inputs cannot always be performed. Convolutional neural networks show great potential in achieving flexibility, which has been the subject of research in many papers such as [15], where the main contribution of work was finding an algorithm that can be used with a dataset that has a high diversity in classes.

This paper uses videos that capture a person performing a specific physical activity as the input data. One of the significant goals of this study was to achieve high system flexibility regarding input data without the loss of performance. The input video can be recorded with a mobile phone, camera or similar device, by a person performing the exercise or by professionals such as sports coaches or medical staff. The system developed in this paper is not sensitive to the input content, therefore the video can be recorded in various environments either outdoors (in nature, on a street and similar) or indoors (in a house, gym and similar). To achieve the highest possible accuracy of the system, it is recommended that the person of interest be the only person in the input video. Further, the system is not sensitive to the appearance of the person being examined, therefore the system performance is not reduced depending on height, weight, gender or other basic characteristic of the person that is performing the physical activity. The observed person can also appear in any clothing combination of any color, with a slight preference given to tight clothing that does not blend with the environment and gives clear body contours to make it easier to detect key points of the body. The type of physical activity, its speed of performance and other elements of the exercise can be arbitrary. The shooting angle must coincide with the observed activity in order to perform estimation for body points and angles that are of interest for the given exercise.

The proposed model offers high flexibility in terms of processing of the input videos. The input diversity can be divided into two categories:

- video content diversity and
- video format diversity.

Three key elements can be highlighted by which the video content can vary: the appearance of the person being filmed, the environment in which they are located, and the perspective in which the video was shot. The model developed in this research aims to achieve flexibility in terms of the diversity of video content that represents input data. As mentioned previously, the model is not sensitive in terms of the environment and the appearance of the person in the video. Flexibility of the perspective is achieved by creating three different modules. Each module represents one perspective:

- front view,
- right side view and
- left side view.

For each perspective, different landmark and angle segmentation is performed which is presented in the next chapter. This improves the visibility of the detected posture in the video.

Videos can also vary in terms of format, dimensions, FPS and other format and quality determinants. The proposed model can accept various video format extensions such as avi, mp4, mov and similar. All output components are scaled to the different dimension of the videos to make the output visible and the display design proportional, clear and user-friendly. This aspect of input data is further elaborated in the design scaling chapter.

2.2 MediaPipe Pose

Pose detection is preformed using MediaPipe Pose Landmark Model. MediaPipe is an open-source framework that offers fully extensible and customizable ML solutions that work across Android, iOS, desktop/cloud, web and Internet of Things [16]. It contains a number of solutions for body and face landmark extraction, object detection including real-time 3D object detection, object segmentation as well as object and motion tracking. MediaPipe Pose outperforms current state-of-the-art approaches that rely primarily on powerful desktop environments for inference, and it achieves real-time performance on most modern mobile phones, desktop and laptop computers, in Python and even on the web [16].

MediaPipe Pose is based on the BlazePose convolutional neural network for human body pose perception. BlazePose utilizes a two-step detector-tracker ML pipeline, in which it first locates the region-of-interest (ROI) within the frame first and subsequently predicts the pose landmarks and segmentation mask within the ROI using the ROI-cropped frame as input [17]. The ROI is a person or persons pose which is captured in

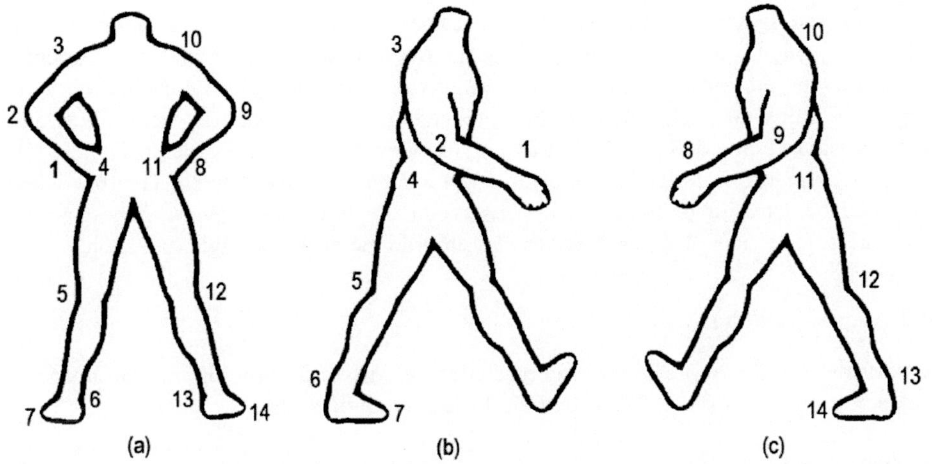

Fig. 1. Extracted landmarks for each perspective (1 - right wrist, 2 - right elbow, 3 - right shoulder, 4 - right hip, 5 - right knee, 6 - right ankle, 7 - right foot index toe, 8 - left wrist, 9 - left elbow, 10 - left shoulder, 11 - left hip, 12 - left knee, 13 - left ankle, 14 - left foot index toe); (a) front view; (b) right view; (c) left view

the input video, which brings great challenges in detection due to the diversity of the environment in which the person is, the appearance of the persons being observed as well as the clothes they wear, the movements they perform and the perspectives from which they are filmed.

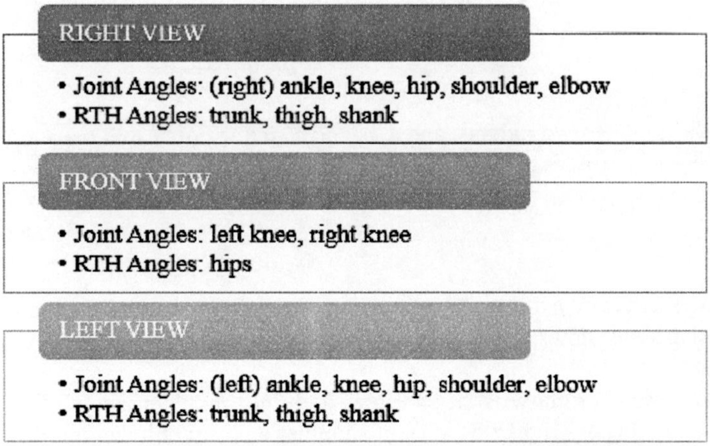

Fig. 2. Joint angles and angles relative to horizontal (RTH)

In this paper, Pose Landmark Model is used and adjusted for multi-perspective landmark detection and extended with joint angles and angles relative to horizontal detection. The developed solution includes three modules with different perspectives:

front, left and right. Depending on the observed perspective, different pose landmarks are extracted, which is presented in Fig. 1. For the front view 14 landmarks were extracted, while for the left and right view 7 landmarks were extracted. For the right side view right foot index toe, right ankle, right knee, right hip, right shoulder, right elbow and right wrist are being extracted. For the left side view left foot index toe, left ankle, left knee, left hip, left shoulder, left elbow and left wrist are being extracted. The front view included the detection of all landmarks observed in the left and right views. These points are crucial for each of the perspectives and provide the basis for angle detection.

2.3 Angle Detection

For a simple and comprehensive interpretation of correct posture, the measurement of key angles was performed. Measured angles can be divided into two groups:

- joint angles and
- angles relative to the horizontal.

For each perspective, an estimation of a different set of angles was performed depending on which key points on the body were isolated and which body angles were relevant for determining body posture. The overview of the extracted angles depending on the observed perspective and the type of the angle is presented in the Fig. 2.

To determine the angles of the joints, previously detected body landmarks are used as sets of points that make up the observed angle. Each angle is calculated based on three previously detected landmarks, where the central point is the point where the observed angle is calculated, and the remaining two points are the closest points in relation to the observed angle. This calculation is performed using the following equation:

$$\alpha_{(rad)} = arctg\frac{c_y - b_y}{c_x - b_x} - arctg\frac{a_y - b_y}{a_x - b_x} \tag{1}$$

where α is the angle size in radians, and a, b, and c are the points that make up the angle. In order to convert radians to degrees, the following equation can be used:

$$\alpha_{(deg)} = \left|\frac{\alpha_{(rad)} * 180}{\pi}\right| \tag{2}$$

The angles detected in this work and the corresponding points from which each angle is calculated are as follows:

- right elbow angle – right wrist, right elbow and right shoulder landmark;
- right shoulder angle – right elbow, right shoulder and right hip landmark;
- right hip angle – right shoulder, right hip and right knee landmark;
- right knee angle – right hip, right knee and right ankle landmark;
- right ankle angle – right knee, right ankle and right foot index toe landmark;
- left elbow angle – left wrist, left elbow and left shoulder landmark;
- left shoulder angle – left elbow, left shoulder and left hip landmark;
- left hip angle – left shoulder, left hip and left knee landmark;

- left knee angle – left hip, left knee and left ankle landmark;
- left ankle angle – left knee, left ankle and left foot index toe landmark.

Figure 3 illustrates all the detected angles with the landmarks that intersect them. The view of the angles is divided into three perspectives, and it can be noted that the angles detected in the left and right view are collectively estimated in the front view. The figure clearly shows the points used to determine the size of each angle.

A similar method is used for determining angles relative to horizontal, where the central point between the two landmarks detected in the previous stage of the model is primarily defined. For the trunk point estimation, the middle point between the hip and shoulder is used; for the thigh point estimation, the middle point between the hip and knee is used; for the shank point estimation, the middle point between the knee and ankle is used; for the hips point estimation, the middle point between the left and right hip is used. The next step is defining a horizontal line in the central point and extracting the point from the horizontal line for estimating the angle of the observed point.

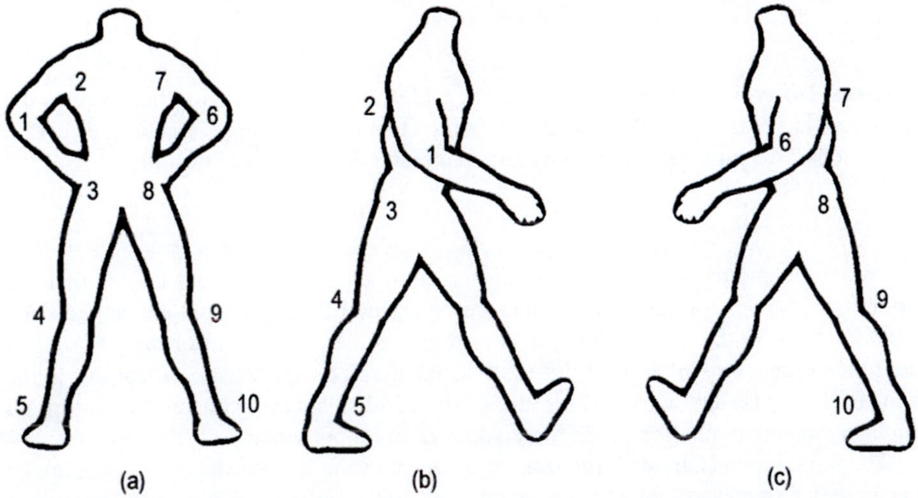

Fig. 3. Extracted angles for each perspective (1 - right elbow, 2 - right shoulder, 3 - right hip, 4 - right knee, 5 - right ankle, 6 - left elbow, 7 - left shoulder, 8 - left hip, 9 - left knee, 10 - left ankle); (a) front view; (b) right view; (c) left view

Figure 4 shows the angles relative to the horizontal that are included in this work. The figure represents the angles for each perspective individually. In the front perspective, one angle relative to horizontal is defined. It indicates if the hips are in line with the horizontal. This is particularly significant because it shows if the body is leaning towards one side when performing the exercise. This can lead to a greater load on the left or the right side of the body, which can cause numerous injuries, especially in the knee area. Ideally, this angle should be equal to zero. In the case of the left and right perspective, three angles relative to horizontal are observed: trunk, thigh and shank. These angles help determine the angle at which the observed body parts are bent when performing the

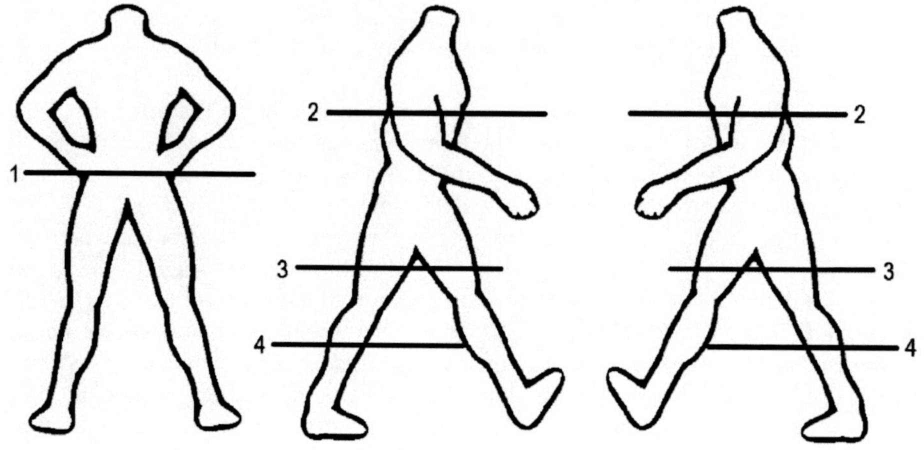

Fig. 4. Extracted angles relative to horizontal for each perspective (1 - hips, 2 -trunk, 3 - thigh, 4 - shank); (a) front view; (b) right view; (c) left view

exercises. Different exercises have different preferred angles and with these values the supervisor can establish if the trainee is moving at an incorrect angle, make a correction and thereby prevent potential injuries and maximize the benefits of training.

2.4 Design Scaling

The proposed model is detecting the aforementioned landmarks and angles that are further used for generating a graphical representation of the body pose with angles. The graph is added to the input video, which represents the final output of the system. As input videos can vary in terms of dimensions, the graph design must be adaptable to the input for better readability and visibility of the results. This is achieved by scaling all display elements in relation to the dimensions of the input video.

The input video is divided into frames and each frame is viewed as a separate image that is being processed for posture estimation and joint angle detection. The image processing results in a graph that follows the body pose. Each graph consists of joint points, lines that connect them and angle values. Line thickness is determined using the following equation:

$$lt = max\left(\frac{\min(w, h)}{500}, 1\right) \tag{3}$$

Where w represents image width and h represents image height. The dimension of the landmarks is obtained by using the formula (1) and multiplying the result with the number 3.

An additional component over which dimensional scaling has been applied is the visual representation of angle values. Text size scale is calculated using the equation:

$$ts = \frac{\min(w, h)}{2500} \tag{4}$$

In addition to the font size scaling of the text that shows the values of the angles according to formula (2), scaling was performed over the background that visually distinguishes this segment of the display. The starting coordinates of the background softbox are determined using the following formula:

$$s = \left(p_x + lt * 6, p_y + \frac{th}{2}\right) \tag{5}$$

where p is the point denoting the joint for which the angle estimation is made, while th is the height of the text contained in the softbox. The ending coordinates of the softbox are established using the following formula:

$$e = \left(s_x + tw + \frac{bl}{2} + \max\left(\frac{lt}{2}, 1\right) * 2, s_y - th\right) \tag{6}$$

where tw represents text width and bl is text baseline. The width of the text contained in the softbox can vary depending on the angle size.

2.5 Hardware and Software Requirements

The proposed software for pose estimation and joint angle detection was developed in Python (version 3.8) programming language using Jupyter Notebook as the development environment. The model integrates the functions of TensorFlow software library for ML and AI, MediaPipe library, ffmpeg library for managing video formats, opencv library for computer vision tasks, NumPy comprehensive library for mathematical functions and glob and pathlib libraries for file path management.

The software is implemented and tested in the local environment but with further development it can be applied in web-based environments. The hardware configuration used in this research includes NVIDIA GeForce GTX 1650 Ti graphics processing unit, AMD Ryzen 5 4600H 3.00 GHz central processing unit and 8 GB of installed physical memory (RAM). Thus, 6 s long video with 1920 frame width and 1080 frame height requires less than 30 s for complete processing and exporting. Consequently, it can be concluded that with the improvement of the processing configuration, there can be a significant improvement in the processing time, and additional optimization could lead to real time processing and display of posture estimation and joint angle detection during physical activity.

3 Results

In this research, using the previously described methodology, a software for recognizing body posture and determining joint angles was created. A video containing a person performing physical activity is used as input. The first step is to choose the perspective of the person shown in the video. The video is segmented into frames on which the key landmark and angle estimation is performed. Applied ML algorithms are trained to identify key points on the body in a flexible environment. Based on the detected points the angles are estimated, and landmarks, their connections and angles are combined into a connected graph. The graph is plotted over the original image, and the specified process is repeated for each frame of the initial input. After processing individual images, conversion to the video format is performed, which is the final product of this software.

The software provides the ability to process three different perspectives: left, right and front. Depending on the chosen perspective, different points on the body are detected and a custom view is generated. Also, in addition to the estimation of the pose, the measurement of joint angles is performed. This system is the implementation of a comprehensive model that combines several different methods for evaluating posture and thus provides complete and highly accurate information to the person monitoring physical movements. This system can be widely used in medicine during physiotherapy and similar activities, as well as in fitness during exercise supervision. Therefore, coaches and medical staff are assisted in correcting the improper posture of the supervised person during physical activity, which prevents potential injuries and leads to the better overall health of the individual.

Another benefit of this paper is reflected in the flexibility of the system in terms of input. The software is applicable to inputs of a variety of content and formats, both in terms of reliable landmark prediction and in terms of processing and displaying the resulting output. The result is a fast, flexible and comprehensive software that detects key points on the body and provides insight into body alignment for computer-aided posture estimation. The software has been tested on numerous inputs, including various examinees, perspectives, physical activities and other elements of the video. Figure 5 shows one example of video frame before and after processing.

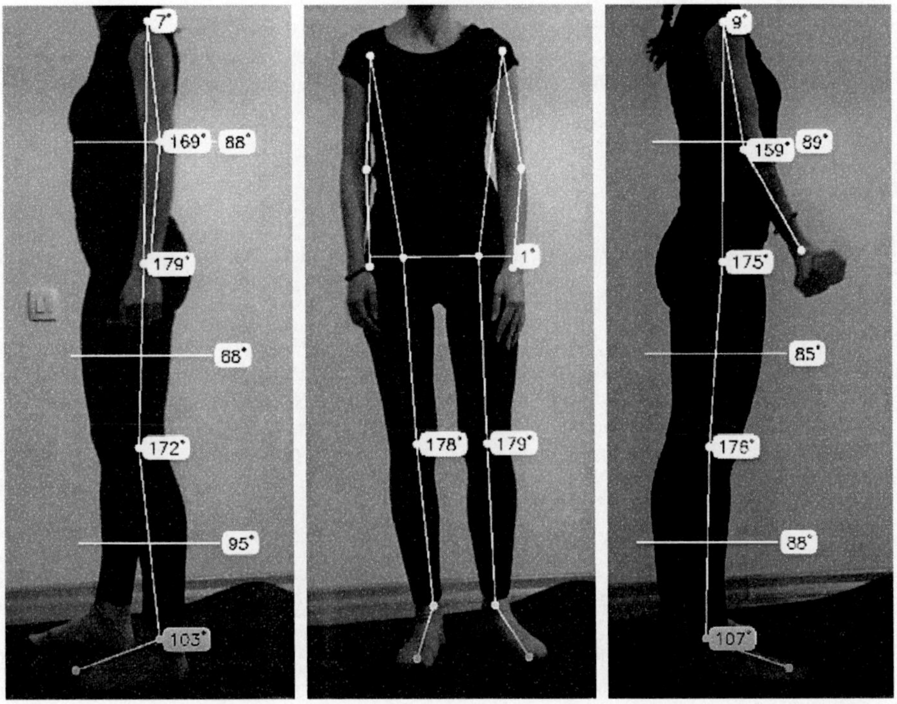

Fig. 5. A frame extracted from input video after processing: left, front and right perspective

4 Conclusions

This paper addresses one of the important elements that affect human health. Physical activity as a mechanism for maintaining health has numerous advantages, but due to the improper performance of these activities, serious injuries can occur. Various methods for the prevention of injuries of this type have been developed during previous decades.

In this paper, a system that combines several different methods for injury prevention is presented. It applies the most modern techniques of AI in order to provide reliable information about the posture of a person during exercise. In this study, the integration of posture estimation and joint angle detection was performed in order to identify potential problems that could lead to injuries. In addition, the detection of key points on the body in relation to the horizontal is performed, which provides valuable information about body alignment.

The proposed software was developed locally using an average hardware configuration. In the following stages of development, the system can be web-based and adapted to real-time video processing in order to provide feedback on body posture during physical activity without delays. In addition, integration with the incorrect exercise performance detection can be done, which implies indicating the critical points and providing advice for correction. This would enable the system to provide complete assistance without the supervision of professionals.

Acknowledgments. This study was supported by the Ministry of Education, Science and Technological Development of the Republic of Serbia, and these results are parts of the Grant No. 451–03-68/2022–14/200132 with University of Kragujevac - Faculty of Technical Sciences Čačak.

References

1. Vuori, I.: Does physical activity enhance health? Patient Educ. Couns. **33**, S95–S103 (1998)
2. Health Education Authority: Sports Council. Belmont Press, Allied Dunbar Fitness Survey. London (1992)
3. Schmidt, S.C., Tittlbach, S., Bös, K., Woll, A.: Different types of physical activity and fitness and health in adults: an 18-year longitudinal study. BioMed Res. Int. (2017)
4. Mendonça, L.D., Ley, C., Schuermans, J., Wezenbeek, E., Witvrouw, E.: How injury prevention programs are being structured and implemented worldwide: an international survey of sports physical therapists. Phys. Ther. Sport **53**, 143–150 (2022)
5. Mitrović, K., Milošević, D.: Classification of malaria-infected cells using convolutional neural networks. In: 15th International Symposium on Applied Computational Intelligence and Informatics (SACI), pp. 323–328. IEEE (2021)
6. Narayanan, B.N., Ali, R., Hardie, R.C.: Performance analysis of machine learning and deep learning architectures for malaria detection on cell images. Appl. Mach. Learn. SPIE **11139**, 240–247 (2019)
7. Agarwal, S., et al.: FitMe: a fitness application for accurate pose estimation using deep learning. In: 2nd International Conference on Secure Cyber Computing and Communications, pp. 232–237. IEEE (2021)
8. Chen, S., Yang, R.R.: Pose trainer: correcting exercise posture using pose estimation. ArXiv preprint arXiv:2006.11718 (2020)
9. Zou, J., et al.: Intelligent fitness trainer system based on human pose estimation. In: Sun, S., Fu, M., Xu, L. (eds.) Signal and Information Processing, Networking and Computers. ICSINC 2018. Lecture Notes in Electrical Engineering, vol. 550, pp. 593–599. Springer, Singapore (2019). https://doi.org/10.1007/978-981-13-7123-3_69
10. Jeon, H., Yoon, Y., Kim, D.: Lightweight 2D human pose estimation for fitness coaching system. In: 36th International Technical Conference on Circuits/Systems, Computers and Communications, pp. 1–4. IEEE (2021)
11. Eivindsen, J.E., Kristensen, B.Y.: Human Pose Estimation Assisted Fitness Technique Evaluation System, Master's thesis, NTNU (2020)
12. Sivakumar, S., Gopalai, A.A., Lim, K.H., Gouwanda, D., Chauhan, S.: Joint angle estimation with wavelet neural networks. Sci. Rep. **11**(1), 1–15 (2021)
13. Mundt, M., Johnson, W.R., Potthast, W., Markert, B., Mian, A., Alderson, J.: A comparison of three neural network approaches for estimating joint angles and moments from inertial measurement units. Sensors **21**(13), 4535 (2021)
14. Koch, P., et al.: sEMG-based hand movement regression by prediction of joint angles with recurrent neural networks. In: 43rd Annual International Conference of the IEEE Engineering in Medicine & Biology Society, pp. 6519–6523 (2021)
15. Mitrović, K., Milošević, D.: Flower classification with convolutional neural networks. In: 23rd International Conference on System Theory, Control and Computing (ICSTCC), pp. 845–850 (2019)
16. MediaPipe Pose. https://google.github.io/mediapipe/. Accessed 23 July 2022
17. On-device, Real-time Body Pose Tracking with MediaPipe BlazePose. https://ai.googleblog.com/2020/08/on-device-real-time-body-pose-tracking.html. Accessed 23 July 2022

Application of AI in Histopathological Image Analysis

Jelena Štifanic[1](✉), Daniel Štifanić[1], Ana Zulijani[2], and Zlatan Car[1]

[1] Faculty of Engineering, University of Rijeka, 58 Vukovarska Street, 51000 Rijeka, Croatia
{jmusulin,dstifanic,car}@riteh.hr

[2] Department of Oral Surgery, Clinical Hospital Center of Rijeka, 40 Krešimirova Street, 51000 Rijeka, Croatia

Abstract. Over the past decade, improvements in image analysis methods and substantial advancements in the processing power have allowed the development of powerful computer-aided analytical approaches to medical data. Tissue histology slides can now be scanned and preserved in digital form, thanks to the recent introduction of entire slide digital scanners. In such a form, they can serve as input data for Artificial Intelligence (AI) algorithms that can speed up standard procedures for histology analysis with high accuracy and precision. This research aimed to create an automated system based on AI for histopathological image analysis. The first step was normalizing H&E-stain images and then using them as input to the convolutional neural network. The best results are achieved using ResNet50 with the highest AUC value of 0.98 ($\pm\sigma = 0.02$). Such an approach proved to be successful in analyzing histopathological images.

Keywords: Artificial Intelligence · Convolutional Neural Network · Histopathological analysis · Oral squamous cell carcinoma

1 Introduction

Image data is extremely important in healthcare. Lately, the massive accumulation of digital images has increased the demand for their analysis, such as computer-aided diagnosis using Artificial Intelligence algorithms. Medical image analysis is one of the areas where histological tissue patterns are combined with computer-aided image analysis to improve the detection and classification of diseases [1]. There is also the possibility of automating and speeding up processes that take a long time to complete manually. AI-based models may learn to recognize specific traits in these images, making the diagnostic procedure much faster and more accurate.

The dataset used in this research consists of histopathological images of the oral squamous cell carcinoma (OSCC) region, which contains abnormalities. Oral squamous cell carcinoma is the most common histological neoplasm of head and neck cancers, and while it is located in an easily visible area and can be detected early, this does not always eventuate [2]. Over the past decade, the incidence of oral cancer has increased, especially among young adults. The cause of oral squamous cell carcinoma is multifactorial, and the

N. Filipovic (Ed.): AAI 2022, LNNS 659, pp. 121–131, 2023.
https://doi.org/10.1007/978-3-031-29717-5_9

consumption of tobacco and alcohol have been well-established as significant risk factors for the development of oral cancer [3]. Despite advances in therapeutic approaches, the morbidity and mortality rates from OSCC have not improved significantly over the last 30 years. The 5-year survival rate for patients with OSCC ranges between 40% and 50% [4]. The most prevalent reasons why OSCC is detected in advanced stages include an incorrect initial diagnosis, and ignorance from the patient or the attending physician.

Clinical examination, conventional oral examination (COE), and histological evaluation following biopsy are procedures for detecting oral cancer. These procedures can detect cancer in the stage of established lesions with significant malignant changes. However, the subjective component of the examination, respectively inter- and intra-observer variability, is the fundamental difficulty in employing histopathological examination for tumor differentiation. Moreover, from the pathologist's point of view, providing exact histological identification in the context of multi-class grading is crucial. According to numerous studies, the World Health Organization (WHO) classification system is not a reliable prognostic and predictive factor for patent outcomes. This could be partly due to the fact that grading of OSCC is a subjective process that depends on the area of tumor samples and the evaluation criteria of the individual pathologist. The majority of OSCCs show histological heterogeneity, and in these cases, the highest grade should be recorded [5]. For this reason, a combination of AI-based approaches with a clinical perspective could reduce inter- and intra-observer variability as well as assist pathologists in terms of reducing the load of manual inspection in a shorter time [2].

1.1 Related Work

Nowadays, there is interest in leveraging AI tools, and pathology departments have several commercial digital pathology platforms available for diagnostic work. Several papers have demonstrated the viability of developing AI-based algorithms to analyze histopathology images.

Pantanowitz et al. (2020) developed an AI-based algorithm using H&E-stained slides of prostate core needle biopsies (CNBs). Their paper describes the successful development, validation, and deployment in clinical practice of an AI-based algorithm for accurately detecting, grading, and evaluating clinically relevant findings in digitized slides of prostate CNBs [6]. Song et al. (2020) used a deep learning algorithm trained on H&E-stained whole slide images of gastric cancer to create a clinically applicable system. They show that the system could help pathologists improve diagnostic accuracy and avoid misdiagnoses [7]. On the other hand, Chen et al. (2020) demonstrate that a deep learning algorithm could be used to assist pathologists in the classification of histopathology H&E images in liver cancer [8].

Literature reveals that most researchers have applied AI-based algorithms in order to develop computer-assisted diagnostic tools to evaluate biopsies and improve diagnostic accuracy.

2 Materials and Methods

This section provides a detailed description of the dataset used for the classification of oral squamous cell carcinoma, as well as a brief overview of AI-based models.

2.1 Dataset Description

For this research 127 histopathological H&E-stained images with 768x768-pixel size have been used to create the dataset. Haematoxylin and eosin (H&E) staining is one type of tissue staining that is of particular interest to pathologists. This is because the H stain highlights nuclei in blue against a pink cytoplasmic background (and other issue regions). This allows a pathologist to quickly identify and examine tissue, which is a labor-intensive operation. As mentioned in the Introduction, the gold standard for diagnosis of oral cancer is tissue biopsy with routine histopathological examination. According to the WHO, OSCC can be graded as well-differentiated, moderately differentiated and poorly differentiated, based on the degree of resemblance to the normal squamous epithelium and the amount of keratin production [9].

The OSCC samples were retrieved from the archives of the Clinical Department of Pathology and Cytology, Clinical Hospital Center in Rijeka. Two unbiased pathologists analyzed sample slides and classified them according to the 4th edition of the WHO classification of Head and Neck malignancies and the 8th edition of the American Joint Committee on Cancer (AJCC) Cancer Staging Manual. Following the aforementioned classification, images have been divided into two classes, well- and moderately differentiated OSCC as shown in Fig. 1.

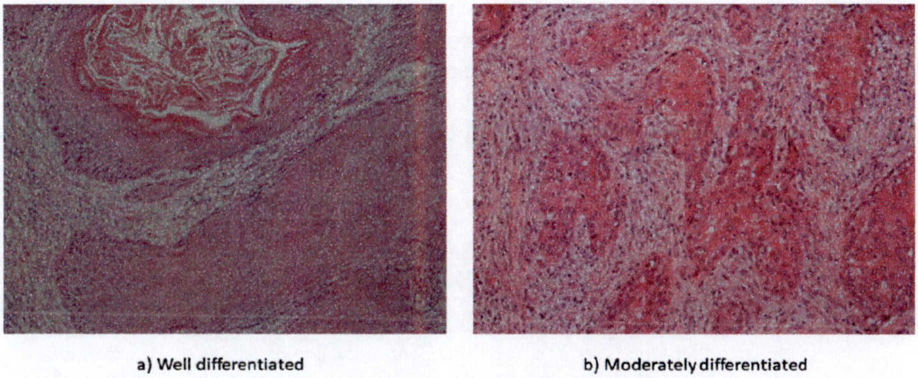

Fig. 1. OSCC group of well- and moderately differentiated OSCC with magnification x10

Hematoxylin-eosin (HE) staining is most used method in the histopathological examination of tissue sections. For HE-staining, 4 μm thick sections were deparaffinized with xylene and rehydrated in a graded ethanol series and stained with HE according to standard protocol. HE stained sections were captured using the light microscope (Olympus BX51, Olympus, Japan) equipped with a digital camera (DP50, Olympus, Japan) and transmitted to a computer by CellF software (Olympus, Japan). Images were captured with x10 objective lenses.

Since fields like medical image analysis rarely have access to a large number of samples, whilst AI models rely on a large number of samples to achieve good performance and avoid overfitting, it is required to use augmentation techniques to significantly improve the amount and quality of the data [10]. Geometrical transformations used for the augmentation procedure are horizontal flip, horizontal flip combined with 90 degrees anticlockwise rotation, vertical flip, and vertical flip combined with 90 degrees anticlockwise rotation, 90 degrees anticlockwise rotation, 180 degrees anticlockwise rotation and 270 degrees anticlockwise rotation. The augmentation process is used only for the development of training samples, as newly generated data are variants of the original data. Testing samples are not augmented.

2.2 Convolutional Neural Network Architectures

Convolutional neural networks (CNN) have emerged as the most prominent strain of neural networks in research in recent years [11]. They have revolutionized computer vision, achieving cutting-edge results in many fundamental tasks while also making significant advances in natural language processing, reinforcement learning, and many other areas. In this research, for classification purposes, we are using three deep CNN architectures.

VGG-16
Simonyan & Zisserman in their paper investigates how the CNN depth affects its accuracy in the large-scale image recording setting. Their main contribution is a thorough evaluation of networks of increasing depth using an architecture with very small convolution filters, which demonstrates that increasing the depth to 16–19 weight layers results in a significant improvement over prior-art configurations [12]. The ImageNet database was used to train the VGG-16 network. Because of the extensive training that the VGG-16 network has received, it provides excellent accuracies even with small image data sets [13]. A detailed description of VGG-16 layers is provided in Table 1.

Table 1. VGG-16 architecture representation.

Layer	Number of kernels	Kernel size
Conv1_x/2	64	3 × 3/1
Maxpool		2 × 2/2
Conv2_x/2	128	3 × 3/1
Maxpool		2 × 2/2
Conv3_x/3	256	3 × 3/1
Maxpool		2 × 2/2
Conv4_x/3	512	3 × 3/1
Maxpool		2 × 2/2
Conv5_x/3	512	3 × 3/1

ResNet50
Due to the well-known vanishing gradient problem, deep neural networks become increasingly difficult to train. For that reason, He et al. (2016) propose a residual network

(ResNets) to aid in the training of deep neural networks. They refined the residual block as well as the pre-activation variant of the residual block, allowing vanishing gradients to flow unhindered to any previous layer via the shortcut connections. ResNet50 architecture replaces every 2-layer block in the 34-layer network with a 3-layer bottleneck block, which results in 50 layers, as shown in Table 2 [14].

Table 2. ResNet50 architecture representation.

Layer	Output	50-layer
Conv1	112 × 112	7 × 7, 64, stride 2
		3 × 3 max pool, stride 2
Conv2_x	56 × 56	$\begin{pmatrix} 1\times1, 64 \\ 3\times3, 64 \\ 1\times1, 256 \end{pmatrix} \times 3$
Conv3_x	28x28	$\begin{pmatrix} 1\times1, 128 \\ 3\times3, 128 \\ 1\times1, 512 \end{pmatrix} \times 4$
Conv4_x	14x14	$\begin{pmatrix} 1\times1, 256 \\ 3\times3, 256 \\ 1\times1, 1024 \end{pmatrix} \times 6$
Conv5_x	7 × 7	$\begin{pmatrix} 1\times1, 512 \\ 3\times3, 512 \\ 1\times1, 2048 \end{pmatrix} \times 3$
	1 × 1	Flatten, 3-d fully connected, Softmax

This architecture can be used for image classification, object localization, and object detection in computer vision tasks. This framework can also be applied to non-computer vision tasks to provide the benefit of depth while also reducing computational expenses [15].

InceptionResNetv2

Szegedy et al. proposed several techniques for optimizing the network to loosen the constraints for easier model adaptation in an InceptionV3 architecture, including factorized convolutions, regularization, dimension reduction, and parallelized computations [16]. Furthermore, because the Inception architecture has been demonstrated to be successful at a low computational cost, Szegedy et al. introduce the InceptionResNetv2, which combines the Inception architecture with residual connections, as shown in Fig. 2. This type of architecture improved both recognition performance and training speed [17].

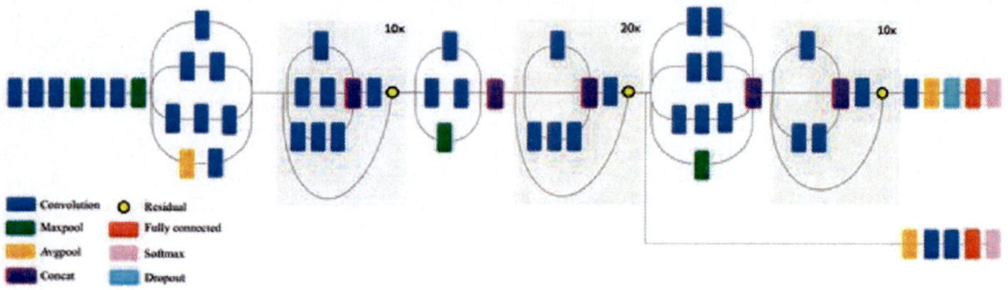

Fig. 2. Graphical representation of InceptionResNetv2 architecture [18]

The network has 164 layers and can classify images into 1000 different object categories; as a result, the network has learned rich feature representations for a wide range of images. It employs a sophisticated architecture to retrieve key information from the images and it was hence our choice of CNN for the image classification [19].

3 Results and Discussion

An automated H&E-stain histopathological analysis could assist the pathologist in discovering new informative features and in analyzing the tumor microenvironment. In order to perform automated image analysis, H&E-stained images need to be normalized as shown in Fig. 3.

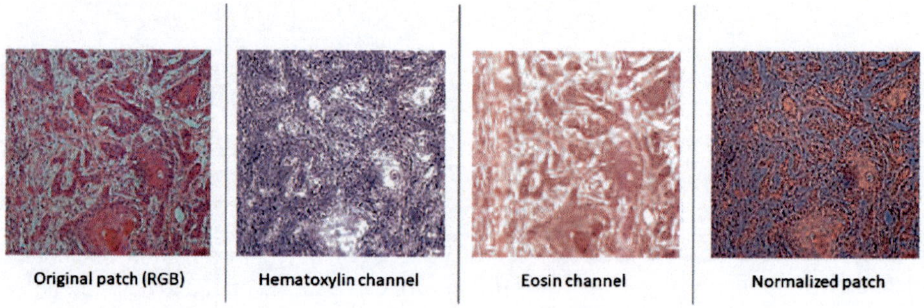

Fig. 3. Visual representation of H&E-stained normalization

This is due to the large color variations in images caused by sample preparation and imaging settings. In our research, we used the Macenko approach [20] where the Singular Value Decomposition (SVD) geodesic method is used for obtaining stain vectors. The first step is to convert the RGB color vector to their corresponding optical density (OD) values and then remove data with OD intensity less than β. A threshold value of $\beta = 0.15$ was found to provide the most robust results while removing as little data as possible. The next step is to calculate singular value decomposition (SVD) on the OD tuples and then create a plane from the SVD directions corresponding to the two largest singular values. After projecting data onto the plane and normalizing to the unit length, we calculate the angle of each point regarding the first SVD direction. The final step is to convert extreme values back to OD space. Figure 4 shows images before and after normalization.

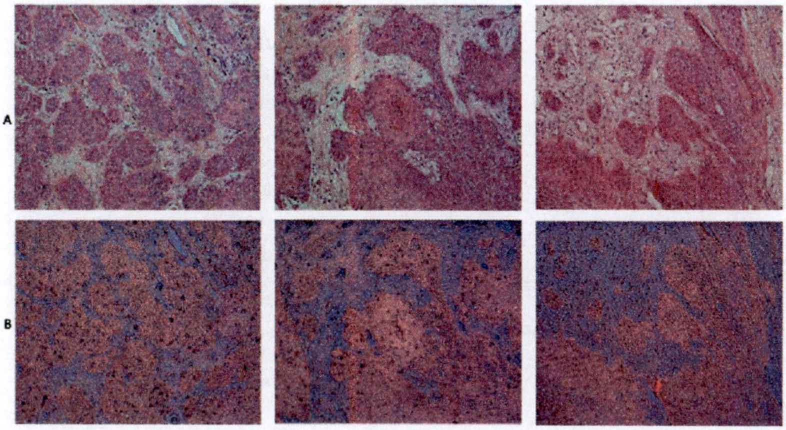

Fig. 4. Visual representation of A) H&E-stained images and B) normalized H&E-stained images

Normalized H&E-stained images are then used as input for deep CNN architectures. The first experimental results are achieved with VGG-16, ResNet50 and InceptionResNetv2 which are pretrained on ImageNet. Stratified 5-fold cross-validation is used to estimate the performance of the AI-based model while Area Under the ROC Curve (AUC) is used as an evaluation metric. In our contribution, the pretrained CNN networks were fine-tuned to suit the mentioned classification task based on the normalized H&E images. This fine-tuning was accomplished based on additional layers at the end of the aforementioned architectures. The first added layer was global average pooling, and the second was the fully connected layer, also known as the output layer.

In the case of VGG-16, the highest AUC value of 0.98 is achieved in the fifth fold as shown in Fig. 5. However, the same architecture achieves the lowest values in the fourth fold. The mean AUC value of five-fold cross-validation is 0.93 along with a standard deviation (σ) of 0.03.

On the other hand, by using ResNet50 architecture the highest AUC value of 1.00 is achieved in the third fold, while the fourth fold achieves the lowest value as shown in Fig. 6. The mean AUC and standard deviation values of 5-fold cross-validation are $0.98 \pm \sigma = 0.02$.

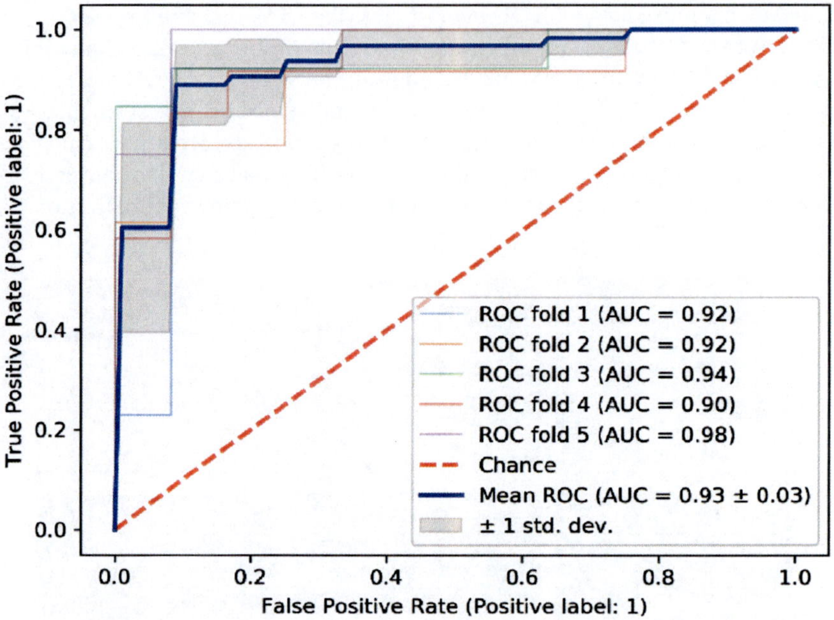

Fig. 5. AUC values of 5-fold cross-validation utilizing VGG-16 architecture

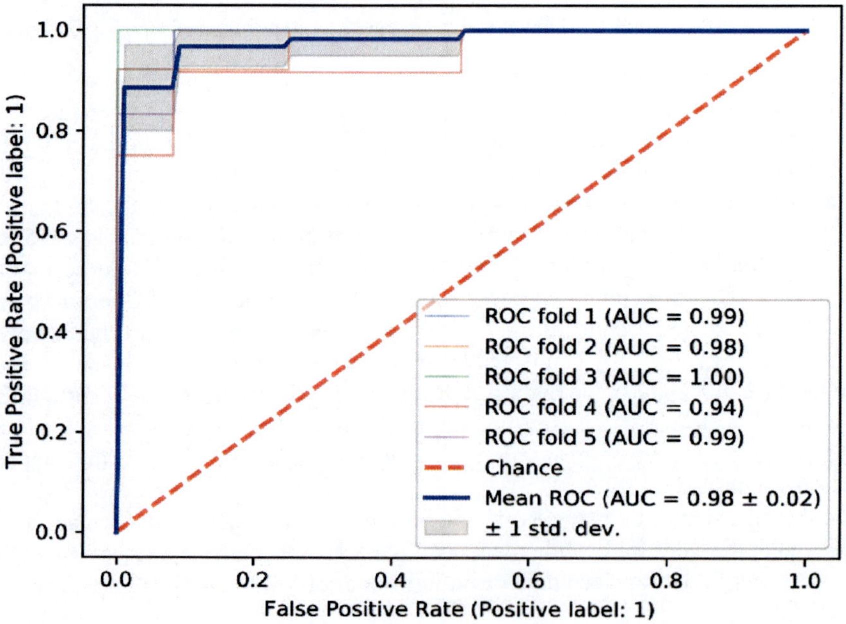

Fig. 6. AUC values of 5-fold cross-validation utilizing ResNet50 architecture

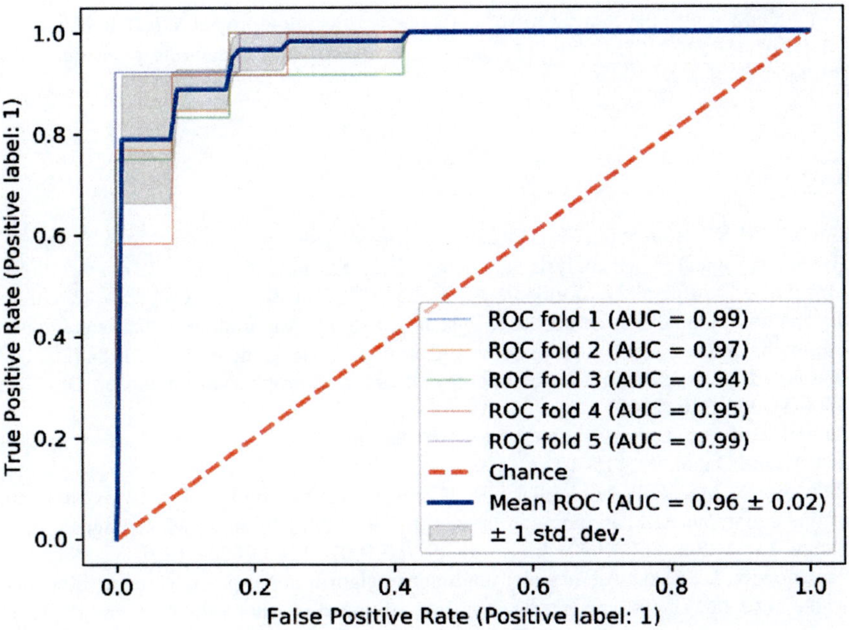

Fig. 7. AUC values of 5-fold cross-validation utilizing InceptionResNetv2 architecture

Figure 7 shows that InceptionResNetv2 architecture achieves the overall highest AUC value of 0.99 in the first and the last fold. Moreover, the lowest value is achieved in the third fold. The mean AUC value of five-fold cross-validation is 0.96 along with a standard deviation of 0.02.

After image normalization ResNet50 resulted in the highest classification value of 0.98 ($\sigma \pm 0.02$) AUC. Figures 5, 6 and 7 represent the ROC metric to evaluate classifier output quality using cross-validation. ROC curves are shown for each of the 5-fold cross-validation and the overall average ROC curve (blue), based on VGG-16, ResNet50 and InceptionResNetv2 predictions.

4 Conclusions

Obtained results reveal that the application of AI-based algorithms along with preprocessing methods, such as image normalization for image analysis, has great potential in the diagnosis of OSCC. Integration of preprocessing method along with the convolutional neural network resulted in 0.98 ($\sigma \pm 0.02$) AUC. However, data availability was a limitation of the research so future work should use a dataset with more histopathology images to create a more robust system. The presented approach is the first step in automating histopathological image analysis, therefore, in future work, we plan to integrate more preprocessing methods with other AI classification algorithms.

Acknowledgments. This research has been (partly) supported by the CEEPUS network CIII-HR-0108, European Regional Development Fund under the grant KK.01.1.1.01.0009 (DATACROSS), project CEKOM under the grant KK.01.2.2.03.0004, Erasmus+ project WICT under the grant 2021–1-HR01-KA220-HED-000031177 and University of Rijeka scientific grant uniri-tehnic-18–275-1447.

References

1. Gurcan, M.N., Boucheron, L.E., Can, A., Madabhushi, A., Rajpoot, N.M., Yener, B.: Histopathological image analysis: a review. IEEE Rev. Biomed. Eng. **2**, 147–171 (2009)
2. Musulin, J., Štifanić, D., Zulijani, A., Ćabov, T., Dekanić, A., Car, Z.: An enhanced histopathology analysis: an AI-based system for multiclass grading of oral squamous cell carcinoma and segmenting of epithelial and stromal tissue. Cancers **13**(8), 1784 (2021)
3. Warnakulasuriya, S.: Global epidemiology of oral and oropharyngeal cancer. Oral Oncol. **45**(4–5), 309–316 (2009)
4. Zanoni, D.K., et al.: Survival outcomes after treatment of cancer of the oral cavity (1985–2015). Oral Oncol. **90**, 115–121 (2019)
5. Rahman, N., MacNeill, M., Wallace, W., Conn, B.: Reframing histological risk assessment of oral squamous cell carcinoma in the era of UICC 8th edition TNM staging. Head Neck Pathol. **15**(1), 202–211 (2020). https://doi.org/10.1007/s12105-020-01201-8
6. Pantanowitz, L., et al.: An artificial intelligence algorithm for prostate cancer diagnosis in whole slide images of core needle biopsies: a blinded clinical validation and deployment study. Lancet Digital Health **2**(8), e407–e416 (2020)
7. Song, Z., et al.: Clinically applicable histopathological diagnosis system for gastric cancer detection using deep learning. Nat. Commun. **11**(1), 1–9 (2020)
8. Chen, M., et al.: Classification and mutation prediction based on histopathology H&E images in liver cancer using deep learning. NPJ Precis. Oncol. **4**(1), 1–7 (2020)
9. El-Naggar, A.K., Chan, J.K.C., Rubin Grandis, J., Takata, T., Slootweg, P.J.: International Agency for Research on Cancer. WHO classification of head and neck tumours. World Health Organization classification of tumours. 4. Lyon: International Agency for Research on Cancer (2017)
10. Musulin, J., et al.: Automated grading of oral squamous cell carcinoma into multiple classes using deep learning methods. In: 2021 IEEE 21st International Conference on Bioinformatics and Bioengineering (BIBE), pp. 1–6 (2021)
11. Štifanić, D., Musulin, J., Car, Z., Čep, R.: Use of convolutional neural network for fish species classification. Pomorski zbornik **59**(1), 131–142 (2020)
12. Simonyan, K., Zisserman, A.: Very deep convolutional networks for large-scale image recognition. arXiv preprint arXiv:1409.1556 (2014)
13. Theckedath, D., Sedamkar, R.R.: Detecting affect states using VGG16, ResNet50 and SE-ResNet50 networks. SN Comput. Sci. **1**(2), 1–7 (2020)
14. He, K., Zhang, X., Ren, S., Sun, J.: Proceedings of the IEEE Conference on Computer Vision and Pattern Recognition (2016)
15. Tian, X., Chen, C.: Modulation pattern recognition based on Resnet50 neural network. In: 2019 IEEE 2nd International Conference on Information Communication and Signal Processing (ICICSP), pp. 34–38 (2019)
16. Szegedy, C., Vanhoucke, V., Ioffe, S., Shlens, J., Wojna, Z.: Rethinking the inception architecture for computer vision. In: Proceedings of the IEEE Conference on Computer Vision and Pattern Recognition, pp. 2818–2826 (2016)
17. Szegedy, C., Ioffe, S., Vanhoucke, V., Alemi, A.A.: Inception-v4, inception-resnet and the impact of residual connections on learning. In: Thirty-first AAAI Conference on Artificial Intelligence (2017)

18. Mahdianpari, M., Salehi, B., Rezaee, M., Mohammadimanesh, F., Zhang, Y.: Very deep convolutional neural networks for complex land cover mapping using multispectral remote sensing imagery. Remote Sens. **10**(7), 1119 (2018)
19. Bhatia, Y., Bajpayee, A., Raghuvanshi, D., Mittal, H.: Image captioning using Google's inception-resnet-v2 and recurrent neural network. In: 2019 Twelfth International Conference on Contemporary Computing (IC3), pp. 1–6. IEEE (2019)
20. Macenko, M., et al.: A method for normalizing histology slides for quantitative analysis. In: 2009 IEEE International Symposium on Biomedical Imaging: From Nano to Macro, pp. 1107–1110 (2009)

The Projects Evaluation and Selection by Using MCDM and Intuitionistic Fuzzy Sets

Aleksandar Aleksić(✉), Snežana Nestić, and Danijela Tadić

Faculty of Engineering, University of Kragujevac, 6 Sestre Janjić Street, 34000 Kragujevac, Serbia
{aaleksic,s.nestic,galovic}@kg.ac.rs

Abstract. The process of project evaluation and selection is essential in many companies, especially in information technology (IT) organizations. The objective of this research is to propose a two-stage multi-attribute decision-making (MADM) model combined with intuitive fuzzy sets (IFs) to evaluate and select projects concerning predefined criteria. In the first stage, the fuzzy pairwise comparison matrix of the criteria's relative importance is constructed. The fuzzy weights vector of criteria is determined by using AHP extended with IFs. In the second step, the MABAC extended with IFs is proposed and applied for ranking the considered alternatives, in this case, projects. The proposed methodology enables rejection of the considered project proposals if they do not meet any defined criteria. In this way, the rank of the treated project proposals is determined at the level of each group of treated criteria, so the decision-makers can have the final call on the selection of project funding and execution. The model is tested on real-life data from one IT company operating in the Republic of Serbia.

Keywords: project management · intuitionistic fuzzy set · IF-AHP · F-MABAC

1 Introduction

In the IT sector, most companies design their organization to be project-based. The functioning of these companies is determined by ongoing project selection and their execution, so their employees are seeking new project opportunities constantly. In this situation, there can be an accumulated long list of potential projects that can be realized but it should be mentioned that each organization has limited resources, so they need to choose the projects wisely as they usually strive for a lean management approach [1]. During the process of project assessment and selection, some companies also perform vulnerability of projects [2]. The Project Portfolio Selection can be very complex [3]. Also, the decision brought today will determine the allocation of resources in the future so the project selection can be analyzed from the perspective of the companies' sustainability [4]. Inevitably, there is a conclusion that it is much more convenient to develop an appropriate project portfolio than to randomly select projects for their execution.

The goal of this research is to provide a reliable model for the project evaluation and selection. In the past few years, the level of uncertainty has increased in business

N. Filipovic (Ed.): AAI 2022, LNNS 659, pp. 132–150, 2023.
https://doi.org/10.1007/978-3-031-29717-5_10

terms all over the world caused of pandemic and geopolitical issues. These new conditions seek to employ different models in business that can handle uncertainty which is also applicable to project management issues [5]. The motivation of this research is to enhance the pool of models that are used for the evaluation and selection of projects that are capable of handling uncertainties efficiently. The intuitionistic fuzzy sets theory was introduced by Atanassov [6] which allows vagueness to be represented fairly quantitatively. In this manuscript, uncertainties into the relative importance of criteria and their values are modelled by the intuitionistic fuzzy numbers (IFNs) which present the special type of the intuitionistic fuzzy set [7]. In general, membership function and non-membership function can have different shapes. In the literature, triangular intuitionistic fuzzy numbers (TIFNs) [8], and trapezoidal intuitionistic fuzzy numbers (TrIFNs) are mostly used for handling various uncertainties [9, 10].

Many authors suggest that decision-maker (DM) can make a more precise evaluation of the relative importance criteria if each pair of criteria is evaluated separately [11–13]. The elements of the fuzzy pair-wise comparison matrix of the relative importance of criteria are described by TIFs [9, 13] and TrIFNs [8], as in this research. Assessment consistency of DMs is checked by applying the eigenvector method as suggested in conventional AHP [14]. Transformation of the fuzzy pair-wise comparison matrix into a pair-wise comparison matrix can be realized in different ways [15]. In the scope of this research, it is achieved by using the defuzzification procedure [16]. The overall weights vector is given by using the fuzzy geometric operator, as suggested by Dutta and Guha, [17]. In this way, according to fuzzy algebra rules [6], the weights of criteria are described by TrIFNs.

The rank of projects can be obtained by using the various multi-criteria attribute method (MADM) which are based on the different theoretical foundations [18]. Choosing these MADMs can be considered a problem on its own. It is important to mention that if the project does not fulfill only one criterion it is not taken into further consideration. Concerning the nature of the problem, the authors suggest that by applying MABAC we can come to the best possible solution. In the literature, we can find many papers in which MABAC is extended with the intuitionistic fuzzy sets theory [19–23]. Determining the distance between the current values of the decision matrix and elements of BAA [20] is based on a procedure developed by Mishra et al. [21]. Belonging to BAA [23] is based on using the weighting function which is proposed in this paper. New procedures for determining the distance between two IFNs are proposed in [22, 24]. Belonging to BAA is determined in a way similar to Mishra et al. [20]. It should be noted that the choice of distance presents the main difference between analyzed papers.

2 Problem Statement

The project evaluation and selection represent one of the most strategic management tasks involving different managers as the decision-makers. The number of projects that are going to be selected to be adjoined to the project's portfolio is presented by a set of indices $\{1, \ldots, i, \ldots, I\}$. The total number of identified investment projects is denoted as I, and i, $i = 1, \ldots, I$ is the index of the project. In the general case, investment projects can be estimated concerning criteria groups, which can be formally represented by a set of indices $\{1, \ldots, k, .., K\}$. The total number of criteria groups is denoted as K. The index of the criteria group is k, $k = 1, \ldots, K$. These criteria groups consist of many criteria which can be presented by a set of indices $\{1, \ldots, j, \ldots, J_k\}$. The total number of criteria under criteria group k, $k = 1, .., K$ is denoted as J_k, and j, $j = 1, \ldots, J_k$ is the index of criterion.

In this research, the assessment is performed in compliance with the criteria defined by Pinto [25] which are presented in Table 1. As there is a lot of uncertainty regarding the discussed criteria, fuzzy sets are employed within the proposed models as in already conducted research [26, 27].

In the problem, the assumption is introduced: (i) the relative importance of criteria under each criterion group does not have equal importance and it is presented by a fuzzy pair-wise comparison matrix, (ii) the criteria under criterion groups are a benefit type and cost type; criteria values are assessed by DMs who use natural language expressions.

Table 1. Criteria in project evaluation and selection

Risk–unpredictability to the firm ($k = 1$)	Technical ($j = 1$) Financial ($j = 2$) Safety ($j = 3$) Quality ($j = 4$) Legal exposure ($j = 5$)
Commercial–market potential ($k = 2$)	Expected return on investment ($j = 1$) Payback period (j = 2) Potential market share ($j = 3$) Long-term market dominance ($j = 4$) Initial cash outlay ($j = 5$) Ability to generate future business/new markets ($j = 7$)
Internal operating–changes in firm operations ($k = 3$)	Need to develop/train employees ($j = 1$) Change in workforce size or composition ($j = 2$) Change in physical environment ($j = 3$) Change in manufacturing or service operations ($j = 4$)
Additional ($k = 4$)	Patent protection ($j = 1$) Impact on company's image ($j = 2$) Strategic fit ($j = 3$)

2.1 Modelling of Uncertainties

Uncertainties that exist in the model such as relative importance criteria and their values are assessed by DMs. Respecting the nature of human thinking, it can be argued that DMs express their assessments better by using natural language words instead of precise numbers. In this paper, modelling of predefined linguistic statements is based on TrIFNs [6].

All the considered criteria for evaluating investment projects have different relative importance which can be considered unchangeable during the considered period of time. The relative importance of the criteria is assessed by the decision-making team (chief executive officer, marketing manager, chief operating officer, chief information officer). The elements of the fuzzy pairwise comparison matrix of the relative importance of criteria are described by pre-defined linguistic expressions and their corresponding TrIFNs are presented:

Equal importance (EW) - ([1,1,1,1];0.95,0.05)
Very low importance (VLW) - ([1,1,1.5,5.5];0.85,0.1)
Low importance (LW) - ([1,3.5,4.5,7];0.65,0.3)
Medium importance (MW) - ([2.5,5,6,8.5];0.55,0.4)
High importance (HW) - ([3,5.5,6.5,9];0.7,0.25)
Very high value (VHW) - ([4.5,8.5,9,9];0.75,0.2)

It is supposed that, the criteria values could be adequately described by using 5 linguistic expressions which are modelled by TrIFNs as presented:

Very low value (VLV) - ([1,1,2,3.5];0.9,0)
Low value (LV) - ([1, 2, 3, 4];0.8,0.1)
Medium value (MV) - ([3,4.5,5.5,7];0.5,0.4)
High value (HV) - ([6,7,8,9];0.7,0.3)
Very high value (VHV) - ([7.5,8,9,9];0.8,0.1)

The domains of these TrIFNs are defined into real line intervals [1–9]. The value 1 indicates the lowest relative importance and value of criteria, respectively.Similarly, the value 9 indicates the highest relative importance and value of criteria, respectively. The values of membership function and non-membership function have been established according to the rating of DMs. They base their assessments on knowledge and experience. The overlap of TrIFNs describing the relative importance of criteria is large. This indicates a lack of knowledge of DMs about the importance of the considered criteria.

Granularity is defined as the number of fuzzy numbers assigned to the relative importance criteria, as well as their values, and depend on the problem and type of size as well as estimation of DMs.

All the considered criteria for evaluating investment projects have different relative importance which can be considered unchangeable during the considered period. The relative importance of the criteria is assessed by DMs. The decision-making team is in charge of the assessment of the criteria values, too. The elements of the fuzzy pairwise comparison matrix of the relative importance of criteria are described by five different pre-defined linguistic expressions corresponding to TrIFNs. The overlap of TrIFNs describing the relative importance of criteria is large. This indicates a lack of knowledge of DMs about the importance of the considered criteria.

3 Methodology

The proposed algorithm can be summarized as presented in Fig. 1. The proposed model consists of two stages. A further description of the model is applied to each group of the treated criteria.

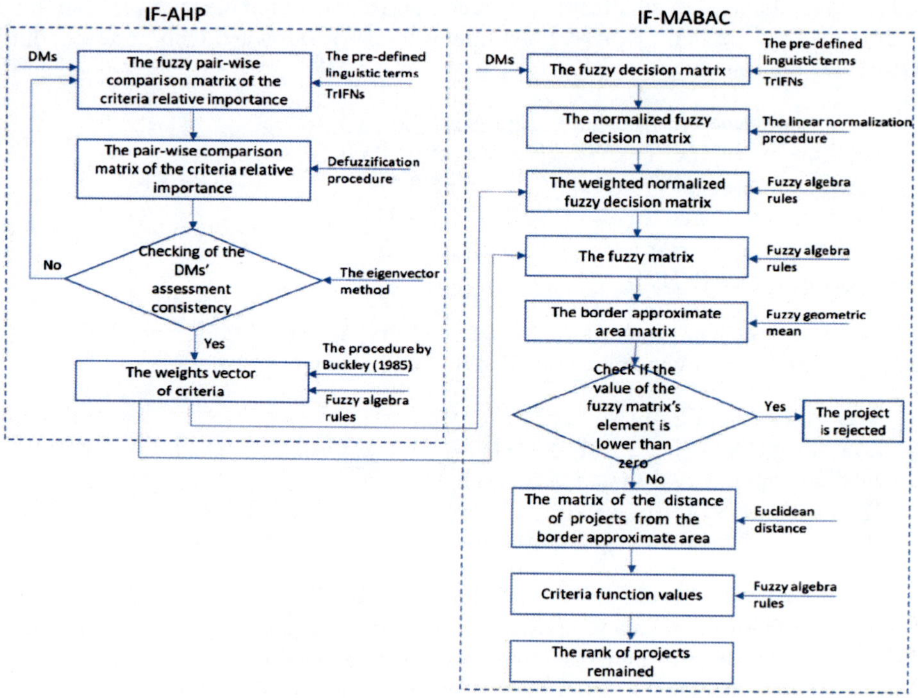

Fig. 1. The proposed two-stage research model

The first stage starts with defining a fuzzy pairwise comparison matrix. This matrix is transformed into a pairwise comparison matrix by using the defuzzification procedure [28]. Checking of the DMs' assessment consistency is performed by using the Eigenvector method [14]. Determining the weights vector of criteria is based on fuzzy algebra rules [6] and the procedure defined by Buckley [29].

The second stage is used for determining the project's rank by using the extended MABAC method [30] with TrIFNs. The proposed method is executed in more steps compared to the conventional due to the specificity of TrIFNs mathematical operations. The treated criteria are cost and benefit type, so the normalization procedure is applied. The elements of the weighted normalized fuzzy decision matrix are calculated as a product of the criteria weight and the normalized value. The fuzzy matrix is constructed as the sum of weights vector and weighted normalized fuzzy decision matrix. The new procedure is developed to check alternatives' belonging to the border approximate areas. If the value of the fuzzy matrix's element is lower than zero, then the project is rejected.

The rest of the projects are further considered. The value of the criteria function is calculated as a sum of the distance between the two TrIFNs. Those TrIFNs describe the values of fuzzy matrix elements and the values of the border approximate area matrix. In the first place in the rank, there is a project with the highest value adjoined.

3.1 The Proposed Algorithm

Step 1. Fuzzy rating of the relative importance of criteria j over j', j, j' = j = 1, ..., J_k; $j \neq j'$ is performed by DMs at the level of criteria group k, $k = 1, .., K$

$$\tilde{W}_{jj'}^{k} = \left(\left[a_{jj'}^{k}, b_{jj'}^{k}, c_{jj'}^{k}, d_{jj'}^{k}\right]; \mu_{jj'}^{k}, \vartheta_{jj'}^{k}\right) \tag{1}$$

Step 2. Transform the pair-wise comparison matrix into the pair-wise comparison matrix by using the [31]:

$$\left[W_{jj'}^{k}\right] \tag{2}$$

Values of the pair-wise comparison matrix, $W_{jj'}^{k}$, $j, j' = 1, \ldots, J_k$; $k = 1, .., K$ are given by using the defuzzification procedure presented in [32]. The consistency of the estimates of DMs is performed by using the eigenvector method. If C.I. is less than or equal to 0.1 it can be considered that the assessment errors made do not significantly affect the accuracy of the solution.

Step 3. Determination of weights vector of criteria under criteria group k, $\overset{\sim}{\omega}_j^{k}$, $j = 1, .., J_k$; $k = 1, .., K$ is based on the procedure which is proposed by Bruckley [30]. So that:

$$\overset{\sim}{\omega}_j^{k} = \left(\left[\begin{array}{c} \dfrac{\sqrt[J_k]{\prod_{j'=1}^{J_k} a_{jj'}^{k}}}{\sum_{j=1,\ldots,J_k}\sqrt[J_k]{\prod_{j'=1}^{J_k} d_{jj'}^{k}}}, \dfrac{\sqrt[J_k]{\prod_{j'=1}^{J_k} b_{jj'}^{k}}}{\sum_{j=1,\ldots,J_k}\sqrt[J_k]{\prod_{j'=1}^{J_k} c_{jj'}^{k}}}, \\ \dfrac{\sqrt[J_k]{\prod_{j'=1}^{J_k} c_{jj'}^{k}}}{\sum_{j=1,\ldots,J_k}\sqrt[J_k]{\prod_{j'=1}^{J_k} b_{jj'}^{k}}}, \dfrac{\sqrt[J_k]{\prod_{j'=1}^{J_k} d_{jj'}^{k}}}{\sum_{j=1,\ldots,J_k}\sqrt[J_k]{\prod_{j'=1}^{J_k} a_{jj'}^{k}}} \\ \min_{j=1,\ldots,J_k} \mu_j, \max_{=1,\ldots,J_k} \vartheta_j \end{array}\right];\right) \tag{3}$$

Step 5. The fuzzy decision matrix at the level criteria group k, $k = 1, \ldots, K$ is stated:

$$\left[\tilde{x}_{ij}^{k}\right]_{I\times J_k} \tag{4}$$

Where $\tilde{x}_{ij}^{k}$ is TrIFN which describes the value criterion j, $j = 1, \ldots, J_k$ for project i, $i = 1, .., I$ at the level criterion group k, $k = 1, .., K$

$$\tilde{x}_{ij}^{k} = \left(\left[l_{ij}^{k}, m_{ij}^{k}, n_{ij}^{k}, p_{ij}^{k}\right]; \mu_i, \vartheta_i\right) \tag{5}$$

Step 6. The normalized fuzzy decision matrix is constructed by using the linear normalization procedure:

$$\left[\tilde{r}_{ij}^{k}\right]_{\mathrm{I}\times J_k} \tag{6}$$

where:

Benefit type

$$\tilde{r}_{ij}^{k}=\left(\left[\frac{l_{ij}^{k}}{p^{*}},\frac{m_{ij}^{k}}{p^{*}},\frac{n_{ij}^{k}}{p^{*}},\frac{p_{ij}^{k}}{p^{*}}\right];\mu_i,\vartheta_i\right) \tag{7}$$

$$p^{*}=\max_{i=1,..,I}p_{ij}^{k},j=1,..,J_k;\,k=1,..,K \tag{8}$$

Cost type

$$\tilde{r}_{ij}^{k}=\left(\left[\frac{l^{-}}{p_{ij}^{k}},\frac{l^{-}}{n_{ij}^{k}},\frac{l^{-}}{m_{ij}^{k}},\frac{l^{-}}{l_{ij}^{k}}\right]\right) \tag{9}$$

$$l^{-}=\min_{i=1,..,I}l_{ij}^{k},j=1,..,J_k;\,k=1,..,K \tag{10}$$

Step 7. The weighted fuzzy decision matrix at the level of each criteria group k, $k=1,\ldots,K$ is constructed:

$$\left[\tilde{z}_{ij}^{k}\right]_{\mathrm{I}\times J_k} \tag{11}$$

where:

$$\tilde{z}_{ij}^{k}=\tilde{\omega}_{\mathrm{k}}\cdot\tilde{x}_{ij}^{k}=\left[L_{ij}^{k},M_{ij}^{k},N_{ij}^{k},P_{ij}^{k}\right];\min\left(\mu_i,\mu_j\right),\max\left(\vartheta_i,\vartheta_j\right) \tag{12}$$

Step 8. Construct to the fuzzy matrix, $\left[\tilde{v}_{ij}^{k}\right]_{I\times J_k}$, $i=1,\ldots,I;j=1,..,J_k;k=1,\ldots,K$. The elements of this matrix are given by applying the fuzzy algebra rules [33]:

$$\tilde{v}_{ij}^{k}=\tilde{z}_{ij}^{k}+\tilde{\omega}_{j}^{k} \tag{13}$$

Step 9. The column matrix, $\left[\tilde{g}_{j}^{k}\right]_{1\times J_k}$, $j=1,..,J_k;\mathrm{k}=1,\ldots,\mathrm{K}$ is constructed. The elements of the border approximate area matrix are given:

$$\tilde{g}_{j}^{k}=\sqrt[I]{\prod_{i=1,..,I}\tilde{v}_{ij}^{k}} \tag{14}$$

where:

$$\tilde{g}_{j}^{k}=\left[\alpha_{j}^{k},\beta_{j}^{k},\gamma_{ij}^{k},\delta_{ij}^{k}\right];\min\left(\mu_{i}^{k}\right),\max\left(\vartheta_{i}^{k}\right) \tag{15}$$

Step 10. Affiliation/belonging of the project i, $i = 1, .., I$ to the approximate area are defined in conventional MABAC [30]:

$$i \in \{ \begin{matrix} G^{+} \; if \; \tilde{v}_{ij}^{k} > \tilde{g}_{j}^{k} \\ G \; if \; \tilde{v}_{ij}^{k} = \tilde{g}_{j}^{k} \\ G^{-} \; if \; \tilde{v}_{ij}^{k} < \tilde{g}_{j}^{k} \end{matrix} \tag{16}$$

The comparison of fuzzy numbers is based on the procedure introduced by De and Das [34]. It should be emphasized that the project which belongs to G^{-}, it is not considered in the future.

Step 11. Criteria function values by the project are calculated according to the procedure proposed in conventional MABAC [29]:

$$S_i = \sum_{j=1,.,J_k} d\left(\tilde{v}_{ij}^{k}, \tilde{g}_{j}^{k}\right) \tag{17}$$

where: $d\left(\tilde{v}_{ij}^{k}, \tilde{g}_{j}^{k}\right)$ is the distance between to TrIFNs [34].

Step 12. The rank of projects under criteria group k, $k = 1, .., K$ is given by respecting criteria function values, S_i. In the first place in the rank, there is the project which is associated with the highest value S_i.

4 An Illustrative Example

Project evaluation and selection models [25] can be numeric or non-numeric with features such as realism, cost-effectiveness, comparability, ease of use, etc. In this example, the proposed numeric model is tested on real-life data. The company from the IT sector that operates in Central Serbia has conducted the process of project selection for the next period. The data needed for model testing is obtained by taking into account the evidence data and the decision-making team that brought their decisions by reaching a consensus. The decision-making team has consisted of three DMs: chief executive officer, marketing manager, chief operating officer, and chief information officer. The seven different project proposals have been taken into the consideration.

According to the proposed Algorithm (Step 1 to step 2), the fuzzy pair-wise comparison matrix at the level of each criteria group k is stated:

Risk–unpredictability to the firm (***k*** = 1)

$$\begin{bmatrix} - & \frac{1}{VLW} & LW & LW & MW \\ & - & MW & LW & VHW \\ & & - & \frac{1}{LW} & MW \\ & & & - & LW \\ & & & & - \end{bmatrix}, C.I. = 0.04$$

Commercial–market potential ($k = 2$)

$$\begin{bmatrix} - & VLW & HW & HW & LW & VHW \\ & - & EW & LW & \frac{1}{LW} & MW \\ & & - & VLW & \frac{1}{MW} & LW \\ & & & - & \frac{1}{LW} & HW \\ & & & & - & MW \\ & & & & & - \end{bmatrix}, C.I. = 0.06$$

Internal operating–changes in firm operations ($k = 3$)

$$\begin{bmatrix} - & VLW & \frac{1}{MW} & \frac{1}{LW} \\ & - & \frac{1}{LW} & \frac{1}{VLW} \\ & & - & LW \\ & & & - \end{bmatrix}, C.I. = 0.07$$

Additional ($k = 4$)

$$\begin{bmatrix} - & MW & LW \\ & - & \frac{1}{VLW} \\ & & - \end{bmatrix}, C.I. = 0.08$$

The weights vector for each criteria group is calculated by procedure (step 3 of the proposed Algorithm).

Risk–unpredictability to the firm ($k = 1$)

$$\tilde{\omega}_1^1 = ([0.07, 0.27, 0.40, 0.96]; 0.55, 0.4)$$

$$\tilde{\omega}_2^1 = ([0.13, 0.34, 0.50, 1.39]; 0.55, 0.4)$$

$$\tilde{\omega}_3^1 = ([0.03, 0.07, 0.10, 0.36]; 0.55, 0.4)$$

$$\tilde{\omega}_4^1 = ([0.04, 0.11, 0.17, 0.63]; 0.55, 0.4)$$

$$\tilde{\omega}_5^1 = ([0.02, 0.03, 0.04, 0.15]; 0.55, 0.4)$$

Commercial–market potential ($k = 2$)

$$\tilde{\omega}_1^2 = ([0.14, 0.34, 0.48, 0.86]; 0.55, 0.4)$$

$$\tilde{\omega}_2^2 = ([0.05, 0.13, 0.18, 0.44]; 0.55, 0.4)$$

$$\tilde{\omega}_3^2 = ([0.04, 0.07, 0.10, 0.29]; 0.55, 0.4)$$

$$\tilde{\omega}_4^2 = ([0.03, 0.06, 0.09, 0.27]; 0.55, 0.4)$$

$$\tilde{\omega}_5^2 = ([0.08, 0.22, 0.32, 0.88]; 0.55, 0.4)$$

$$\tilde{\omega}_6^2 = ([0.01, 0.02, 0.06, 0.12]; 0.55, 0.4)$$

Internal operating–changes in firm operations ($\boldsymbol{k} = 3$)

$$\tilde{\omega}_1^3 = ([0.04, 0.08, 0.11, 0.49]; 0.55, 0.4)$$

$$\tilde{\omega}_2^3 = ([0.03, 0.10, 0.15, 0.40]; 0.55, 0.4)$$

$$\tilde{\omega}_3^3 = ([0.14, 0.48, 0.70, 1.81]; 0.55, 0.4)$$

$$\tilde{\omega}_4^3 = ([0.07, 0.16, 0.25, 0.99]; 0.55, 0.4)$$

Additional ($\boldsymbol{k} = 4$)

$$\tilde{\omega}_1^4 = ([0.21, 0.60, 0.81, 1.81]; 0.55, 0.44)$$

$$\tilde{\omega}_2^4 = ([0.04, 0.11, 0.16, 0.34]; 0.55, 0.44)$$

$$\tilde{\omega}_3^4 = ([0.08, 0.01, 0.20, 0.82]; 0.55, 0.4)$$

The input data are presented in Appendix 2 (step 5 of the proposed algorithm).

The proposed Algorithm (Step 6 to step 12) is illustrated by data for criteria group 2. The values criterion ($j = 1$) for the project ($i = 1$) is:

$$\tilde{r}_{11}^2 = \left(\left[\frac{3}{9}, \frac{4.5}{9}, \frac{5.5}{9}, \frac{7}{9}\right]; 0.5, 0.4\right) = \tilde{r}_{11}^2 = ([0.33, 0.50, 0.61, 0.78]; 0.8, 0.1)$$

The values criterion ($j = 2$) for the project ($i = 1$):

$$([6, 7, 8, 9]; 0.7, 0.3)$$

$$\tilde{r}_{12}^2 = \left(\left[\frac{1}{9}, \frac{1}{8}, \frac{1}{7}, \frac{1}{6}\right]; 0.7, 0.3\right) = ([0.11, 0.12, 0.14, 0.17]; 0.7, 0.3)$$

The weighted normalized criterion value, $\tilde{z}_{11}^2$ and fuzzy value, $\tilde{v}_{11}^2$ for the project ($i = 1$) under criterion ($j = 1$) of criterion group ($k = 2$) is:

$$\tilde{\omega}_1^2 = ([0.14, 0.34, 0.48, 0.86]; 0.55, 0.4)$$

$$\begin{aligned}\tilde{z}_{11}^2 &= ([0.14, 0.34, 0.48, 0.86]; 0.55, 0.4) \cdot ([0.33, 0.50, 0.61, 0.78]; 0.8, 0.1) \\ &= ([0.05, 0.17, 0.29, 0.67]; 0.55, 0.4)\end{aligned}$$

$$\begin{aligned}\tilde{v}_{11}^2 &= ([0.05, 0.17, 0.29, 0.67]; 0.55, 0.4) + ([0.14, 0.34, 0.48, 0.86]; 0.55, 0.4) \\ &= ([0.19, 0.51, 0.77, 1.53]; 0.55, 0.4)\end{aligned}$$

Table 2. The fuzzy matrix under criterion group ($k = 2$)

	$j = 1$	$j = 2j = 2$
$i = 1$	$([0.19, 0.51, 0.77, 1.53]; 0.55, 0.4)$	$([0.06, 0.17, 0.27, 0.88]; 0.55, 0.4)$
$i = 2$	$([0.23, 0.61, 0.91, 1.72]; 0.55, 0.4)$	$([0.06, 0.14, 0.20, 0.50]; 0.55, 0.4)$
$i = 3$	$([0.16, 0.41, 0.64, 1.24]; 0.55, 0.4)$	$([0.06, 0.15, 0.22, 0.59]; 0.55, 0.4)$
$i = 4$	$([0.19, 0.51, 0.77, 1.53]; 0.55, 0.4)$	$([0.06, 0.15, 0.22, 0.59]; 0.55, 0.4)$
$i = 5$	$([0.19, 0.51, 0.77, 1.53]; 0.55, 0.4)$	$([0.06, 0.17, 0.27, 0.88]; 0.55, 0.4)$
$i = 6$	$([0.16, 0.41, 0.64, 1.24]; 0.55, 0.4)$	$([0.06, 0.15, 0.22, 0.59]; 0.55, 0.4)$
$i = 7$	$([0.23, 0.61, 0.91, 1.72]; 0.55, 0.4)$	$([0.06, 0.19, 0.36, 0.88]; 0.55, 0.4)$
	$j = 3$	$j = 4$
$i = 1$	$([0.07, 0.12, 0.19, 0.58]; 0.55, 0.4)$	$([0.05, 0.11, 0.17, 0.54]; 0.55, 0.4)$
$i = 2$	$([0.05, 0.10, 0.16, 0.52]; 0.55, 0.4)$	$([0.05, 0.11, 0.18, 0.54]; 0.55, 0.4)$
$i = 3$	$([0.04, 0.09, 0.13, 0.47]; 0.55, 0.4)$	$([0.05, 0.11, 0.17, 0.54]; 0.55, 0.4)$
$i = 4$	$([0.07, 0.13, 0.58, 0.58]; 0.55, 0.4)$	$([0.05, 0.11, 0.17, 0.54]; 0.55, 0.4)$
$i = 5$	$([0.07, 0.12, 0.19, 0.58]; 0.55, 0.4)$	$([0.05, 0.11, 0.18, 0.54]; 0.55, 0.4)$
$i = 6$	$([0.05, 0.10, 0.16, 0.52]; 0.55, 0.4)$	$([0.04, 0.09, 0.14, 0.48]; 0.55, 0.4)$
$i = 7$	$([0.07, 0.12, 0.19, 0.58]; 0.55, 0.4)$	$([0.05, 0.11, 0.17, 0.54]; 0.55, 0.4)$
	$j = 5$	$j = 6$
$i = 1$	$([0.10, 0.29, 0.48, 1.76]; 0.55, 0.4)$	$([0.02, 0.04, 0.12, 0.24]; 0.55, 0.4)$
$i = 2$	$([0.10, 0.29, 0.48, 1.76]; 0.55, 0.4)$	$([0.02, 0.04, 0.12, 0.24]; 0.55, 0.4)$
$i = 3$	$([0.09, 0.26, 0.39, 1.17]; 0.55, 0.4)$	$([0.02, 0.04, 0.12, 0.24]; 0.55, 0.4)$
$i = 4$	$([0.17, 0.25, 0.37, 1.03]; 0.55, 0.4)$	$([0.02, 0.04, 0.12, 0.24]; 0.55, 0.4)$
$i = 5$	$([0.10, 0.29, 0.48, 1.76]; 0.55, 0.4)$	$([0.02, 0.04, 0.12, 0.24]; 0.55, 0.4)$
$i = 6$	$([0.10, 0.33, 0.64, 1.76]; 0.55, 0.4)$	$([0.01, 0.02, 0.08, 0.17]; 0.55, 0.4)$
$i = 7$	$([0.10, 0.29, 0.48, 1.76]; 0.55, 0.4)$	$([0.02, 0.04, 0.12, 0.24]; 0.55, 0.4)$

In a similar way, the element values of the fuzzy matrix, are calculated and presented in Table 2.

The border approximate area matrix, $\tilde{g}_j^2$ is given:

$$\begin{bmatrix} ([0.19, 0.50, 0.77, 1.49]; 0.55, 0.4) \\ ([0.06, 0.14, 0.21, 0.53]; 0.55, 0.4) \\ ([0.06, 0.11, 0.23, 0.54]; 0.55, 0.4) \\ ([0.05, 0.11, 0.15, 0.53]; 0.55, 0.4) \\ ([0.11, 0.28, 0.47, 1.45]; 0.55, 0.4) \\ ([0.02, 0.04, 0.11, 0.24]; 0.55, 0.4) \end{bmatrix}^T$$

Determine the belonging of the project ($i = 3$) to BAA which is evaluated according to criterion ($j = 1$) within the criteria group ($k = 2$).($k = 2$).

$$V_\mu\left(\tilde{v}_{11}^2\right) = \frac{0.16 + 2 \cdot (0.41 + 0.64) + 1.24}{6} \cdot 0.55 = 0.3208$$

$$V_\vartheta\left(\tilde{v}_{11}^2\right) = \frac{0.16 + 2 \cdot (0.41 + 0.64) + 1.24}{6} \cdot (1 - 0.4) = 0.2333$$

$$A_{\mu}\left(\tilde{v}_{31}^{2}\right)=\frac{1.24-0.16-2\cdot(0.41-0.64)}{3}\cdot 0.55=0.2567$$

$$A_{\vartheta}\left(\tilde{v}_{31}^{2}\right)=\frac{1.24-0.16-2\cdot(0.41-0.64)}{3}\cdot(1-0.4)=0.3080$$

$$V\left(\tilde{v}_{31}^{2}\right)=\frac{0.3208+0.2333}{2}=0.2770$$

$$A\left(\tilde{v}3_{1}^{2}\right)=\frac{0.2567+0.3080}{2}=0.2823$$

$$R\left(\tilde{v}_{31}^{2}\right)=V\left(\tilde{v}_{31}^{2}\right)+A\left(\tilde{v}_{31}^{2}\right)=0.2770+0.2823=0.5593$$

Similarly, we calculate $R(\tilde{g}_{1}^{2})=0.7320$.

If $R(\tilde{v}_{31}^{2})$ is lower than $R(\tilde{g}_{1}^{2})$, it can be said that project $(i=3)$ belongs to lower BAA, according to the rules defined in the conventional MABAC. Respecting the nature of the problem, project $(i=3)$ is not considered under criterion group 2 in the future. Similarly, the belonging of the rest projects under the criterion $(j=1)$ of the treated criterion group are given:

$R(\tilde{v}_{11}^{2})=0.7667$, $R(\tilde{v}_{21}^{2})=0.7889$, $R(\tilde{v}_{41}^{2})=0.7667$, $R(\tilde{v}_{51}^{2})=0.7667$, $R(\tilde{v}_{61}^{2})=0.5593$, and $Rv712=0.7889$ and presented in Fig. 2.

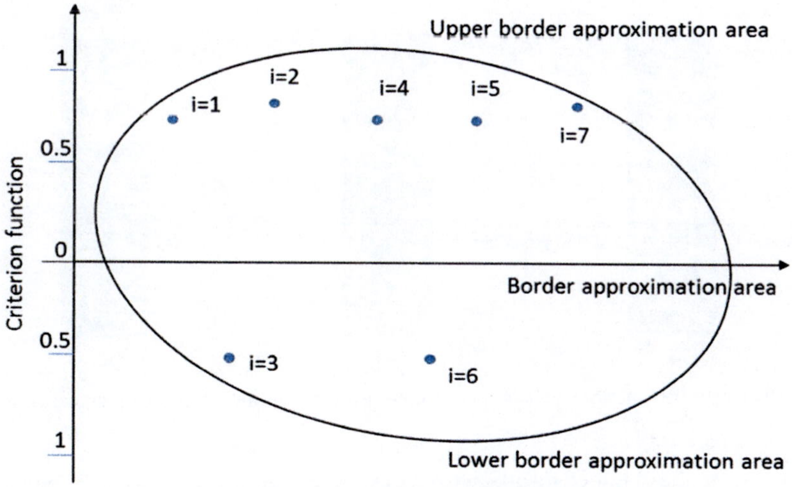

Fig. 2. Projects belonging to the BAA

In a similar way, the determination of the rest project belonging to BAA at the level of each criterion and criteria group ($k = 2$) and presented in Table 3.

Table 3. Project belonging to BAA at the level of criteria group ($k = 2$)

	$j = 1$	$j = 2$	$j = 3$	$j = 4$	$j = 5$	$j = 6$
$i = 1$	G^+	G^+	G^+	G^+	G^+	G^+
$i = 2$	G^+	G^+	–	G^+	G^+	G^+
$i = 3$	–	–	–	G^+	–	G^+
$i = 4$	G^+	–	G^+	G^+	–	G^+
$i = 5$	G^+	G^+	G^+	G^+	G^+	G^+
$i = 6$	–	–	–	–	G^+	–
$i = 7$	G^+	G^+	G^+	G^+	G^+	G^+

Concerning all criteria under criterion group 2, criteria function values are determined by applying the proposed Algorithm (Step 11) and presented in Fig. 3.

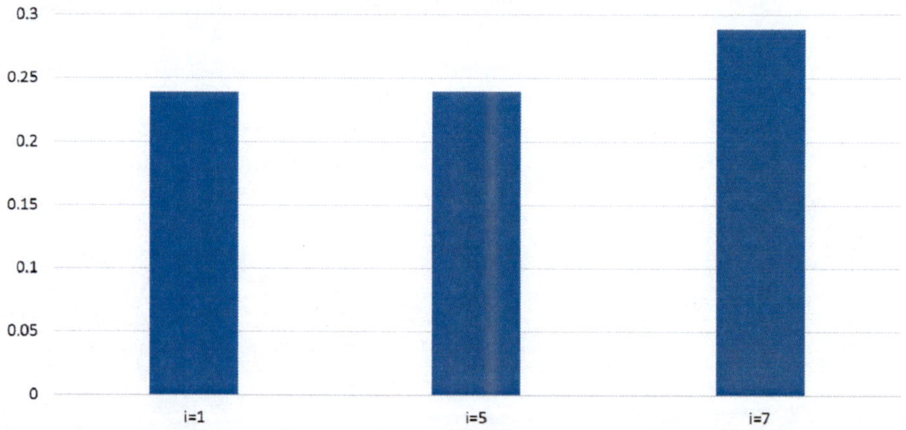

Fig. 3. The criteria functions values for selected projects within the criterion group ($k = 2$)

According to the obtained results, within the criteria group $k = 2$, it can be concluded that the best project is projected $i = 7$. It should be emphasized that the projects $i = 1$ and $i = 5$ can also be considered.

In a similar way, we can also evaluate the projects under other criteria groups. The obtained results are presented in Fig. 4.

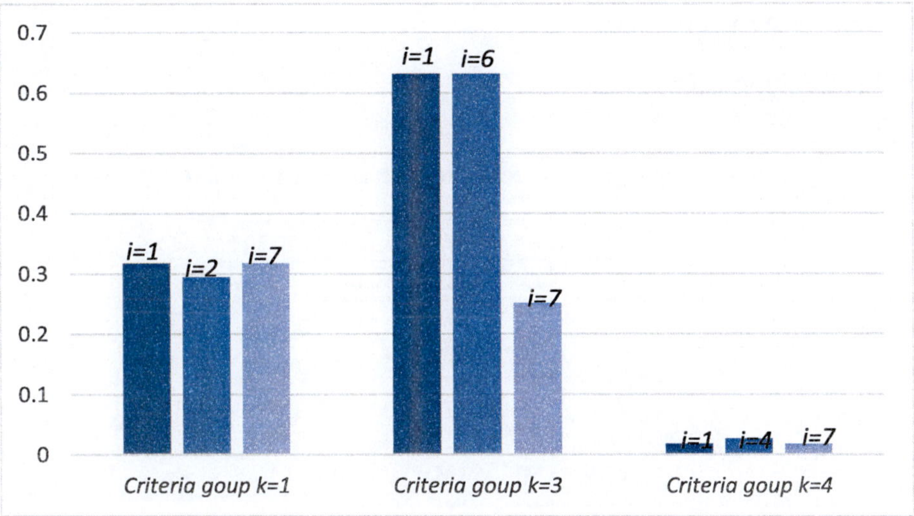

Fig. 4. The criteria functions values for selected projects within the other criterion groups

Respecting the obtained results (see Fig. 4), the projects ($i = 1$), ($i = 4$), and ($i = 7$) are close to the border approximation area. This information is main for DMs to take more into account when designing projects so that the results of projects can be patented and contribute more to the company's image. According to the results which are presented in Fig. 3 and the Fig. 4 it can be seen that the two projects ($i = 1$) and ($i = 7$) have satisfied all criteria of the considered criteria groups. By applying the proposed method, we do not get the answer to which project is the best but we determine a set of projects that decision-makers can further consider. Choice of projects can be based on the assessment of DMs or these projects need to be further analyzed.

Also, if the available budget is sufficient DMs can implement both projects.

5 Conclusions

The proposed model is tested on the real data from the company operating in the IT sector in the Republic of Serbia. The input data is obtained through collaboration with the decision-makers team from the company as suggested in the description of the two-stage model. At the level of each considered group of the treated criteria, the rank of the projects is obtained. As the rank is obtained at the level of each four criteria groups, the decision-makers should determine which project will be selected for the realization.

The main theoretical contribution comes from the modification of the MABAC method in terms of checking if the considered alternative belongs to the border approximate areas. Due to the nature of the considered problem, if the alternative belongs to the lower border approximate area, this alternative is rejected.

The future research will include the analysis of different methods for determining the most suitable project for funding taking into account the considered criteria groups.

Appendix A: Input data

See Tables 4, 5, 6 and 7

Table 4. Criteria values for Risk–unpredictability to the firm ($k = 1$)

	$j=1$	$j=2$	$j=3$	$j=4$	$j=5$
$i=1$	VLV	LV	LV	LV	VLV
$i=2$	LV	LV	MV	LV	LV
$i=3$	MV	MV	LV	HV	VHV
$i=4$	HV	LV	VHV	VHV	LV
$i=5$	LV	VLV	VLV	MV	HV
$i=6$	VHV	MV	MV	VHV	VLV
$i=7$	VLV	LV	LV	LV	VLV

Table 5. Criteria values for Commercial–market potential ($k = 2$)

	$j=1$	$j=2$	$j=3$	$j=4$	$j=5$	$j=6$
$i=1$	MV	LV	MV	HV	LV	VHV
$i=2$	HV	VLV	LV	VHV	LV	HV
$i=3$	LV	MV	VLV	HV	MV	HV
$i=4$	MV	MV	MV	HV	HV	HV
$i=5$	MV	LV	VHV	VHV	LV	VHV
$i=6$	LV	MV	MV	MV	VLV	LV
$i=7$	HV	VLV	VHV	HV	LV	HV

Appendix B: Preliminaries

Definition 1. An intuitionistic fuzzy set A in the universe of discourse X is defined with the form [6]:

$$\tilde{A} = \left(x, \mu_{\tilde{A}}(x), \vartheta_{\tilde{A}}(x) | x \in X\right)$$

Table 6. Criteria values for Internal operating–changes in firm operations ($k = 3$)

	$j=1$	$j=2$	$j=3$	$j=4$
$i=1$	LV	VLV	LV	VLV
$i=2$	LV	VHV	VHV	LV
$i=3$	MV	MV	HV	LV
$i=4$	HV	HV	VHV	VLV
$i=5$	LV	VHV	MV	VLV
$i=6$	MV	VLV	LV	VLV
$i=7$	LV	LV	MV	VLV

Table 7. Criteria values for Additional ($k = 4$)

	$j=1$	$j=2$	$j=3$
$i=1$	*VHV*	*VHV*	*HV*
$i=2$	*HV*	*VHV*	*MV*
$i=3$	*VHV*	*HV*	*HV*
$i=4$	*VHV*	*VHV*	*VHV*
$i=5$	*HV*	*HV*	*HV*
$i=6$	*HV*	*HV*	*LV*
$i=7$	*VHV*	*VHV*	*HV*

where:

the numbers $\mu_{\tilde{A}}(x) \to [0,\ 1]$ and $\vartheta_{\tilde{A}}(x) \to [0,\ 1]$ denote the membership degree and non-membership degree.

with the condition

$$0 \leq \mu_{\tilde{A}}(x) + \vartheta_{\tilde{A}}(x) \leq 1, \forall x \in X$$

for each intuitionistic fuzzy set $\tilde{A}$ from set X, the following holds:

$$\pi_{\tilde{A}}(x) = 1 - \mu_{\tilde{A}}(x) - \vartheta_{\tilde{A}}(x)$$

$$0 \leq \pi_{\tilde{A}}(x) \leq 1, \forall x \in X$$

The value of $\pi_{\tilde{A}}(x)$ is called the degree of indeterminacy (or hesitation). The smaller $\pi_{\tilde{A}}(x)$, more certain $\tilde{A}$.

Definition 2. An IFS $\tilde{A} = \left(x,\ \mu_{\tilde{A}}(x),\ \vartheta_{\tilde{A}}(x) | x \in X\right)$ of the real line is called an intuitionistic fuzzy number (IFN) whose membership function and non-membership function are defined as follows [35].

$$\vartheta_{\tilde{A}}(x) = \begin{pmatrix} \frac{b-x+(x-a_1)\cdot\vartheta_{\tilde{A}}(x)}{b-a_1} & if\ a_1 \le x < b \\ \vartheta_{\tilde{A}}(x) & if\ \ b < x \le c \\ \frac{x-c+(d_1-x)\cdot\vartheta_{\tilde{A}}(x)}{d_1-c} & if\ c < x \le d_1 \\ 1 & else \end{pmatrix}$$

where a, b, c, d, a_1, d_1 are real numbers, and $a_1 \le a \le b \le c \le d \le d_1$.

If $[a, b, c, d] = [a_1, b, c, d_1]$, then the TrIFN can be simply denoted as:

$$\tilde{A} = \left([a, b, c, d];\ \mu_{\tilde{A}}(x), \vartheta_{\tilde{A}}(x)\right)$$

If $a \ge 0$ and one of the three values, b, c, d is not equal to 0, then the TrIFN $\tilde{A} = \big([a, b, c, d];\ \mu_{\tilde{A}}(x),\ \vartheta_{\tilde{A}}(x)\big)$ is called a positive TrIFN.

Definition 3. Let $\tilde{A} = \big([a_1, b_1, c_1, d_1];\ \mu_{\tilde{A}}(x),\ \vartheta_{\tilde{A}}(x)\big)$ and $\tilde{B} = \big([a_2, b_2, c_2, d_2];\ \mu_{\tilde{B}}(x),\ \vartheta_{\tilde{B}}(x)\big)$ be two positive TrIFNs. And λ is a real number. The operations of these TrIFNs are [30]:

$$\begin{aligned}\widetilde{A+B} =& ([a_1 - a_2, b_1 - b_2, c_1 - c_2, d_1 \\ & -d_2];\ \min\left(\mu_{\tilde{A}}(x), \mu_{\tilde{B}}(x)\right), \max\left(\vartheta_{\tilde{A}}(x), \vartheta_{\tilde{B}}(x)\right))\end{aligned}$$

$$\begin{aligned}\widetilde{A-B} =& ([a_1 - d_2, b_1 - c_2, c_1 - b, d_1 \\ & -a_2];\ \min\left(\mu_{\tilde{A}}(x), \mu_{\tilde{B}}(x)\right), \max\left(\vartheta_{\tilde{A}}(x), \vartheta_{\tilde{B}}(x)\right))\end{aligned}$$

$$\begin{aligned}\tilde{A}\cdot\tilde{B} =& ([a_1 \cdot a_1, b_1 \cdot b_1, c_1 \cdot c_1, d_1 \\ & \cdot d_2] \min\left(\mu_{\tilde{A}}(x), \mu_{\tilde{B}}(x)\right), \max\left(\vartheta_{\tilde{A}}(x), \vartheta_{\tilde{B}}(x)\right))\end{aligned}$$

$$\lambda \cdot \tilde{A} = \left([\lambda \cdot a_1, \lambda \cdot b_1, \lambda \cdot c_1, \lambda \cdot d_1];\ \mu_{\tilde{A}}(x), \vartheta_{\tilde{A}}(x)\right)$$

$$(\tilde{A})^{-1} = \left(\left[\frac{1}{d_1}, \frac{1}{c_1}, \frac{1}{b_1}, \frac{1}{a_1}\right]; \mu_{\tilde{A}}(x), \vartheta_{\tilde{A}}(x)\right)$$

Definition 4. Let $\tilde{A} = \big([a_1, b_1, c_1, d_1];\ \mu_{\tilde{A}}(x),\ \vartheta_{\tilde{A}}(x)\big)$ and $\tilde{B} = \big([a_2, b_2, c_2, d_2];\ \mu_{\tilde{B}}(x),\ \vartheta_{\tilde{B}}(x)\big)$ be two positive TrIFNs. The Euclidean distance between two TrIFNs is defined [36]:

$$d\left(\tilde{A}, \tilde{B}\right) = \frac{1}{2} \cdot \sqrt{\begin{array}{c}(a_1 - a_2)^2 + (b_1 - b_2)^2 + (c_1 - c_2)^2 + (d_1 - d_2)^2 + \\ \max\left\{\left(\mu_{\tilde{A}}(x) - \mu_{\tilde{B}}(x)\right)^2, \left(\vartheta_{\tilde{A}}(x) - \vartheta_{\tilde{B}}(x)\right)^2\right\}\end{array}}$$

References

1. Sousa, P., Tereso, A., Alves, A.A., Gomes, L.: Implementation of project management and lean production practices in a SME Portuguese innovation company. Procedia Comput. Sci. **138**, 867–874 (2018)
2. Aleksić, A., Puskarić, H., Tadić, D., Stefanović, M.: Project management issues: vulnerability management assessment. Kybernetes **46**(7), 1171–1188 (2017)
3. Gutiérrez, E., Magnusson, M.: Dealing with legitimacy: a key challenge for project portfolio management decision makers. Int. J. Project Manage. **32**, 30–39 (2014)
4. Shaukat, M.B., Latif, K.F., Sajjad, A., Eweje, G.: Revisiting the relationship between sustainable project management and project success: the moderating role of stakeholder engagement and team building. Sustain. Dev. **30**(1), 58–75 (2022)
5. Ma, J., Harstvedt, J.D., Jaradat, R.R., Smith, B.: Sustainability driven multi-criteria project portfolio selection under uncertain decision-making environment. Comput. Ind. Eng. **140**(2), 106236 (2020)
6. Atanassov, K.: Intuitionistic Fuzzy Sets: Theory and Applications. Physica-Verlag, Wyrzburg, Germany (1999)
7. Ye, J.: Multicriteria group decision-making method using vector similarity measures for trapezoidal intuitionistic fuzzy numbers. Group Decis. Negot. **21**(4), 519–530 (2012)
8. Wan, S.P.: Power average operators of trapezoidal intuitionistic fuzzy numbers and application to multi-attribute group decision making. Appl. Math. Model. **37**, 4112–4126 (2013)
9. Mirghafoori, S.H., Izadi, M.R., Daei, A.: Analysis of the barriers affecting the quality of electronic services of libraries by VIKOR, FMEA and entropy combined approach in an intuitionistic-fuzzy environment. J. Intell. Fuzzy Syst. **34**(4), 2441–2451 (2018)
10. Saini, N., Bajaj, R.K., Gandotra, N.: Dwivedi RP, Multi-criteria decision making with triangular intuitionistic fuzzy number based on distance measure & parametric entropy approach. Proc. Comput. Sci. **125**, 34–41 (2018)
11. Liu, H.C., You, J.X., You, X.Y., Shan, M.M.: A novel approach for failure mode and effects analysis using combination weighting and fuzzy VIKOR method. Appl. Soft Comput. **28**, 579–588 (2015)
12. Sakthivel, G., Saravanakumar, D., Muthuramalingam, T.: Application of failure mode and effect analysis in manufacturing industry - an integrated approach with FAHP-fuzzy TOPSIS and FAHP-fuzzy VIKOR. Int. J. Prod. Qual. Manage. **24**(3), 398 (2018)
13. Gojković, R., Đurić, G., Tadić, D., Nestić, S., Aleksić, A.: Evaluation and selection of the quality methods for manufacturing process reliability improvement – intuitionistic fuzzy sets and genetic algorithm approach. Mathematics **9**(13), 1531 (2021)
14. Saaty, T.L.: The modern science of multicriteria decision making and its practical applications: the AHP/ANP approach. Oper. Res. **61**(5), 1101–1118 (2013)
15. Tadić, D., et al.: The evaluation and enhancement of quality, environmental protection and seaport safety by using FAHP. Nat. Hazard. **17**(2), 261–275 (2017)
16. Wan, S.P., Wang, Q.Y., Dong, J.Y.: The extended VIKOR method for multi-attribute group decision making with triangular intuitionistic fuzzy numbers. Knowl.-Based Syst. **52**, 65–77 (2013)
17. Dutta, B., Guha, D.: Preference programming approach for solving intuitionistic fuzzy AHP. Int. J. Comput. Intell. Syst. **8**(5), 977–991 (2015)
18. Tadić, D., Arsovski, S., Aleksić, A., Stefanović, M., Nestić, S.: A fuzzy evaluation of projects for business processes' quality improvement. In: Kahraman, C., Çevik Onar, S. (eds) Intelligent Techniques in Engineering Management. Intelligent Systems Reference Library, p. 87 (2015)

19. Xue, Y.X., You, J.X., Lai, X.D., Liu, H.C.: An interval-valued intuitionistic fuzzy MABAC approach for material selection with incomplete weight information. Appl. Soft Comput. **38**, 703–713 (2016)
20. Mishra, A.R., Chandel, A., Motwani, D.: Extended MABAC method based on divergence measures for multi-criteria assessment of programming language with interval-valued intuitionistic fuzzy sets. Granular Comput. **5**(1), 97–117 (2018). https://doi.org/10.1007/s41066-018-0130-5
21. Mishra, A.R., Rani, P., Jain, D.: Information measures based TOPSIS method for multicriteria decision making problem in intuitionistic fuzzy environment. Iran. J. Fuzzy Syst. **14**(6), 41–63 (2017)
22. Liang, R.-X., He, S.-S., Wang, J.-Q., Chen, K., Li, L.: An extended MABAC method for multi-criteria group decision-making problems based on correlative inputs of intuitionistic fuzzy information. Comput. Appl. Math. **38**(3), 1–28 (2019). https://doi.org/10.1007/s40314-019-0886-5
23. Zhao, M., Wei, G., Chen, X., Wei, Y.: Intuitionistic fuzzy MABAC method based on cumulative prospect theory for multiple attribute group decision making. Int. J. Intell. Syst. **36**(11), 6337–6359 (2021)
24. Verma, R.: On intuitionistic fuzzy order-α divergence and entropy measures with MABAC method for multiple attribute group decision-making. J. Intell. Fuzzy Syst. **40**(1), 1191–1217 (2021)
25. Pinto, J.K.: Project Management: Achieving Competitive Advantage, 4th Edition. Pearson education, Pennsylvania State University – Erie (2016)
26. Aleksić, A., Runić Ristić, M., Komatina, N., Tadić, D.: Advanced risk assessment in reverse supply chain processes: a case study in republic of serbia. Adv. Prod. Eng. Manag. **14**(4), 421–434 (2019)
27. Nestić, S., Lampón, J.F., Aleksić, A., Cabanelas, P., Tadić, D.: Ranking manufacturing processes from the quality management perspective in the automotive industry. Expert. Syst. **36**(6), 1–16 (2019)
28. Atanassov, K.T.: On Intuitionistic Fuzzy Sets Theory, vol. 283. Springer, Cham (2012)
29. Buckley, J.J.: Fuzzy hierarchical analysis. Fuzzy Sets Syst. **17**(3), 233–247 (1985)
30. Pamučar, D., Ćirović, G.: The selection of transport and handling resources in logistics centers using Multi-Attributive Border Approximation area Comparison (MABAC). Expert Syst. Appl. **42**(6), 3016–3028 (2015)
31. Atanassov, K.T., et al.: Novel Developments in Uncertainty Representation and Processing Advances in Intuitionistic Fuzzy Sets and Generalized Nets – Proceedings of 14th International Conference on Intuitionistic Fuzzy Sets and Generalized Nets Advances in Intelligent Systems and Computing AISC, vol. 401. Springer Cham (2016). https://doi.org/10.1007/978-3-319-26211-6
32. Awasthi, A., Govindan, K., Gold, S.: Multi-tier sustainable global supplier selection using a fuzzy AHP-VIKOR based approach. Int. J. Prod. Econ. **195**, 106–117 (2018)
33. Dinagar, D.S., Thiripurasundari, K.: A navel method for solving fuzzy transportation problem involving intuitionistic trapezoidal fuzzy numbers. Int. J. Curr. Res. **6**(6), 7038–7041 (2014)
34. De, P.K., Das, D.: A study on ranking of trapezoidal intuitionistic fuzzy numbers. Int. J. Comput. Inf. Syst. Ind. Manage. Appl. **6**, 437–444 (2014)
35. Hao, Y., Chen, X., Wang, X.: A ranking method for multiple attribute decision-making problems based on the possibility degrees of trapezoidal intuitionistic fuzzy numbers. Int. J. Intell. Syst. **34**(1), 24–38 (2018)
36. Grzegorzewski, P.: Distances between intuitionistic fuzzy sets and/or interval-valued fuzzy sets based on the Hausdorff metric. Fuzzy Sets Syst. **148**(2), 319–328 (2004). https://doi.org/10.1016/j.fss.2003.08.005

Application of MCDM DIBR-Rough Mabac Model for Selection of Drone for Use in Natural Disaster Caused by Flood

Duško Z. Tešić[1](✉), Darko I. Božanić[1], and Boža D. Miljković[2]

[1] Military Academy, University of Defense in Belgrade, 33 Veljka Lukića Kurjaka Street, 11042 Belgrade, Serbia
tesic.dusko@yahoo.com, dbozanic@yahoo.com

[2] Faculty of Education, University of Novi Sad, 4 Podgorička Street, 25101 Sombor, Serbia
bole@ravangrad.net

Abstract. Natural disasters around the world have resulted in enormous casualties and economic damage. Floods, as one of the natural disasters caused by climate change and inadequate human attitude towards nature, often create major problems for countries on all continents. Although preventive action is one of the ways to prevent the occurrence of floods, we are witnessing that they continue to happen, so the elimination of the consequences of floods and saving lives is given great attention. With the advancement of technology, modern means, machines and devices are increasingly used to rescue people from flooded areas, both to provide assistance to the endangered and their evacuation, as well as to monitor and reconnoiter flood-affected locations. The paper presents the application of the MCDM model DIBR-Rough MABAC in the selection of drones, based on the characteristics of drones, for use during floods, i.e., for surveying flooded areas and delivery of necessary materials, food and water. The presented model was successfully applied, where the optimal one was selected from a set of seven alternatives. Validation of the model was performed by analyzing its sensitivity to changes in weight coefficients. The results obtained by sensitivity analysis indicate that the mentioned model is stable to the change of the weight coefficients of the criteria, that is, that the correlation coefficients tend towards an ideal positive correlation.

Keywords: MCDM · Rough MABAC · DIBR · Flood · Drone

1 Introduction

The unpredictability of climate change, worldwide, significantly affects the increase in the number of natural disasters, especially those of meteorological and hydrological origin. Natural disasters are phenomena that disrupt the normal course of life [1] resulting in casualties, causing great damage to property or causing loss of property, damage to infrastructure and greatly endangering the environment [2], where the community does not have the ability to repair the losses and damage without someone's help [3].

N. Filipovic (Ed.): AAI 2022, LNNS 659, pp. 151–169, 2023.
https://doi.org/10.1007/978-3-031-29717-5_11

Natural disasters are caused by natural forces, and are manifested through earthquakes, fires, floods, droughts, avalanches, storms, landslides and landslides, hurricane winds, volcanic eruptions, etc. Of course, man, with his attitude towards nature, also contributes to the fact that natural disasters have greater consequences than in previous years, by not regulating the riverbeds, that at high water levels leads to the formation of floods. Floods "represent the overflow of water beyond natural or artificial boundaries, i.e., leaving their beds by flooding larger or smaller areas, endangering people and material goods." [4] posing a threat to all elements of the state and society.

The occurrence of large floods often causes the problem of physical access to flooded areas, and for the purpose of reconnaissance of these places, finding people who need help, delivering the necessary medicines, food, water and other necessary materials, drones can be used. Drone (Unmanned Aircraft Vehicle), is a synthesis of unmanned aerial vehicle and devices necessary to control it [5], i.e., it is an aircraft that can fly without a human operator in it [6]. Depending on their purpose, drones are equipped with different types of sensors and cameras, and their application in different areas indicates the importance of their existence. The basic and essential characteristics of drones are: "weight, payload, endurance and range, speed, wing loading, cost, engine type and power" [6–8].

This paper presents the MCDM model DIBR-Rough MABAC for the selection of drones for use during floods, i.e., for surveying flooded areas and the delivery of necessary materials, food and water, based on the characteristics of drones, through two goals. The first goal is to use the DIBR method to obtain weight coefficients, which will clearly reflect the importance of each of the criteria. The second goal refers to the selection of drones, using the Rough MABAC method, with successful and quality treatment of imprecisions and uncertainties.

The selection of drones using different MCDM methods has been presented in many papers: for "last mile" delivery using the interval-valued inferential fuzzy TOPSIS method [9], with methods AHP, COPRAS I TOPSIS [10], for the specific needs of farmers using the software Expert Choice (AHP method) [11], agriculture drone for little forming space using the methods AHP and TOPSIS [12], in the defence field with methods AHP and TOPIS [13] etc. Drones' applications for supporting disaster management, has been the subject of research, for example, in papers [14–19].

2 Description of Model and Used Methods

The DIBR-Rough MABAC MCDM model consists of three phases. The appearance of the model is presented in Fig. 1.

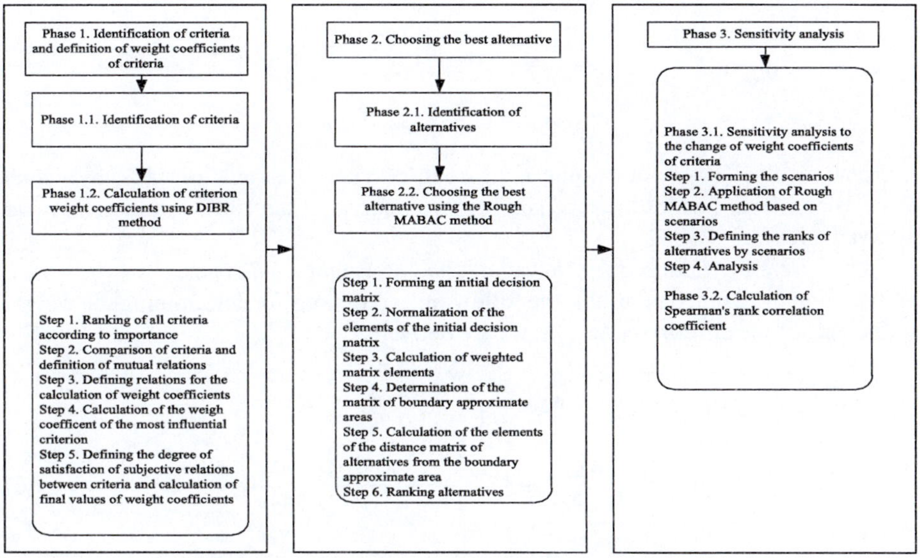

Fig. 1. DIBR-Rough MABAC MCDM model

In the following, a description of the DIBR method, rough numbers and Rough MABAC method is given.

2.1 DIBR Method

There are different methods for determining the weight coefficients of the criteria [20–24] etc., and one of them is the DIBR method [25]. This method is based on defining the relationship between ranked criteria, i.e., it considers the relationships between adjacent criteria, and this method consists of five steps presented below [25]:

Step 1. Ranking of criteria according to significance.

On a defined set of n criteria $C = \{C_1, C_2, ..., C_n\}$ the criteria are ranked according to significance as $C_1 > C_2 > C_3 > ... > C_n$, where n represents the total number of criteria in the set C.

Step 2. Comparison of criteria and definition of mutual relations.

When comparing the criteria, values $\lambda_{12}, \lambda_{13}, ..., \lambda_{n-1,n}$ and λ_{1n} are obtained, that is, when the criterion C_1 is compared with C_2, the value λ_{12} is obtained etc., and these values should satisfy the condition that it is $\lambda_{n-1,n}, \lambda_{1n} \in [0, 1]$. Based on the previously defined conditions, the following relationships between the criteria are reached:

$$w_1 : w_2 = (1 - \lambda_{12}) : \lambda_{12} \quad (1)$$

$$w_2 : w_3 = (1 - \lambda_{23}) : \lambda_{23} \quad (2)$$

$$\ldots$$

$$w_{n-1} : w_n = (1 - \lambda_{n-1,n}) : \lambda_{n-1,n} \tag{3}$$

$$w_1 : w_n = (1 - \lambda_{1,n}) : \lambda_{1,n} \tag{4}$$

Relationships (1)–(4) and value $\lambda_{n-1,n}$ can be viewed as relationships, by which the decision maker divides the total significance interval of the 100% criterion into two observed criteria.

Step 3. Defining relations for the calculation of weight coefficients.

Based on the defined relations, the following expressions for determining the weight coefficients of the criteria $w_2, w_3, \ldots, w_n$ are derived:

$$w_2 = \frac{\lambda_{12}}{(1 - \lambda_{12})} w_1 \tag{5}$$

$$w_3 = \frac{\lambda_{23}}{(1 - \lambda_{23})} w_2 = \frac{\lambda_{12}\lambda_{23}}{(1 - \lambda_{12})(1 - \lambda_{23})} w_1 \tag{6}$$

$$\ldots$$

$$w_n = \frac{\lambda_{n-1,n}}{\left(1 - \lambda_{n-1,n}\right)} w_{n-1} = \frac{\lambda_{12}\lambda_{23} \cdot \ldots \cdot \lambda_{n-1,n}}{(1 - \lambda_{12})(1 - \lambda_{23}) \cdot \ldots \cdot \left(1 - \lambda_{n-1,n}\right)} w_1$$
$$= \frac{\prod_{i=1}^{n-1} \lambda_{i,i+1}}{\prod_{i=1}^{n-1} \left(1 - \lambda_{i,i+1}\right)} w_1 \tag{7}$$

Step 4. Calculation of the weight coefficient of the most influential criterion.

Based on expressions (5)–(7) and satisfying the condition that $\sum_{j=1}^{n} w_j = 1$, the following relation is defined:

$$w_1 \left(1 + \frac{\lambda_{12}}{(1 - \lambda_{12})} + \frac{\lambda_{12}\lambda_{23}}{(1 - \lambda_{12})(1 - \lambda_{23})} + \ldots + \frac{\prod_{i=1}^{n-1} \lambda_{i,i+1}}{\prod_{i=1}^{n-1} \left(1 - \lambda_{i,i+1}\right)}\right) = 1 \tag{8}$$

Based on expression (8), the expression for the calculation of the weight coefficient of the most influential criterion is defined:

$$w_1 = \frac{1}{1 + \frac{\lambda_{12}}{(1-\lambda_{12})} + \frac{\lambda_{12}\lambda_{23}}{(1-\lambda_{12})(1-\lambda_{23})} + \ldots + \frac{\prod_{i=1}^{n-1} \lambda_{i,i+1}}{\prod_{i=1}^{n-1} (1-\lambda_{i,i+1})}} \tag{9}$$

Based on the value of the weighting coefficient of the most influential criterion, expression (9), the weighting coefficients of other criteria are calculated $w_2, w_3, \ldots, w_n$.

Step 5. Defining the degree of satisfaction of subjective relations between criteria.

Based on expression (4), the value of the weight coefficient of the criterion w_n is defined:

$$w_n = \frac{\lambda_{1n}}{(1 - \lambda_{1n})} w_1 \tag{10}$$

Expression (4) is a relation for controlling expression (7), which is intended to check the satisfaction of the decision maker's preference, and from which the value $\lambda'_{1,n}$, expression (11)

$$\lambda'_{1,n} = \frac{w_n}{w_1 + w_n} \tag{11}$$

If the values λ_{1n} and $\lambda'_{1,n}$ are approximately equal, then it can be concluded that the preference of the DM decision is satisfied. If they differ, it is necessary to first check the relationship for λ_{1n}. If the decision maker considers that the relationship λ_{1n} is well defined, the relationship between the criteria should be redefined and the weighting of the criteria should be recalculated. If that is not the case, it is necessary to redefine the relationship λ_{1n}. It is necessary that the deviation of the value λ_{1n} and $\lambda'_{1,n}$ be up to a maximum of 10%. If this is not the case, it is necessary to redefine the relationships between the criteria in order to achieve this condition.

2.2 Rough Numbers

Treating inaccuracies and uncertainties in decision-making is one of the most common challenges in multi-criteria decision-making. Coarse set theory is a tool that successfully treats this area, where their application does not assume inaccuracies, as in fuzzy numbers, and the advantage of rough numbers is based on the fact that all parameters of this theory are based on the data set and no additional information is needed. Also, inaccuracy is expressed by approximations, which represent the basic concept of rough numbers [26, 27].

Rough numbers in combination with some of the MCDM methods have found their application in solving a large number of problems in many spheres of society: to evaluate design concepts under subjective environment in MCDM model AHP-VIKOR [28], for design concept evaluation under uncertain environments with methods AHP and TOPSIS [29], selection of optimal direction for the creation of a temporary military route in a hybrid IR-DEMATEL-COPRAS multi-criteria model [30], the selection of the optimal landing operations point for overcoming water obstacles in MCDM model IVFRN-MAIRCA [31], for the selection of the location for construction, reconstruction and repair of flood defence facilities with MAIRCA method [32], for evaluation and supplier selection based on fuzzy PIPRECIA–IRSAW model [33] etc.

In the following text, a description of rough numbers is given, as follows [27, 34]:

Suppose that U is a universe containing all objects, that Y is a boundary object of the universe U, and R is a set containing elements $(H_1, H_2, H_3, ...H_t)$ which encompass all objects in the universe U. If these elements are arranged in an array as $H_1 < H_2 < H_3 < ... < H_t$ then $\forall Y \in U$, $H_q \subset R$, $1 \leq q \leq t$, , on the basis of which the upper $(\underline{Apr}(H_q))$ and lower approximations $(\overline{Apr}(H_q))$ and the boundary region $(Bnd(H_q))$ of the element H_q are defined, according to the following:

$$\underline{Apr}(H_q) = \cup\{Y \in U/R(Y) \leq H_q\} \tag{12}$$

$$\overline{Apr}(H_q) = \cup\{Y \in U/R(Y) \geq H_q\} \tag{13}$$

$$Bnd(H_q) = \cup\{Y \in U/R(Y) \neq H_q\} = \{Y \in U/R(Y) > H_q\} \cup \{Y \in U/R(Y) < H_q\} \tag{14}$$

Then the element G_q can be represented as a rough number ($RN(H_q)$)which is defined by its lower limit ($\underline{Lim}(H_q)$) and upper limit ($\overline{Lim}(H_q)$), where is

$$\underline{Lim}(H_q) = \frac{1}{M_L}\sum R(Y)|Y \in \underline{Apr}(H_q) \tag{15}$$

$$\overline{Lim}(H_q) = \frac{1}{M_U}\sum R(Y)|Y \in \overline{Apr}(H_q) \tag{16}$$

$$RN(H_q) = \left[\underline{Lim}(H_q), \overline{Lim}(H_q)\right] \tag{17}$$

where M_L and M_U represent the number of objects contained in $\underline{Apr}(H_q)$ and $\overline{Apr}(H_q)$ respectively. The difference between them is a rough boundary interval ($IRBnd(H_q)$):

$$(IRBnd(H_q) = \overline{Lim}(H_q) - \underline{Lim}(H_q)) \tag{18}$$

The rough boundary interval indicates the uncertainty of the element H_q, with a higher number indicating greater inaccuracy, while a lower number indicates better precision, and only then can the subjective information be denoted by a rough number.

When manipulating rough numbers, it is necessary to know arithmetic operations. Let two rough numbers $RN(A) = \left[\underline{Lim}(A), \overline{Lim}(A)\right]$ and $RN(B) = \left[\underline{Lim}(B), \overline{Lim}(B)\right]$ be given, as well as a constant μ where $\mu \neq 0$, then [22, 29]:

$$RN(A)x\mu = \left[\underline{Lim}(A), \overline{Lim}(A)\right]x\,\mu = \left[\underline{\mu\, x\, Lim}(A), \overline{\mu\, x\, Lim}(A)\right] \tag{19}$$

$$\begin{aligned} RN(A) + RN(B) &= \left[\underline{Lim}(A), \overline{Lim}(A)\right] + \left[\underline{Lim}(B), \overline{Lim}(B)\right] \\ &= \left[\underline{Lim}(A) + \underline{Lim}(B), \overline{Lim}(A) + \overline{Lim}(B)\right] \end{aligned} \tag{20}$$

$$\begin{aligned} RN(A) - RN(B) &= \left[\underline{Lim}(A), \overline{Lim}(A)\right] - \left[\underline{Lim}(B), \overline{Lim}(B)\right] \\ &= \left[\underline{Lim}(A) - \underline{Lim}(B), \overline{Lim}(A) - \overline{Lim}(B)\right] \end{aligned} \tag{21}$$

$$\begin{aligned} RN(A)\,x\,RN(B) &= \left[\underline{Lim}(A), \overline{Lim}(A)\right]x\left[\underline{Lim}(B), \overline{Lim}(B)\right] \\ &= \left[\underline{Lim}(A)x\underline{Lim}(B), \overline{Lim}(A)x\overline{Lim}(B)\right] \end{aligned} \tag{22}$$

$$\begin{aligned} RN(A)/RN(B) &= \left[\underline{Lim}(A), \overline{Lim}(A)\right]/\left[\underline{Lim}(B), \overline{Lim}(B)\right] \\ &= \left[\underline{Lim}(A)/\underline{Lim}(B), \overline{Lim}(A)/\overline{Lim}(B)\right] \end{aligned} \tag{23}$$

In order to convert a rough number $RN(A) = \left[\underline{Lim}(A), \overline{Lim}(A)\right]$ into a crisp value, the following methodology is used, i.e., the conversion is done by using expressions from (24) to (26) [35]:

$$RN(A_i) = \left[\underline{Lim}(A_i), \overline{Lim}(A_i)\right] = \left\{ \begin{array}{l} \underline{Lim}(A_i) = \dfrac{\underline{Lim}(A_i) - \min\limits_i\left\{\underline{Lim}(A_i)\right\}}{\max\limits_i\left\{\overline{Lim}(A_i)\right\} - \min\limits_i\left\{\underline{Lim}(A_i)\right\}} \\ \overline{Lim}(A_i) = \dfrac{\overline{Lim}(A_i) - \min\limits_i\left\{\underline{Lim}(A_i)\right\}}{\max\limits_i\left\{\overline{Lim}(A_i)\right\} - \min\limits_i\left\{\underline{Lim}(A_i)\right\}} \end{array} \right\} \tag{24}$$

where $\underline{Lim}(A_i)$ and $\overline{Lim}(A_i)$ represent the upper and lower limits of the rough number $RN(A_i)$, respectively.

After normalization, the total normalized crisp value was obtained, expression (14):

$$A_i^N = \frac{\underline{Lim}(A_i)x\left\{1 - \underline{Lim}(A_i)\right\} + \overline{Lim}(A_i)x\overline{Lim}(A_i)}{1 - \underline{Lim}(A_i) + \overline{Lim}(A_i)} \tag{25}$$

The final crisp value A_i^{crisp} for the rough number $RN(A)$ was obtained by applying the expression (15):

$$A_i^{crisp} = \min_i\left\{\underline{Lim}(A_i)\right\} + A_i^N x\left[\max_i\left\{\overline{Lim}(A_i)\right\} - \min_i\left\{\underline{Lim}(A_i)\right\}\right] \tag{26}$$

2.3 Rough MABAC Method

The MABAC (Multi-Attributive Border Approximation area Comparison) method was developed in 2015 [36]. The improvement of the MABAC method with rough numbers is shown, for example, in the following papers [37–42].

The mathematical formulation of the Rough MABAC method consists of 6 steps [36, 37, 43, 44].

Step 1. Forming an initial decision matrix (X).

The first step is to evaluate m alternatives according to n criteria.

$$X = \begin{array}{c} \\ A_1 \\ A_2 \\ A_3 \\ \ldots \\ A_m \end{array} \begin{array}{c} \begin{array}{cccc} C_1 & C_1 & \ldots & C_1 \end{array} \\ \begin{pmatrix} RN(x_{11}) & RN(x_{12}) & \ldots & RN(x_{1n}) \\ RN(x_{21}) & RN(x_{22}) & \ldots & RN(x_{2n}) \\ RN(x_{31}) & RN(x_{32}) & \ldots & RN(x_{3n}) \\ \ldots & \ldots & & \ldots\ldots \\ RN(x_{m1}) & RN(x_{m2}) & \ldots & RN(x_{mn}) \end{pmatrix} \end{array} \tag{27}$$

m denotes the finite number of alternatives, and n the finite number of criteria.

The decision matrix can also be written as $X = \left[\underline{x_{ij}}, \overline{x_{ij}}\right]_{mxn}$, where $i = 1,2,\ldots, m$, a $j = 1,2,\ldots,m$.

Step 2. Normalization of the elements of the initial decision matrix (X).

$$N = \left[\underline{t_{ij}}, \overline{t_{ij}}\right]_{mxn} \tag{28}$$

Matrix elements (N) are defined using the following expressions [37]:

a) Benefit criteria

$$\underline{t_{ij}} = \frac{\underline{x_{ij}} - x_j^-}{x_j^+ - x_j^-}, \quad \overline{t_{ij}} = \frac{\overline{x_{ij}} - x_j^-}{x_j^+ - x_j^-} \tag{29}$$

b) Cost criteria

$$\underline{t_{ij}} = \frac{\underline{x_{ij}} - x_j^+}{x_j^- - x_j^+}, \quad \overline{t_{ij}} = \frac{\overline{x_{ij}} - x_j^-}{x_j^- - x_j^+} \tag{30}$$

where:

$$x_j^+ = \left\{ \max_{1 \le i \le m} \left(\overline{x_{ij}}\right),\ \textit{for benefit type criteria},\ \min_{1 \le i \le m} \left(\underline{x_{ij}}\right),\ \textit{for}\ \cos t\ \textit{type criteria} \right. \tag{31}$$

$$x_j^+ = \left\{ \min_{1 \le i \le m} \left(\underline{x_{ij}}\right),\ \textit{for benefit type criteria},\ \max_{1 \le i \le m} \left(\overline{x_{ij}}\right),\ \textit{for}\ \cos t\ \textit{type criteria} \right. \tag{32}$$

Step 3. Calculation of weighted matrix elements (V).

$$V = \left[\underline{v_{ij}}, \overline{v_{ij}}\right]_{m\,x\,n} \tag{33}$$

The elements of the matrix (*V*) are obtained by applying the expression:

$$\underline{v_{ij}} = \underline{w_i}\,x\left(\underline{t_{ij}} + 1\right),\ \overline{v_{ij}} = \overline{w_i}\,x\left(\overline{t_{ij}} + 1\right) \tag{34}$$

where $\left[\underline{w_i}, \overline{w_i}\right]$ represent the weight coefficient of the criteria.

Step 4. Determination of the matrix of boundary approximate areas (G).

The matrix of boundary approximate domains is formed in the format *n x 1*, ie

$$G = [g_1, g_2, \ldots, g_n]\ \textit{where}\ g_j = \left[\underline{g_j}, \vec{g_j}\right] \tag{35}$$

Boundary Approximate Area (BAA) for each criterion is obtained by applying the expression (36):

$$\underline{g_j} = \left(\prod\nolimits_{i=1}^{m} \underline{v_{ij}}\right)^{1/m},\ \overline{g_j} = \left(\prod\nolimits_{i=1}^{m} \overline{v_{ij}}\right)^{1/m} \tag{36}$$

Step 5. Calculation of the elements of the matrix of distance of alternatives from the boundary approximate area (Q)

$$Q = \left(\left[\underline{q_{ij}}, \overline{q_{ij}}\right]\right)_{m\,x\,n} \tag{37}$$

where:

$$q_{ij} = \left\{ \begin{array}{l} d_E\left(v_{ij}, g_j\right),\ \textit{if}\ RN\left(v_{ij}\right) > RN\left(g_j\right) \\ -d_E\left(v_{ij}, g_j\right),\ \textit{if}\ RN\left(v_{ij}\right) < RN\left(g_j\right) \end{array} \right\} \textit{for benefit type criteria} \tag{38}$$

$$q_{ij} = \begin{Bmatrix} -d_E(v_{ij}, g_j), \ if\ RN(v_{ij}) > RN(g_j) \\ d_E(v_{ij}, g_j), \ if\ RN(v_{ij}) < RN(g_j) \end{Bmatrix} for\ \cos t\ type\ criteria \tag{39}$$

$$d_E(v_{ij}, g_j) = \begin{Bmatrix} \sqrt{\left(\underline{v_{ij}} - \overline{g_j}\right)^2 + \left(\overline{v_{ij}} - \underline{g_j}\right)^2}, \ for\ benefit\ type\ criteria \\ \sqrt{\left(\underline{v_{ij}} - \underline{g_j}\right)^2 + \left(\overline{v_{ij}} - \overline{g_j}\right)^2}, \ for\ \cos t\ type\ criteria \end{Bmatrix} \tag{40}$$

where $\left[\underline{g_j}, \overline{g_j}\right]$ is the boundary approximate area for the criterion C_j $(j = 1, 2, ..., n)$.

Alternative A_i may belong to the boundary approximate area (G), the upper approximate area (G^+) or the lower approximate area (G^-). The upper approximate area (G^+) is the area where the ideal alternative is located (A^+), while the lower approximate area (G^-) is the area where the anti-ideal alternative (A^-) is located (Fig. 2).

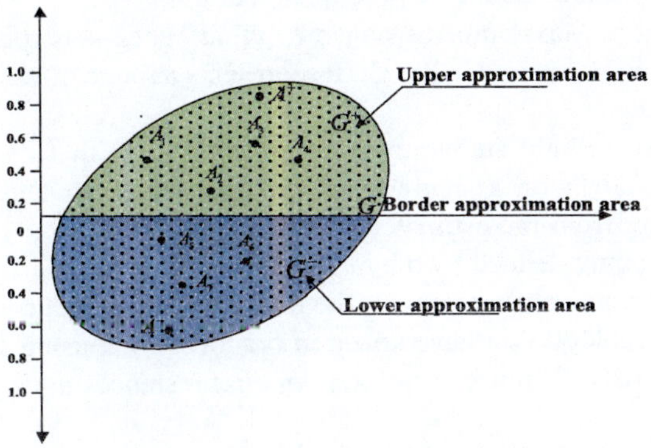

Fig. 2. Approximate areas [31]

Step 6. Ranking alternatives.

By summing the elements of the matrix Q, by rows, the final values of the criterion functions of the alternatives are obtained:

$$S_i = \sum_{j=1}^{n} q_{ij}, \ j = 1, 2, ..., n, \ i = 1, 2, ..., m. \tag{41}$$

3 Application of MCDM Model and Results

By analyzing the available literature [6–8] and taking into account the specifics of the research problem, the following criteria for the selection of drones are defined and their description is given:

Criterion 1 (C_1) – Cost – represents the monetary value of the drone on the market in euros (€), i.e., the price range of available drones.

Criterion 2 (C_2) – Range – represents the distance measured in kilometers (km), from the place where the drone takes off to the maximum distance to which it can be controlled, measured in a straight line horizontally, i.e., in the shortest possible direction. Distance is the range of values, from the estimated maximum range to the maximum declared range defined by the manufacturer.

Criterion 3 (C_3) – Load capacity – represents the maximum load capacity in grams, and is presented as a range of values, from the estimated maximum load capacity to the maximum declared load capacity defined by the manufacturer.

Criterion 4 (C_4) – Flight speed – represents the maximum speed of the drone in km / h. Speed is presented as a range of values, from the estimated maximum speed to the maximum declared speed defined by the manufacturer.

Criterion 5 (C_5) – Maximum flight altitude – represents the distance measured in meters, from the place where the drone takes off to the maximum remote place to which it is possible to control it, measured vertically, i.e., in the shortest possible direction. Distance represents the range of values, from the estimated maximum altitude to the maximum declared altitude defined by the manufacturer.

Criterion 6 (C_6) – Maximum resistance to wind speed – is the resistance of the aircraft to wind speed, measured in km / h. It represents a range of values, and is defined by the manufacturer.

Criterion 7 (C_7) – Flight autonomy – refers to the time that the drone can spend continuously in the air, with a single charge of the battery or tank. The time value represents the range, from the estimated maximum flight autonomy to the maximum declared flight autonomy defined by the manufacturer.

Estimated maximum values represent the declared values reduced by 10% to 20%, depending on the problems that have arisen in practice when using specific drones, in order to reduce the risk of failure or unforeseen circumstances during the operation of the asset.

When choosing, all drones under consideration have cameras with a minimum resolution of 20MP and infrared cameras.

The criteria are ranked in order of importance, from the most important (C_1) to the least important (C_7) for a specific research problem and all are of the benefit type, except for criterion C_1 which is a cost type.

Following the phases and steps of the MCDM model, presented in Fig. 1, the calculation of the weight coefficients of the criteria is approached, using the DIBR method.

Based on the defined criteria, a set of seven criteria $C_1, C_2, \ldots, C_7$ was determined, which are ranked in order of importance as $C_1 > C_2 > C_3 > C_4 > C_5 > C_6 > C_7$.

Based on the rank of the criteria, the values $\lambda_{12}, \lambda_{13}, \ldots, \lambda_{67}$ and λ_{17} are defined, according to the following: $\lambda_{12} = 0.49$, $\lambda_{23} = 0.46$, $\lambda_{34} = 0.47$, $\lambda_{45} = 0.46$, $\lambda_{56} = 0.49$, $\lambda_{67} = 0.47$ and $\lambda_{17} = 0.34$.

Based on the defined values $\lambda_{n-1,n}$, the following relations between the criteria are defined by applying relations (1)–(4): $w_1 : w_2 = 0.51 : 0.49$; $w_2 : w_3 = 0.54 : 0.46$; $w_3 : w_4 = 0.53 : 0.47$; $w_4 : w_5 = 0.54 : 0.46$; $w_5 : w_6 = 0.51 : 0.49$; $w_6 : w_7 = 0.53 : 0.47$; $w_1 : w_7 = 0.66 : 0.34$.

Based on the previous relations, using the expressions (5)–(7), the expressions for the values of the weight coefficients of the criteria are defined: $w_2 = 0.961w_1; w_3 = 0,852w_2 = 0,818w_1$; $w_4 = 0,887w_3 = 0,726w_1; w_5 = 0,839w_4 = 0,609w_1; w_6 = 0,961w_5 = 0,585w_1$ and $w_7 = 0,887w_6 = 0,519w_1$.

Based on condition $\sum_{j=1}^{7} w_j = 1$ and expression (9) follows that

$$w_1 = \frac{1}{1 + 0,961 + 0,818 + 0,726 + 0,609 + 0,585 + 0,519} = 0.1916$$

Using expressions (5) - (7), the weight coefficients of other criteria are calculated $w_2 = 0.1841$; $w_3 = 0.1569$; $w_4 = 0.1391$; $w_5 = 0.1167$; $w_6 = 0.1121$ i $w_7 = 0.0994$.

Using expression (11), the control value $\lambda'_{1,7}$ is calculated.

$$\lambda'_{1,7} = \frac{w_7}{w_1 + w_7} = \frac{0.09494}{0.1916 + 0.0994} = 0.3416$$

Since $\lambda_{17} \approx \lambda'_{1,7}$, ie $\lambda'_{1,7} = 0.2701$ and $\lambda_{17} = 0.27$, it is concluded that expert preferences are well defined, i.e., that the transitive relations that define the significance of the criteria are met.

By applying the previously explained steps of the DIBR method, the following weight coefficients of the criteria are obtained (Table 1):

Table 1. The values of weight coefficients of criteria

Criterion	The value of the weight coefficient
C_1	0.1916
C_2	0.1841
C_3	0.1569
C_4	0.1391
C_5	0.1167
C_6	0.1121
C_7	0.0994

After obtaining the weight coefficients of the criteria, the ranking of 7 alternatives, which represent 7 different models of drones available on the market, is approached, using the Rough MABAC method.

The first step in applying this method is to form an initial decision matrix:

Step 1. Forming an initial decision matrix (X).

$$X = \begin{array}{c} \\ A_1 \\ A_2 \\ A_3 \\ A_4 \\ A_5 \\ A_6 \\ A_7 \end{array} \begin{array}{ccccccc} C_1 & C_2 & C_3 & C_4 & C_5 & C_6 & C_7 \\ \hline [1300, 1450] & [5.4, 6] & [1700, 2000] & [60.88, 68.4] & [5400, 6000] & [26, 32] & [0.5, 0.55] \\ [1300, 1350] & [4.5, 5] & [425, 500] & [44.28, 54] & [3560, 4000] & [35, 41] & [0.49, 0.6] \\ [1280, 1300] & [5.1, 6] & [900, 1000] & [48.6, 54] & [4100, 5000] & [48, 52] & [0.45, 0.5] \\ [1250, 1420] & [5.1, 6] & [1620, 1800] & [45.05, 53] & [4500, 5000] & [10.5, 13] & [0.54, 0.6] \\ [1245, 1355] & [5.4, 6] & [1275, 1500] & [45, 50] & [5340, 6000] & [25, 33] & [0.54, 0.6] \\ [1300, 1580] & [14.95, 16.8] & [720, 800] & [44.28, 54] & [5340, 6000] & [39, 49] & [0.44, 0.52] \\ [1350, 1491] & [5.1, 6] & [510, 600] & [64.8, 72] & [5400, 6000] & [32, 41] & [0.47, 0.53] \end{array}$$

Step 2. Normalization of the elements of the initial decision matrix.
By applying expressions (29) and (30) we obtain a normalized matrix (N):

$$N = \begin{array}{c} \\ A_1 \\ A_2 \\ A_3 \\ A_4 \\ A_5 \\ A_6 \\ A_7 \end{array} \begin{array}{ccccccc} C_1 & C_2 & C_3 & C_4 & C_5 & C_6 & C_7 \\ \hline [0.39, 0.84] & [0.88, 0.93] & [0, 0.19] & [0.13, 0.4] & [0, 0.25] & [0.48, 0.63] & [0.32, 0.66] \\ [0.69, 0.84] & [0.96, 1] & [0.95, 1] & [0.65, 1] & [0.82, 1] & [0.27, 0.41] & [0, 0.68] \\ [0.84, 0.9] & [0.88, 0.95] & [0.63, 0.7] & [0.65, 0.84] & [0.41, 0.78] & [0, 0.1] & [0.63, 0.95] \\ [0.48, 0.99] & [0.88, 0.95] & [0.13, 0.24] & [0.69, 0.97] & [0.41, 0.61] & [0.94, 1] & [0, 0.38] \\ [0.67, 1] & [0.88, 0.93] & [0.32, 0.46] & [0.79, 0.97] & [0, 0.27] & [0.46, 0.65] & [0, 0.38] \\ [0, 0.84] & [0, 0.15] & [0.76, 0.81] & [0.65, 1] & [0, 0.27] & [0.07, 0.31] & [0.51, 1] \\ [0.27, 0.69] & [0.88, 0.95] & [0.89, 0.95] & [0, 0.26] & [0, 0.25] & [0.27, 0.48] & [0.44, 0.81] \end{array}$$

Step 3. Calculation of weighted matrix elements (V).
The elements of the weighted matrix (V) are obtained by applying expression (34):

$$V = \begin{array}{c} \\ A_1 \\ A_2 \\ A_3 \\ A_4 \\ A_5 \\ A_6 \\ A_7 \end{array} \begin{array}{ccccccc} C_1 & C_2 & C_3 & C_4 & C_5 & C_6 & C_7 \\ \hline [0.27, 0.35] & [0.35, 0.35] & [0.16, 0.19] & [0.16, 0.19] & [0.12, 0.15] & [0.17, 0.18] & [0.13, 0.17] \\ [0.32, 0.35] & [0.36, 0.37] & [0.31, 0.31] & [0.23, 0.28] & [0.21, 0.23] & [0.14, 0.16] & [0.1, 0.17] \\ [0.35, 0.36] & [0.35, 0.36] & [0.26, 0.27] & [0.23, 0.26] & [0.16, 0.21] & [0.11, 0.12] & [0.16, 0.19] \\ [0.28, 0.38] & [0.35, 0.36] & [0.18, 0.19] & [0.23, 0.27] & [0.16, 0.19] & [0.22, 0.22] & [0.1, 0.14] \\ [0.32, 0.38] & [0.35, 0.35] & [0.21, 0.23] & [0.25, 0.27] & [0.12, 0.15] & [0.16, 0.19] & [0.1, 0.14] \\ [0.19, 0.35] & [0.18, 0.21] & [0.28, 0.28] & [0.23, 0.28] & [0.12, 0.15] & [0.12, 0.15] & [0.15, 0.2] \\ [0.24, 0.32] & [0.35, 0.36] & [0.3, 0.31] & [0.14, 0.18] & [0.12, 0.15] & [0.14, 0.17] & [0.14, 0.18] \end{array}$$

Step 4. Determination of the matrix of boundary approximate areas (G).
The boundary approximate area for each criterion is determined using expression (36).

After that, a matrix G of format 7 x 1 is formed (where 7 represents the finite number of criteria).

$$G = \begin{array}{ccccccc} C_1 & C_2 & C_3 & C_4 & C_5 & C_6 & C_7 \\ \hline [[0.28, 0.36] & [0.32, 0.33] & [0.23, 0.25] & [0.21, 0.24] & [0.14, 0.17] & [0.15, 0.17] & [0.12, 0.17]]. \end{array}$$

Step 5. Calculation of the elements of the matrix of distance of alternatives from the boundary approximate area (Q)

The distance of alternatives from the boundary approximate area is determined by using expressions (38)–(40).

$$Q = \begin{array}{c} \\ A_1 \\ A_2 \\ A_3 \\ A_4 \\ A_5 \\ A_6 \\ A_7 \end{array} \begin{array}{ccccccc} C_1 & C_2 & C_3 & C_4 & C_5 & C_6 & C_7 \\ \left[\begin{array}{ccccccc} [-0.09, 0.07] & [-0.03, 0.07] & [-0.01, 0.09] & [-0.07, 0.1] & [-0.04, 0.11] & [-0.17, 0.07] & [-0.11, 0.05] \\ [0.18, 0.23] & [0.19, 0.24] & [0.18, 0.24] & [0.18, 0.24] & [0.18, 0.23] & [0.02, 0.09] & [0.18, 0.24] \\ [-0.01, 0.06] & [0.14, 0.19] & [0.09, 0.14] & [0.01, 0.07] & [0.04, 0.11] & [0.11, 0.16] & [0.13, 0.18] \\ [-0.01, 0.07] & [0.06, 0.15] & [0.06, 0.13] & [0.07, 0.15] & [0.08, 0.15] & [0.06, 0.15] & [-0.03, 0.05] \\ [-0.05, 0.02] & [0.05, 0.11] & [0, 0.08] & [0, 0.06] & [-0.05, 0.02] & [-0.05, 0.02] & [-0.05, 0.02] \\ [0, 0.06] & [-0.03, 0.03] & [-0.05, 0] & [0.05, 0.1] & [0, 0.06] & [-0.05, 0.02] & [-0.03, 0.04] \\ [-0.04, 0.04] & [-0.07, 0.04] & [0, 0.07] & [-0.07, 0.01] & [-0.07, 0.01] & [-0.02, 0.07] & [-0.02, 0.06] \end{array}\right] \end{array}$$

Step 6. Ranking alternatives.

The calculation of the values of the criterion functions by alternatives was obtained by applying expression (41). Table 2 presents the values of the criterion functions of the alternatives.

Table 2. Values of criterion functions of alternatives

	S_i
A_1	[−0.02, 0.56]
A_2	[0.314, 0.85]
A_3	[0.263, 0.749]
A_4	[0.162, 0.737]
A_5	[0.143, 0.691]
A_6	[−0.091, 0.599]
A_7	[0.067, 0.634]

By applying expressions (24) to (26), the values of the criterion functions of alternatives that are in the form of rough numbers are converted into crisp values, after which the ranking of alternatives is performed (Table 3):

Based on the results from Table 3, we conclude that alternative A_2 is the most acceptable solution, i.e., it is the best ranked alternative, while alternatives A_1 and A_6 can in no case be chosen as a solution when choosing a drone. The results obtained by applying the Rough MABAC method are expected, considering the values of each alternative in relation to the defined criteria, in the initial decision matrix (X). Given that criteria C_1, C_2 and C_3 have the highest value of weight coefficients, with a total share of significance over 50%, and that the normalized values of alternatives A_2 and A_3 are generally the highest for those criteria, the obtained ranking of alternatives was expected. Also, alternative A_7 has the lowest normalized value according to almost all criteria, so its place as the last ranked was to be expected. All of the above clearly

Table 3. Crisp values of criterion functions of alternatives and their rank

	S_i^{crisp}	*Rank*
A_1	0.229	6
A_2	0.655	1
A_3	0.549	2
A_4	0.477	3
A_5	0.431	4
A_6	0.201	7
A_7	0.339	5

indicates the validity of the proposed mathematical model. A further validation check will be performed by analyzing the sensitivity of the model to the change in the weight coefficients.

4 Sensitivity Analysis

In such a complex process, such as decision-making, it is possible to make mistakes, and it is necessary to perform a sensitivity analysis [45–54]. The sensitivity of the Rough MABAC method to changes in weight coefficients was analyzed in this paper [45], through 18 scenarios. The values of the weight coefficients of the criteria, according to the scenarios, are given in Table 4.

The ranking of alternatives after the application of previously defined scenarios is given in Table 5.

The shaded ranks in the table represent the difference in the rank of the alternative in relation to the initial rank. From the previous table we can conclude that the Rough MABAC method is sensitive to changes in the weight coefficients of the criteria. What is advantageous is the fact that the three top-ranked alternatives do not change their rank when changing the weight coefficients.

One way to check the consistency of MCDM method results is given in [55, 56] and represents Spearman's rank correlation coefficient (S), calculated according to the following expression:

$$S = 1 - \frac{6\sum_{i=1}^{n} D_i^2}{n(n^2 - 1)} \quad (42)$$

where is D_i – the difference between the rank of a given element in the vector w and the rank of the corresponding element in the reference vector, n – number of ranked elements.

Identical ranks of the elements define the value of Spearman's coefficient 1 ("ideal positive correlation"). The value of Spearman's coefficient -1, means that the ranks are

Table 4. Scenarios with different weight coefficients of criteria

	S_1	S_2	S_3	S_4	S_5	S_6	S_7	S_8	S_9
C_1	0.143	0.180	0.169	0.157	0.146	0.134	0.123	0.111	0.100
C_2	0.143	0.186	0.188	0.190	0.192	0.194	0.196	0.198	0.199
C_3	0.143	0.159	0.161	0.163	0.165	0.166	0.168	0.170	0.172
C_4	0.143	0.141	0.143	0.145	0.147	0.149	0.151	0.153	0.154
C_5	0.143	0.119	0.121	0.122	0.124	0.126	0.128	0.130	0.132
C_6	0.143	0.114	0.116	0.118	0.120	0.122	0.124	0.126	0.127
C_7	0.143	0.101	0.103	0.105	0.107	0.109	0.111	0.113	0.115
	S_{10}	S_{11}	S_{12}	S_{13}	S_{14}	S_{15}	S_{16}	S_{17}	S_{18}
C_1	0.088	0.077	0.065	0.054	0.042	0.031	0.019	0.008	0.002
C_2	0.201	0.203	0.205	0.207	0.209	0.211	0.213	0.215	0.217
C_3	0.174	0.176	0.178	0.180	0.182	0.184	0.186	0.188	0.189
C_4	0.156	0.158	0.160	0.162	0.164	0.166	0.168	0.170	0.172
C_5	0.134	0.136	0.138	0.140	0.142	0.144	0.145	0.147	0.149
C_6	0.129	0.131	0.133	0.135	0.137	0.139	0.141	0.143	0.145
C_7	0.117	0.119	0.121	0.122	0.124	0.126	0.128	0.130	0.132

Table 5. Ranking of alternatives by scenarios

	S_1	S_2-S_7	S_8- S_{16}	S_{17}- S_{18}
A_1	7	6	7	7
A_2	1	1	1	1
A_3	2	2	2	2
A_4	3	3	3	3
A_5	4	4	4	5
A_6	6	7	6	6
A_7	5	5	5	4

absolutely opposite ("ideal negative correlation"), and when the value of Spearman's coefficient is 0, the ranks are uncorrelated.

Applying expression (42) gives the value of Spearman's coefficient (S) (Fig. 3). Correlation of ranks was performed in relation to the initial rank, in accordance with the defined scenarios.

From the previous figure we can conclude that the correlation coefficients in the 18 scenarios given in Table 4 tend towards an ideal positive correlation and that the defined MCDM model is stable with respect to the change of weight coefficients.

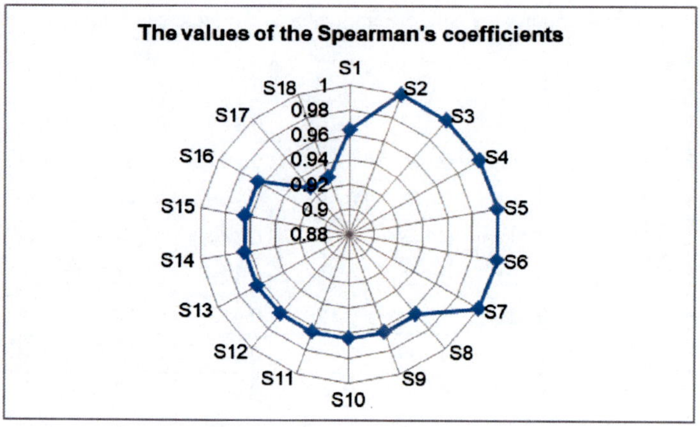

Fig. 3. The values of the Spearman's coefficient

5 Conclusions

This paper presents the application of the new MCDM model DIBR-Rough MABAC on the example of drone selection for the purpose of reconnaissance of flooded areas and delivery of necessary materials, food and water, based on drone characteristics. The paper presents all phases and steps of the proposed MCDM model.

The calculation of weight coefficients was performed by the DIBR method, which is a new method of MCDM, but which has a simple mathematical apparatus and shows great potential for application. The ranking of alternatives was performed using the MABAC method, improved by rough numbers, which further improved the quality of the decision-making process, especially in cases of inaccurate and indeterminate input data.

Also, the sensitivity analysis was performed in the paper, as one of the ways to validate the proposed model. The obtained results indicate that the MCDM model is stable when changing the weights of the criteria, i.e., that the three first-ranked alternatives do not change their rank regardless of changes in their weights, and that the last three ranked alternatives do not become top-ranked in any case. When determining the rank correlation coefficients, all values tend towards an ideal positive correlation, which also indicates the stability of the output results of this model.

This model can be further improved by applying other and more different methods for determining weight coefficients and ranking alternatives, as well as by introducing other sets that treat inaccuracies and uncertainties.

References

1. Degg, M.: Natural disasters: recent trends and future prospects. Geography **77**(3), 198–209 (1992)
2. Cvetković, V.: Zaštita kritične infrastrukture od posledica prirodnih katastrofa. In: 7 međunarodna znastvenostručna konferencija "Dani kriznog upravljanja", 2014, pp. 1281–1295 (2014)

3. Tobin, G.A., Montz, B.E.: Natural Hazards: Explanation and Integration. The Guilford Press, New York (1997)
4. Jovanović, D., Arsić, M.: Logistička operacija pomoći i spasavanja u elementarnim nepogodama. Novi Glasnik, 10–12/2004 (2004)
5. Milić, A., Ranđelović, A., Radovanović, M.: Use of drones in operations in the urban environment. In: 5 th International Scientific Conference Safety and Crisis Management – Theory and Practise Safety for the Future – SecMan 2019, pp. 124–130 (2019)
6. Agbeyangi, A., Odiete, J., Olorunlomerue, A.: Review on UAVs used for aerial surveillance. J. Multidisc. Eng. Sci. Technol. **3**(10), 5713–5719 (2016)
7. Radovanović, M., Milić, A., Ranđelović, A.: The possibility of using drons in the protection of the land security zone (Mogućnost upotrebe dronova u zaštiti kopnene zone bezbednosti). In: 15th International Conference on Risk and Safety Engineering, Kopaonik, Serbia, 2020, pp. 303–311 (2020)
8. Arjomandi, M., Agostino, S., Mammone, M., Nelson, M., Zhou, T.T.: Classification of unmanned aerial vehicles. Report for Mechanical Engineering class, University of Adelaide, Adelaide, Australia (2006)
9. Nur, F., Alrahahleh, A., Burch, R., Babski-Reeves, K., Marufuzzaman, M.: Last mile delivery drone selection and evaluation using the interval-valued inferential fuzzy TOPSIS. J. Comput. Design Eng. **7**(4), 397–411 (2020)
10. Sohaib Khan, M., Ali Shah, S.I., Javed, A., Mumtaz Qadri, N., Hussain, N.: Drone selection using multi-criteria decision-making methods. In: 2021 International Bhurban Conference on Applied Sciences and Technologies (IBCAST), pp. 256–270 (2021)
11. Petkovics, I., Simon, J., Petkovics, Á., Covic, Z.: Selection of unmanned aerial vehicle for precision agriculture with multi-criteria decision making algorithm. In: 2017 IEEE 15th International Symposium on Intelligent Systems and Informatics (SISY), 2017, pp. 151–156 (2017)
12. Rakhade, R.D., Patil, N.V., Pardeshi, M.R., Mhasde, C.S.: Optimal choice of agricultural drone using MADM methods. Int. J. Technol. Innov. Mod. Eng. Sci. (IJTIMES) **7**(4), 2455–2585 (2021)
13. Hamurcu, M., Eren, T.: Selection of unmanned aerial vehicles by using multicriteria decision-making for defence. J. Math. **2020**, 4308756 (2020)
14. Restas, A.: Drone applications for supporting disaster management. World J. Eng. Technol. **3**(03), 316–321 (2015)
15. Sehrawat, A., Choudhury, T.A., Raj, G.: Surveillance drone for disaster management and military security. In: 2017 International Conference on Computing, Communication and Automation (ICCCA), 2017, pp. 470–475 (2017)
16. Lee, S., Har, D., Kum, D.: Drone-assisted disaster management: finding victims via infrared camera and LIDAR sensor fusion. In: 2016 3rd Asia-Pacific World Congress on Computer Science and Engineering (APWC on CSE), pp. 84–89. IEEE (2016)
17. Daud, S.M.S.M., et al.: Applications of drone in disaster management: a scoping review. Sci. Justice **62**(1), 30–42 (2022)
18. Hasan, K.M., Newaz, S.S., Ahsan, M.S.: Design and development of an aircraft type portable drone for surveillance and disaster management. Int. J. Intell. Unmanned Syst. **6**(3), 147–159 (2018)
19. Mishra, B., Garg, D., Narang, P., Mishra, V.: Drone-surveillance for search and rescue in natural disaster. Comput. Commun. **156**, 1–10 (2020)
20. Miljković, B., Žižović, M., Petojević, A., Damljanović, N.: New weighted sum model. Filomat **31**, 2991–2998 (2017)
21. Pamučar, D., Stević, Ž, Sremac, S.: A new model for determining weight coefficients of criteria in MCDM models: full consistency method (FUCOM). Symmetry **10**(9), 393 (2018)

22. Žižović, M., Miljković, B., Marinković, D.: Objective methods for determining criteria weight coefficients: a modification of the CRITIC method. Decis. Making: Appl. Manage. Eng. **3**(2), 149–161 (2020)
23. Žižović, M., Pamučar, D., Miljković, B., Karan, A.: Multiple-criteria evaluation model for medical professionals assigned to temporary SARS-CoV-2 hospitals. Decis. Making: Appl. Manage. Eng. **4**(1), 153–173 (2021)
24. Žižović, M., Pamučar, D., Ćirović, G., Žižović, M.M., Miljković, B.D.: A model for determining weight coefficients by forming a non-decreasing series at criteria significance levels (NDSL). Mathematics **8**(5), 745 (2020)
25. Pamučar, D., Deveci, M., Gokasar, I., Işık, M., Žižovic, M.: Concepts in urban mobility alternatives using integrated DIBR method and fuzzy dombi CoCoSo model. J. Clean. Prod. **323**, 129096 (2021)
26. Pawlak, Z.: Rough Sets-Theoretical Aspects of Reasoning About Data. Kluwer Academic Publishers, Dordrecht (1991)
27. Zhai, L.Y., Khoo, L.P., Zhong, Z.W.: A rough set enhanced fuzzy approach to quality function deployment. Int. J. Adv. Manuf. Technol. **37**, 613–624 (2008)
28. Zhu, G.N., Hu, J., Qi, J., Gu, C.C., Peng, Y.H.: An integrated AHP and VIKOR for design concept evaluation based on rough number. Adv. Eng. Inform. **29**(3), 408–418 (2015)
29. Zhu, G.N., Hu, J., Ren, H.: A fuzzy rough number-based AHP-TOPSIS for design concept evaluation under uncertain environments. Appl. Soft. Comput. **91**, 106228 (2020)
30. Pamučar, D., Božanić, D., Lukovac, V., Komazec, N.: Normalized weighted geometric bonferroni mean operator of interval rough numbers – application in INTERVAL ROUGH DEMATEL-COPRAS MODEL. Facta Univ. Ser.: Mech. Eng. **16**(2), 171–191 (2018)
31. Pamučar, D.S., Ćirović, G., Božanić, D.: Application of interval valued fuzzy-rough numbers in multi-criteria decision making: the IVFRN-MAIRCA model. Yugoslav J. Oper. Res. **29**(2), 221–247 (2019)
32. Božanić, D., Pamučar, D., Tešić, D.: Selection of the location for construction, reconstruction and repair of flood defense facilities by IR-MAIRCA model application. In: Proceedings of the Fifth International Scientific-Profesional Conference Security and Crisis Management–Theory and Practice, SeCMan 2019, pp. 300–308 (2019)
33. Đalić, I., Stević, Ž, Karamasa, C., Puška, A.: A novel integrated fuzzy PIPRECIA – interval rough SAW model: green supplier selection. Decis. Making: Appl. Manage. Eng. **3**(1), 126–145 (2020)
34. Stević, Ž, Pamučar, D., Kazimieras Zavadskas, E., Ćirović, G., Prentkovskis, O.: The selection of wagons for the internal transport of a logistics company: A novel approach based on rough BWM and rough SAW methods. Symmetry **9**(11), 264 (2017)
35. Roy, J., Adhikary, K., Kar, S., Pamucar, D.: A rough strength relational DEMATEL model for analysing the key success factors of hospital service quality. Decis. Making: Appl. Manage. Eng. **1**(1), 121–142 (2018)
36. Pamučar, D., Ćirović, G.: The selection of transport and handling resources in logistics centres using multi-attributive border approximation area comparison (MABAC). Expert Syst. Appl. **42**, 3016–3028 (2015)
37. Roy, J., Chatterjee, K., Bandyopadhyay, A., Kar, S.: Evaluation and selection of medical tourism sites: a rough analytic hierarchy process-based multiattributive border approximation area comparison approach. Expert. Syst. **35**, e12232 (2018)
38. Pamučar, D., Stević, Ž, Zavadskas, E.K.: Integration of interval rough AHP and interval rough MABAC methods for evaluating university web pages. Appl. Soft Comput. **67**, 141–163 (2018)
39. Jia, F., Liu, Y., Wang, X.: An extended MABAC method for multi-criteria group decision making based on intuitionistic fuzzy rough numbers. Expert Syst. Appl. **127**, 241–255 (2019)

40. Sharma, H.K., Roy, J., Kar, S., Prentkovskis, O.: Multi criteria evaluation framework for prioritizing indian railway stations using modified rough AHP-Mabac method. Transp. Telecommun. J. **19**, 113–127 (2018)
41. Chakraborty, S., Dandge, S.S., Agarwal, S.: Non-traditional machining processes selection and evaluation: a rough multi-attributive border approximation area comparison approach. Comput. Ind. Eng. **139**, 106201 (2020)
42. Yazdani, M., Pamučar, D., Chatterjee, P., Chakraborty, S.: Development of a decision support framework for sustainable freight transport system evaluation using rough numbers. Int. J. Prod. Res. **58**(14), 4325–4351 (2020)
43. Božanić, D., Tešić, D., Milićević, J.: Selection of locations for deep draft tank crossing by applying fuzzy Mabac method. In: ICMNEE 2017 the 1st International Conference on Management, Engineering and Environment 2017, pp. 346–358 (2017)
44. Božanić, D., Tešić, D., Milić, A.: Multicriteria decision making model with Z-numbers based on FUCOM and MABAC model. Decis. Making: Appl. Manage. Eng. **3**(2), 19–36 (2020)
45. Pamučar, D., Žižović, M., Biswas, S., Božanić, D.: A new logarithm methodology of additive weights (LMAW) for multi-criteria decision-making: application in logistics. Facta Univ. Ser.: Mech. Eng. **19**(3), 361–380 (2021)
46. Muhammad, L.J., Badi, I., Haruna, A.A., Mohammed, I.A.: Selecting the best municipal solid waste management techniques in Nigeria using multi criteria decision making techniques. Rep. Mech. Eng. **2**(1), 180–189 (2021)
47. Durmić, E., Stević, Ž, Chatterjee, P., Vasiljević, M., Tomašević, M.: Sustainable supplier selection using combined FUCOM–Rough SAW model. Rep. Mech. Eng. **1**(1), 34–43 (2020)
48. Božanić, D., Pamučar, D., Milić, A., Marinković, D., Komazec, N.: Modification of the logarithm methodology of additive weights (LMAW) by a triangular fuzzy number and its application in multi-criteria decision making. Axioms **11**(3), 89 (2022)
49. Pamučar, D.S., Savin, L.M.: Višekriterijumski BWM-COPRAS model za izbor optimalnog terenskog vozila za prevoz putnika. Mil. Tech. Courier **68**(1), 28–64 (2020)
50. Jokić, Ž, Božanić, D., Pamučar, D.: Selection of fire position of mortar units using LBWA and fuzzy MABAC model. Oper. Res. Eng. Sci.: Theory Appl. **4**(1), 115–135 (2021)
51. Božanić, D., Jurišić, D., Erkić, D.: LBWA – Z-MAIRCA model supporting decision making in the army. Oper. Res. Eng. Sci.: Theory Appl. **3**(2), 87–110 (2020)
52. Božanić, D., Ranđelović, A., Radovanović, M., Tešić, D.A.: hybrid LBWA-IR-MAIRCA multi-criteria decision-making model for determination of constructive elements of weapons. Facta Univ. Ser.: Mech. Eng. **18**(3), 399–418 (2020)
53. Pamučar, D., Božanić, D., Kurtov, D.: Fuzzification of the Saaty's scale and a presentation of the hybrid fuzzy AHP-TOPSIS model: an example of the selection of a brigade artillery group firing position in a defensive operation. Mil. Tech. Courier **64**(4), 966–986 (2016)
54. Pamučar, D., Macura, D., Tavana, M., Božanić, D., Knežević, N.: An integrated rough group multicriteria decision-making model for the ex-ante prioritization of infrastructure projects: The Serbian Railways case. Socioecon. Plann. Sci. **79**, 101098 (2022)
55. Srđević, B., Srđević, Z., Suvočarev, K.: Analitički hijerarhijski proces: individualna i grupna konzistentnost donosilaca odluka. Vodoprivreda **41**, 13–21 (2009)
56. Božanić, D., Milić, A., Tešić, D., Salabun, W., Pamučar, D.D.: numbers – FUCOM – Fuzzy RAFSI model for selecting the group of construction machines for enabling mobility. Facta Univ. Ser.: Mech. Eng. **19**(3), 447–471 (2021)

Improving the Low Accuracy of Traditional Earthquake Loss Assessment Systems

Zoran Stojadinović(✉)

Faculty of Civil Engineering, University of Belgrade, Bulevar Kralja Aleksandra 73, 11000 Belgrade, Serbia
joka@grf.bg.ac.rs

Abstract. Rapid earthquake loss assessment implies near real-time prediction of damage to the affected structures and monetizing loss at the community level. The accuracy and speed of prediction are the main quality features of rapid loss assessment systems. The problem with traditional systems is their low accuracy, making them unreliable and unusable in the recovery process, which is their primary purpose. Low accuracy is caused by significant uncertainty in analytical and insufficient data sets to create vulnerability curves in empirical methods. The root cause of low accuracy in analytical methods is assuming *theoretical* vulnerability relations *before* an earthquake. We propose a new kind of rapid earthquake loss assessment system which uses trained assessors to perform on-the-ground observation of *actual* damage on the representative sample *after* an earthquake. Machine learning methods are then used to predict damage for the remaining building portfolio, which is more accurate and still rapid enough. The contributions of this research are: the procedure of representative sampling for creating an informative and sufficient representative set, discovering the minimum building representation that uses only location and building geometry attributes, introducing the soft rule formula for monetizing loss, and combining those elements into a novel, usable framework. Using a building representation without earthquake data eliminates the need for analytical methods, shake maps and robust ground motion sensor networks, making the proposed framework unique and applicable in any region. All findings were verified using the M5.4 2010 Kraljevo earthquake data. Most importantly, a 14% relative error for predicting repair cost was obtained using a 10% representative set. This level of precision is acceptable and significantly better than in traditional systems for loss assessment.

Keywords: earthquake · rapid loss assessment · machine learning · representative sampling

1 Introduction

Rapid earthquake loss assessment implies near real-time prediction of damage to affected structures and monetizing loss at the community level. The purpose of any loss assessment system is to be the starting point for planning the recovery process. Accuracy and speed of prediction are the main quality features of rapid loss assessment systems.

N. Filipovic (Ed.): AAI 2022, LNNS 659, pp. 170–185, 2023.
https://doi.org/10.1007/978-3-031-29717-5_12

The problem with traditional systems is their low accuracy, making them unreliable for use in the recovery process, which is their primary purpose. Low accuracy is caused by the significant uncertainty in analytical and insufficient data sets to create vulnerability curves in empirical methods. The root cause of low accuracy in analytical methods is assuming *theoretical* vulnerability relations *before* an earthquake. The research goal is to overcome the accuracy problem by using a completely different approach – establishing vulnerability relations *after* an earthquake by observing *actual* damage.

We aim to use machine learning for damage prediction, along with an appropriate procedure for selecting buildings into a small set that represents the entire inventory of buildings reasonably well. Further, we will improve monetizing loss by carefully determining repair costs for all combinations of damage states and building types and introducing probabilities in the formulas for monetizing loss. We aim to discover the minimum input data set to represent a building, enhancing the practicality and fostering the implementation of the proposed loss assessment approach. Finally, we will create a framework for implementing the new approach, covering the logistics in each framework step. If the new system is accurate and rapid enough, it would be a significant scientific and practical contribution to the field of rapid loss assessment.

2 The Traditional Approach to Rapid Earthquake Loss Assessment and What's Wrong with It?

When an earthquake occurs, buildings are damaged, and society strives to recover as soon as possible. The first step in the recovery process is determining the damage to each building. On-the-ground surveys have to be conducted to record and verify damage officially, but it takes a long time to visit every building – usually a couple of months. So, a rapid assessment (near real-time prediction of damage and loss) is needed to start planning recovery. The primary goals of a rapid system are to assess safety and occupancy issues (detect higher levels of damage) and to monetize loss at the community level (so the local authorities can plan the budget and other resources). The accuracy and speed of prediction are the main quality features of rapid loss assessment systems.

Rapid earthquake loss assessment consists of two phases: preparing the system (pre-earthquake phase) and activating the system when an earthquake happens to predict loss (co-earthquake phase).

2.1 Pre-earthquake Phase – Preparing the System Before an Earthquake

The preparation of traditional loss assessment systems comprises five steps:

1. Creating the building portfolio - a database of buildings in a city or region. The database is populated with characteristics of the building (e.g., building type, age, and geometry features) and its location (e.g., geo coordinates and soil type). Buildings are grouped into building types (BT) according to their seismic behaviour, depending on structure elements, construction era and methods, materials used and others.

   The problem with this step is that many buildings do not fit the exact building type due to design modifications, improvisations, additions, or omissions. It is not always easy to determine the actual BT, so mislabeling can occur.

2. Defining damage states (DS) for buildings. Various classifications exist globally to describe the severity of the damage, ranging from slight damage to collapse.

   The problem with this step is that actual damage states are not clear-cut and often coexist in a building. Also, DS does not contain information about the quantity of damage, which is critical for budgeting the recovery process.
3. Establishing a Ground Motion Model aims to determine the earthquake effects on particular building locations based on a region-wide estimate of ground motion intensities. Peak Ground Acceleration, Spectral Acceleration, Peak Ground Velocity, or macroseismic intensity are intensity measures shown in shake-maps, which combine estimates of the ground motion intensity field with measurements performed during the earthquake. The most used shake-map generating algorithms are the "USGS ShakeMap® algorithms" [1] and the "Bayesian inference method" [2]. In the pre-earthquake phase, a region needs to build a network of ground motion sensors and establish an organizational unit to be prepared to use the system.

   The problem with ground motion models is in introducing a significant amount of uncertainty in algorithms and measurements of ground motion when generating shake maps.
4. Damage prediction links ground motion and the level of damage. In traditional systems, damage prediction is performed before an earthquake by creating vulnerability curves for each building type. For a particular earthquake and observed ground motion, the system predicts the probability of a building type being in a certain damage state. Methods for predicting damage can be "analytical, empirical, and hybrid" [3].

4.1. In analytical approaches, damage states are predicted using dynamic modelling and analysis [4]. Analytical methods can be based on capacity spectrum, collapse mechanism, or full displacement.

Analytical methods assume that actual buildings can be categorized into building types, which are analyzed in-depth regarding their seismic behavior. Then vulnerability relations are established, connecting PGA values to probabilities of buildings belonging to certain damage classes. Unfortunately, this is a theoretical exercise since actual buildings vary from theoretical models for numerous reasons (diminishing seismic capacity, ageing, deviating from design, low quality of work, low maintenance, climate influences, local soil circumstances, or adding floors/walls). In addition, it is not easy to classify buildings into building types because they differ in geometry and methods, so knowledgeable surveys are required. Approximations are made based on the year of construction, which proved to be a good indicator of building type (building types represent construction eras). Therefore, analytical methods imply introducing much uncertainty to loss assessment.

Machine learning (ML) methods are also used in research to create vulnerability relations but with no significant scientific or practical added value other than to verify already explored methods. Using AI this way does not solve implementation issues on a diverse portfolio.

4.2. In empirical methods, the assessment of damage states is derived from previously recorded damages on different earthquakes. The results are converted into "damage probability matrices or continuous vulnerability curves fitted to the data" [3]. The problem is that obtained data sets rarely are large enough for reliable predictions when implemented locally [5]. Data from a single earthquake has a limited range of intensities, thus producing just a part of the vulnerability curve, so data fitting is required. The empirical procedure relies on combining data from different regions and for various building types, thus introducing another layer of uncertainty. Typical vulnerability curves derived from empirical data are shown in Fig. 1.

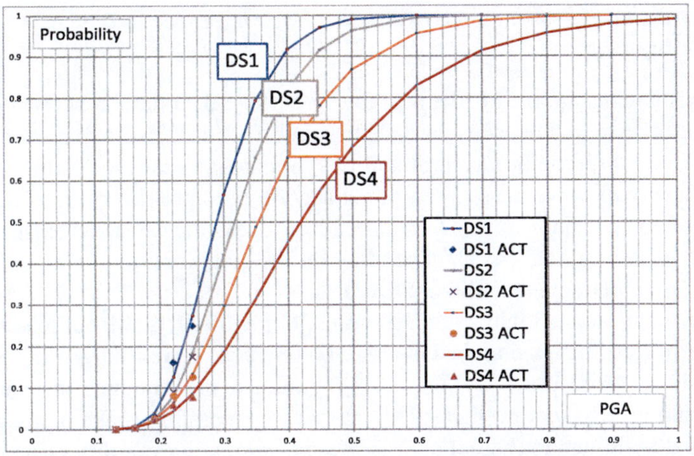

Fig. 1. Vulnerability curves derived from empirical data (Kraljevo 2010 earthquake) showing actual data (dots) and curves fitted to data for different damage states (DS1-DS4)

The fitting procedure was explored in [6], proving that fragility curves created using only a fraction of needed data are unreliable for damage prediction. The PGA values covered a 0.05–0.25 range. The remaining, much bigger part of the curve, for the 0.25–1.00 range, was fitted to data based on mathematical assumptions to mimic the expected curve shape. The prediction accuracy is hard to assess.

5. Loss quantification is the starting point of the recovery process because it gives the local authorities a first assessment of the needed budget. Loss quantification provides costs for repairing all damage states and building types, typically expressed per unit area in a matrix. In the case of collapsed buildings, repair cost turns into replacement cost.

Usually, repair costs are "normalized by the replacement value of the building" [7, 8], which in itself is a source of uncertainty because it implies using ratios (adding more corresponding uncertainty when choosing ratio values). Also, market conditions play a role as replacement values can cause prediction errors too. For example, replacement values in remote rural regions may be lower than the construction cost for a new building, while city centers present opposite challenges.

With repair costs, there is a problem of unknown quantities of repair works, which introduces even more uncertainty into loss assessment. Accordingly, it is possible that repairing lower damage states can cost more than repairing higher damage states. For higher damage states, there is another problem because different construction methods can be used (with corresponding differing costs). Cost matrices need to be up to date, as market conditions can cause significant changes in labour and material prices and replacement values. For a better understanding of the problem, construction management experts should also be involved in upgrading current loss quantification practices.

2.2 Co-earthquake Phase - Activating the System Immediately After an Earthquake

Since all components are set in advance, the co-earthquake phase is simple to perform. When an earthquake occurs, sensors detect ground motions, shake maps are generated and, using vulnerability relations, probabilities for damage states are estimated for each building in near real-time. Loss is computed according to pre-determined vulnerability relations and replacement values. Operational loss assessment systems exist worldwide. They can be local or global [4, 9], depending on the combination of elements to create shake-maps, predict damage and monetize loss.

Although each step is logical and scientifically justified, traditional loss assessment systems are not accurate enough. Uncertainty prevents systems from accomplishing decent accuracy. More precisely, there are significant uncertainties in every step: in algorithms and ground motion measurements when generating shake maps, in approximating actual buildings to theoretical models when creating fragility curves, in fitting empirical data collected from various earthquakes and in repair costs & replacement values when monetizing loss. As pointed out in HAZUS, the total uncertainty is "possibly at best a factor of two or more". This low accuracy was the motivating reason for us to investigate other approaches and try to improve accuracy.

The conclusion is that two problems hamper the traditional approach to loss assessment: analytical methods cannot be accurate because of too many assumptions and approximations, and empirical methods cannot work because they lack large enough data sets to be reliable. Earthquake engineering by itself is stuck, analytical methods cannot be made less uncertain, and there is no solution for inadequate empirical samples.

3 The New Approach and Framework for Rapid Earthquake Loss Assessment

This chapter explains, and the case study verifies, how to overcome deficiencies of traditional loss assessment approaches and improve their accuracy. When faced with uncertainties and limiting small data sets, researchers need to look for multidisciplinary expert knowledge and make higher-level abstractions to enable ML techniques and eliminate unnecessary sources of uncertainty (which are not essential for loss assessment accuracy).

The general idea was to view an earthquake as an event that causes the distribution of various damage states to buildings across a territory. Regardless of the seismic nature of the cause, the hypothesis is that an ML technique could learn such a distribution from a small observed data set. That way, most of the uncertainties could be eliminated from the process. Since obtaining large input data sets will always be challenging, the idea was to make an informative selection that enables even a small data set to teach an ML algorithm.

The first task is to show the capability of ML techniques in predicting damage states. The second is to explore how eliminating earthquake data from the input set affects prediction accuracy. The third task is to improve monetizing loss by creating a repair cost matrix for all BT-DS combinations and including probabilities (inevitable in DS predictions) in formulas for aggregating loss. The fourth is to establish a method for creating a representative sample for a portfolio of buildings. Based on these four elements, a framework is needed to provide logistics enabling the practical implementation of the new approach in earthquake-prone regions. The final task is to explore the relation between the size of the representative sample and corresponding accuracy and find the optimum tradeoff. Only the main findings are presented in this paper, while the complete research is in [10].

3.1 M5.4 2010 Kraljevo Earthquake Data Modeling

The case study relates to the M5.4 2010 Kraljevo earthquake, with almost 6,000 damaged structures, a quarter of which were unsafe to occupy in the aftermath. Two fatalities and one hundred medically treated injuries happened. Due to minor losses recorded in other structures, "loss assessment" in this research relates to residential buildings in the area affected by the earthquake.

It took more than a year to acquire the data set, working with different institutions, which illustrates the difficulty of obtaining large data sets. Finally, a dataset of 1979 buildings (652 damaged) was established, spreading across three typical districts.

According to construction eras, structure elements and construction methods used in the wider region, building types were recognized as follows:

- BT1: Wooden structure with stone foundations (before the 1950s)
- BT2: Masonry structure, old brick format (before 1933)
- BT3: Masonry structure, new brick format (after 1933)
- BT4: Masonry structure with horizontal reinforced concrete ring beams (1963–1975)
- BT5: Masonry structure with both horizontal and vertical reinforced concrete ring beams (1975–1990)
- BT6: Masonry structure with both horizontal and vertical reinforced concrete ring beams (after 1990)

Input attributes representing the *Building* are building type, number of floors, year of construction, and the footprint area.

Input attributes representing the *Location* are GIS x and y coordinates and soil type (a discrete attribute).

The damage states are classified into five categories:

- DS0 - no damage
- DS1 - slight damage
- DS2 - moderate damage

- DS3 - heavy damage
- DS4 - collapse

This classification corresponds with the EMS-98 [11] damage classification, except merging "very heavy" and "destruction" damage states into DS4.

Input attributes representing the *Earthquake* are elastic acceleration response interval [0, 4 s] spectrum values (including the Peak Ground Acceleration) and the distance from the epicenter to the building. The Akkar-Bommer ground motion prediction equation, "often used for seismically active regions in Europe" [12, 13], is used to model the ground motion spatial distribution.

For loss quantification, using construction management methods and local market prices, experts produced a matrix containing repair costs (replacement cost for collapsed buildings) for each BT-DS combination.

3.2 Can ML Be Used for Earthquake Damage Prediction and How?

The Random Forest algorithm was chosen to show the ability of machine learning to discover the relation between the building attributes and damage states. The procedure for testing the prediction of damage states is 10-fold cross-validation using the whole dataset of 1979 buildings. Buildings are represented by all input attributes: *Earthquake* (distance to the epicenter, spectral acceleration values series), *Location* (x and y geo-coordinates, soil type), and *Building* (building type, number of floors, construction year, footprint area). A confusion matrix (Table 1) is a common way to present the prediction performance, aggregating correct and incorrect results from test folds. The results show a high accuracy (85%), but the precision and recall values are not as good. Damage states with many examples (DS0, DS1, and DS3) performed better, while damage states containing smaller numbers of buildings (DS2 and DS4) underperformed. This experiment, fully explained in [10], shows that an ML algorithm can map damage states to building types, but it needs a bigger data set or a data set with a better BT-DS distribution to perform better.

Table 1. Confusion matrix showing the prediction performance (accuracy, precision and recall)

Predicted / Actual	DS0	DS1	DS2	DS3	DS4	Total number of buildings:	**Recall**
DS0	1314	11	0	1	1	1327	0.99
DS1	27	234	38	18	11	328	0.71
DS2	6	64	37	13	7	127	0.29
DS3	7	31	17	69	9	133	0.52
DS4	3	18	14	10	19	64	0.30
total:	1357	358	106	111	47	**Accuracy** 1673/1979=0.85	
Precision	0.97	0.65	0.35	0.62	0.40		

This research is not the only one to explore using machine learning for damage prediction, showing confusion matrices like Table 1 to evaluate prediction performance, i.e. [14]. However, it is essential to point out that researchers make the mistake of using the whole data set acquired long after an earthquake for learning and testing, thus achieving good results (the larger the set, the better the results). Immediately after an earthquake (when rapid assessment takes place), the dataset does not exist! There is only a brief period to gather a small portion of buildings and use that set for prediction. To simulate a real-life situation in this example, we would be obligated to use only a fraction of the buildings dataset (200–300 buildings, not 1979) for learning, not the whole set. The results would be far worse. As we aim to propose a practical and usable framework, we will explore building representations, selecting samples and checking the corresponding accuracy, thus imitating real situations, not theoretical ones.

3.3 Do We Even Need Earthquake Data to Predict Earthquake Damage?

This intriguing hypothesis is very important for the operational versatility of the new approach and would open up various possibilities if proved true. Choosing the right combination of features in a building representation has two goals: to discover the minimum representation that returns acceptable prediction accuracy while using building features that can realistically be obtained. Accordingly, an experiment was conducted to assess the prediction accuracy with different combinations of building attributes. Specifically, we wanted to test the prediction accuracy if earthquake and soil data were excluded from the building representation. The Random Forest ML algorithm was used to train four different damage prediction models: *Earthquake + Location + Building*; *Location + Building*; *Earthquake + Building*; and *XY + Building*.

The 10-fold cross-validation was repeated 100 times. The model using all attributes and other models with lesser representations were compared using a paired t-test [15]. The results from Table 2 show similar results for all models in terms of accuracy. The last two models returned slightly better results in terms of Cohen's Kappa statistics [16], especially knowing that "models with Kappa greater than 0.7 are considered to perform sufficiently well" [17].

Table 2. Comparing accuracy and Kappa for different combinations of building attributes

	Earthquake + Location + Building	Earthquake + Building	Location + Building	XY + Building
Accuracy (%)	85.4	83.6	85.4	85.4
Kappa	0.69	0.68	0.71	0.71

The results (the last two columns in Table 2) proved the starting hypothesis that information about the earthquake can be excluded from the prediction model and obtain even slightly better results. Although surprising at first, this can be explained as follows:

1. Damage states indirectly contain earthquake information and act like rudimentary seismometers.

2. Building types reflect seismic behaviour, and the combination of area and number of floors indicates the building's dynamics.

For further use, the chosen representation is the smallest possible, containing only geo-coordinates, building features and no earthquake data (the last in Table 2).

3.4 Improving Monetizing Loss - Repair Cost Matrix and the "Soft Rule"

Two proposed upgrades can significantly improve the loss of monetization.

Firstly, we developed a repair cost matrix covering all BT-DS combinations with costs expressed in euros per unit footprint area (€/m^2). The matrix was approached differently for each damage state according to its nature. Lower damage states (DS1 and DS2) relate primarily to finishing works to restore the pre-earthquake look. The unit prices are predictable, but the problem is the quantity of work since there was no quantity indicator in the survey forms used in Kraljevo. The problem was approached from a statistical point of view. Heavy damage (DS3) presents a different challenge since repairs include demanding work reinforcing the superstructure according to a permitted specialist design. Since various methods were applied in the reconstruction process, the prevailing one was chosen for further analysis and determining the final matrix values. The cost was calculated by breaking down all needed works and analyzing resource consumption, quantities and unit prices. Finally, the collapsed buildings (DS4) were the easiest to evaluate since the government provided prefabricated wooden homes with commonly known prices per unit area. The repair cost matrix is shown in Table 3. More details regarding the cost matrix are in [10].

Table 3. Mean repair cost matrix (€/m2).

	BT1	BT2	BT3	BT4	BT5	BT6
DS1	3,43	12,04	8,36	10,43	9,14	9,14
DS2	13,40	16,04	15,14	18,86	17,54	18,64
DS3	55,69	46,30	44,15	38,87	29,72	32,75
DS4	350,00	350,00	350,00	350,00	350,00	350,00

Secondly, for calculating the predicted repair cost (PRC) for the whole inventory, a "soft rule" formula is introduced. The building attributes vector **b** represents a building of type t (BT = t) from the set of all buildings B. The damage state (from the DS set) other than DS0 is d. The cost matrix (Table 3) determines the repair cost per unit footprint area $\mathbf{C} = [c_{dt}]$. Building **b** belongs to damage state d with the probability $p_d(\mathbf{b})$. The footprint area of building **b** is $a(\mathbf{b})$. The soft rule formula that calculates PRC is then:

$$PRC_{soft} = \sum_{\mathbf{b}\in B} \sum_{d\in DS} p_d(\mathbf{b})a(\mathbf{b})c_{dt} \tag{1}$$

The soft rule takes into account all damage states with corresponding probabilities, as opposed to the hard rule, where the highest probability determines the damage state of a building. The soft rule's relative error ($|PRC - ARC| / ARC$) was only 5%, while the hard rule performed significantly worse with a relative error of 20% (ARC - actual repair cost). This comparison evidently favours the soft rule and DS probability distributions over discrete DS labels. The confusion matrix from Table 1 offers more insight into the accuracy of the soft rule. DS misclassifications occurred symmetrically to the main diagonal, implying almost equal overestimations and underestimations and enabling a quality PRC estimate. Therefore, for further use, the soft rule will be applied.

3.5 Representative Sampling

Since there is only a limited time to observe a small sample of buildings in the aftermath of an earthquake (if the system is aimed to be rapid), the crucial element of the new approach is pre-selecting buildings into a representative set, appropriately representing the entire inventory. Randomly choosing a representative set would give unsatisfactory prediction results because, inevitably, there would not be enough examples for all BT-DS combinations, thus preventing an ML algorithm from predicting damage states well.

Figure 2 illustrates an algorithm which can select a small set, representing the building portfolio well enough to be used in the pre-earthquake phase. Based on the Chapter 3.3 findings and the minimum building representation debate, a < BT, number of floors > pair was chosen to represent the portfolio. Such a pair of information captures the seismic behavior and the dynamics of buildings. By gradually choosing buildings, the sampling algorithm incrementally determines the distribution of all < BT, number of floors > combinations over a territory.

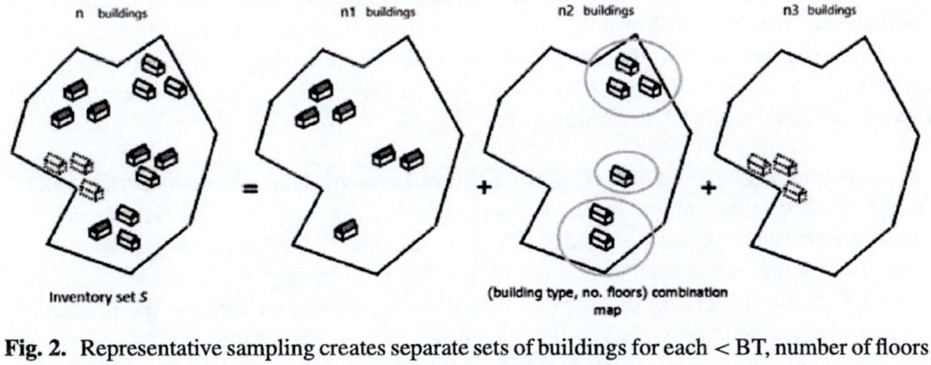

Fig. 2. Representative sampling creates separate sets of buildings for each < BT, number of floors > combination. Circles are showing discovered spatial clusters.

The algorithm divides the inventory into smaller subsets (clusters) according to existing < BT, number of floors > combinations, choosing a proportional number from each subset. The chosen K-means clustering algorithm [18] uses geo-coordinates to discover clusters. It randomly picks buildings to be centroids of starting spatial clusters. The remaining buildings get arranged into clusters according to the distance from the centroids. This procedure repeats until the centroids become stable, and a single building represents every < BT, number of floors > combination for all discovered spatial clusters.

The proposed sampling method selects the building closest to the centroid of its cluster. A more detailed explanation of the sampling algorithm is in [19].

3.6 The Proposed RELA (Rapid Earthquake Loss Assessment) Framework

Clearly, Chapter 2 findings indicate that rapid earthquake loss assessment should be reconsidered and re-formulated in light of new machine learning achievements to improve accuracy. Traditional systems establish vulnerability relations before an earthquake, which inevitably includes too many assumptions that compromise accuracy. The new idea is to learn the BT-DS relationships after the earthquake quickly enough to be considered rapid. The idea is to observe damage states on a small sample and use ML algorithms to predict damage states on the rest of the portfolio.

Two methods exist for gathering damage data after the earthquake: remote sensing and on-the-ground surveys.

Remote sensing involves combining ML methods with image processing of aerial images [20], satellite images [21] or "synthetic aperture radar data" [22] to perform remote damage assessment [23, 24]. But, as stated in [20], "there is a general tendency in remote sensing to underestimate damage". Furthermore, remote sensing struggles to distinguish between slight, moderate and heavy damage.

Since on-the-ground surveys would be easier to implement and must be conducted at least once during recovery, the authors chose to explore such an approach.

As with traditional systems, the proposed RELA framework also consists of two phases:

Pre-earthquake Phase - Preparing the System Before an Earthquake. Form a database of buildings with minimum representation. Form a sample of buildings that represents the portfolio well – a representative set. Train a group of assessors to detect the damage states of buildings. Experts prepare and regularly update a cost matrix with repair costs for each BT-DS combination.

Co-earthquake Phase - Activating the System Immediately After an Earthquake. In the co-earthquake phase, trained assessors visit the representative set, detect the damage state and upload the results. The ML Random Forest algorithm immediately predicts damage states for the rest of the portfolio. Each day, the assessors continue to visit more buildings. As the number of observed buildings increases, the prediction accuracy for the remaining part of the portfolio also increases. After a couple of cycles, the overall prediction process is accurate enough. Since observing the representative set takes a couple of days, the prediction process can be categorized as rapid. The RELA framework for ML-based earthquake loss assessment is shown in Fig. 3.

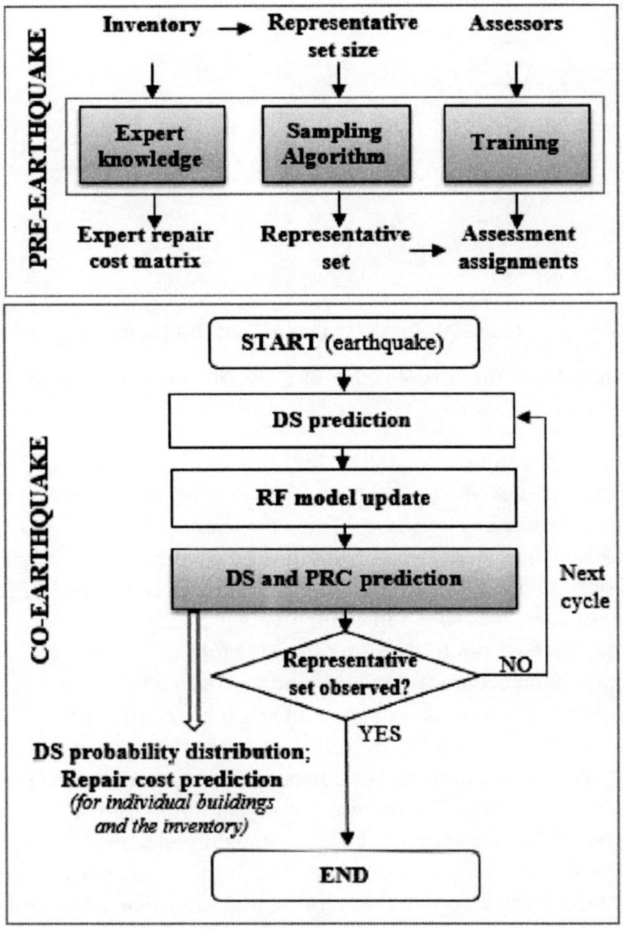

Fig. 3. The RELA framework for ML-based earthquake loss assessment

3.7 Achieved Accuracy Using the RELA Framework

The representative set size (and the accompanying accuracy) is a critical framework feature addressing the issue of small samples, which are inevitable in the co-earthquake phase. The goal is to determine the minimum set size returning satisfactory prediction accuracy. If the set is not small, it takes too much time for assessors to observe damage states, and the whole system is not rapid.

A test calculating the relative error for PRC according to formula (1) was performed to evaluate the prediction accuracy. For 5%, 10%, 15%, and 20% sample sizes, the K-means algorithm was used to create 300 Random Forest models. For future comparisons with other research, it is imperative to recognize that representative sampling was carried out *without* information on the state of damage, simulating the real-life pre-earthquake phase. Figure 4 shows the median values for the relative error for all sample sizes.

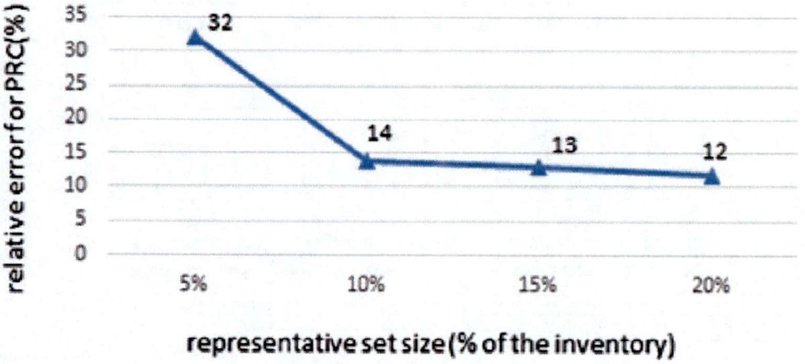

Fig. 4. The relative error of predicting PRC for different representative set sizes

As expected, larger sample sizes deliver higher accuracy. Satisfactory results (relative error of 14%) can be achieved with a sample size of 10%. This precision is significantly better than using traditional systems.

With the assessment speed of 30–40 buildings covered by one assessor daily, 60 assessors could observe nearly 2,000 buildings in one day. In the case of Kraljevo, there are ~ 40.000 houses, so the 10% representative set could be inspected in two days, which can be considered a rapid loss assessment. More on various set sizes delivering corresponding prediction accuracy can be found in [19].

The case study presented in this chapter shows how to solve the problems of traditional damage prediction by combining different kinds of expert knowledge with ML: earthquake engineers for building representation, project management experts for framework logistics and IT experts for representative sampling and ML-based prediction models. The proposed RELA framework can be considered a breakthrough in rapid loss assessment systems since it is more accurate than traditional approaches and far more implementable since it does not need any infrastructure other than a group of trained assessors.

4 Beyond Loss Assessment – The Recovery Process

The purpose of rapid loss assessment is to trigger the essential aspect of an earthquake as a disaster – the recovery process. For efficient management of the recovery process, authorities need information about the total cost (preferably using the soft rule for loss assessment) and the damage distribution since different DS imply different recovery costs and procedures. "Slight" and "Moderate" damage are the most numerous but also the simplest and cheapest. "Heavy" damage implies the longest duration because it includes speciality design, obtaining permits and performing demanding types of work. "Collapsed" buildings carry the most cost, considerable time and a lengthy but straightforward procedure. In the case of the Kraljevo earthquake, Fig. 5 shows the recovery rate for buildings, measured by re-occupancy dates for each damage state.

Repairs of all buildings in the data set were performed simultaneously and completed at approximately the same time, 18 months after the earthquake. It is evident that the

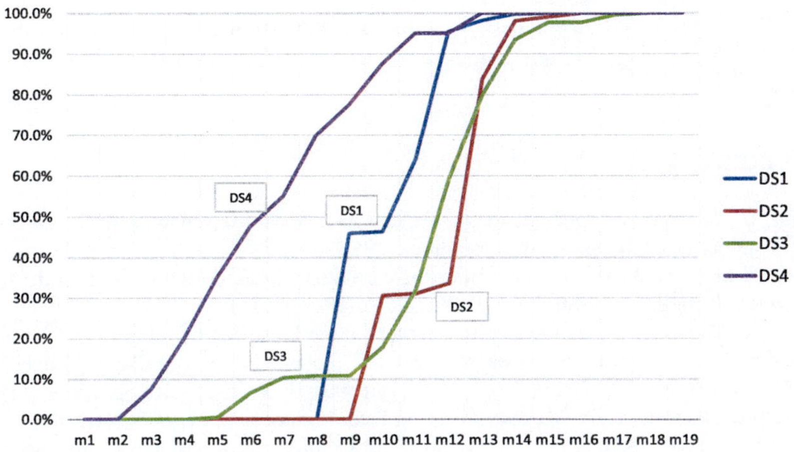

Fig. 5. Rate of recovery of buildings in different damage states [25]

delays in investment prolonged the recovery process. If all of the spent funds were available immediately after the earthquake, the recovery process would have lasted about 14 months and would still be driven by the extensive repairs needed for buildings in DS3. Perhaps, a quick and reliable loss assessment would have helped in that regard.

5 Conclusions

The problem with traditional systems is their low accuracy, making them unreliable and unusable in the recovery process, which is the primary purpose of loss assessment systems. Low accuracy is caused by too much uncertainty in analytical and insufficient data sets to create vulnerability curves in empirical methods. The root cause of low accuracy in analytical methods is caused by assuming *theoretical* vulnerability relations *before* an earthquake. The proposed RELA framework suggests an entirely different approach and uses trained assessors to perform on-the-ground observation of *actual* damage on the representative sample *after* an earthquake. ML methods are then used to predict damage to the remaining building portfolio, which is more accurate and still rapid enough. A valuable contribution is using a building representation without earthquake data which eliminates the need for analytical methods, shake maps and robust ground motion sensor networks, making the proposed framework applicable in any region.

The RELA framework is a result of multidisciplinary expert knowledge.

RELA combines minimum building representation, ML-based damage prediction, representative sampling, on-the-ground surveys performed *after* the earthquake, DS probabilities for monetizing loss (soft rule), and a repair cost matrix in a unique way, making it a new kind of rapid loss assessment system. Furthermore, using buildings as damage sensors opens up the possibility for implementation in disaster scenarios other than earthquakes, using the same representative sample.

The case study relates to the M5.4 2010 Kraljevo earthquake, with almost 6,000 damaged structures, a quarter of which were unsafe to occupy in the aftermath. In the case study, a 10% representative set size delivered predicted repair cost with a relative error of 14%, which is acceptable and significantly better than in traditional loss assessment systems.

References

1. Wald, D., Worden, B., Quitoriano, V., Pankow, K.: ShakeMap manual: technical manual, user's guide, and software guide", Reston: USGS (2005)
2. Gehl, P., Douglas, J., D'Ayala, D.: Inferring earthquake ground-motion fields with bayesian networks. Bull. Seismol. Soc. Am. **107**(6), 2792–2808 (2017)
3. Maio, R., Tsionis, G.: Seismic fragility curves for the European building stock: Review and evaluation of analytical fragility curves, s.l.: JRC Technical Report EUR 27635 EN (2015)
4. Erdik, M., Sesetyan, K., Demircioglu, M.B., Hancılar, U., Zulfikar, C.: Rapid earthquake loss assessment after damaging earthquakes. Soil Dyn. Earthq. Eng. **31**, 247–266 (2011)
5. Eleftheriadou, A., Karabinis, A.I.: Damage probability matrices derived from earthquake statistical data. In: 14th World Conference on Earthquake Engineering, Beijing, (2008)
6. Stojadinović, Z., Kovačević, M., Marinković, D., Stojadinović, B.: Data-driven housing damage and repair cost prediction framework based on the 2010 Kraljevo earthquake data. In: Proceedings of the 16th World Conference on Earthquake Engineering (16WCEE), Chile (2017)
7. Calvi, G., Pinho, R., Magenes, G., Bommer, J., Restrepo-Vélez, L., Crowley, H.: Development of seismic vulnerability assessment methodologies over the past 30 years. J. Earthquake Technol. **43**, 75–104 (2006)
8. FEMA, HAZUS Earthquake Model Technical Manual, Washington, D.C.: Federal Emergency Management Agency (2020)
9. Guerin-Marthe, S., Gehl, P., Negulescu, C., Auclair, S., Fayjaloun, R.: Rapid earthquake response: the state-of-the art and recommendations with a focus on European systems. Int. J. Disaster Risk Reduction, **52**, 101958 (2021)
10. Stojadinović, Z., Kovačević, M., Marinković, D., Stojadinović, B.: Rapid earthquake loss assessment based on machine learning and representative sampling. Earthq. Spectra **38**(1), 152–177 (2022)
11. Grunthal, G.: European Macroseismic Scale, Luxembourg: Chaiers du Centre Européen de Géodynamique et de Séismologie, vol. 15, (1998)
12. Akkar, S., Bommer, J.J.: Empirical equations for the prediction of PGA, PGV, and spectral accelerations in Europe, the Mediterranean region, and the Middle East. Seismol. Res. Lett. **81**(2), 195–206 (2010)
13. Akkar, S., Sandıkkaya, M.A., Bommer, J.J.: Empirical ground-motion models for point- and extended-source crustal earthquake scenarios in Europe and the middle east. Bull. Earthq. Eng. **12**(1), 359–387 (2014)
14. Mangalathu, S., Sun, H., Nweke, C., Yi, Z., Burton, H.: Classifying earthquake damage to buildings using machine learning. Earthq. Spectra **36**(1), 183–208 (2020)
15. Montgomery, D., Runger, C.; Applied Statistics and Probability for Engineers. 6th ed. s.l.: Wiley (2014)
16. Cohen, J.: A coefficient of agreement for nominal scales. Educ. Psychol. Measur. **20**(1), 37–46 (1960)
17. Landis, J., Koch, G.: The measurement of observer agreement for categorical data. Biometrics **33**(1), 159–174 (1977)
18. MacQueen, J.B.: Some methods for classification and analysis of multivariate observations. In: Proceedings of 5-th Berkeley Symposium on Mathematical Statistics and Probability, Berkeley (1967)

19. Kovačević, , Stojadinović, Z., Marinković, D., Stojadinović, B.: Sampling and machine learning methods for a rapid earthquake loss assessment system. In: Proceedings of the 11th National Conference on Earthquake Engineering, paper ID 649, June 25–29, Los Angeles, CA, USA (2018)
20. Booth, E., Saito, K., Spence, R., Madabhushi, G., Eguchi, R.T.: Validating assessments of seismic damage made from remote sensing. Earthq. Spectra **27**, 157-S177 (2011)
21. Tian, T., Nielsen, A., Reinartz, P.: Building damage assessment after the earthquake in Haiti using two post-event satellite stereo imagery and DSMs. Int. J. Image Data Fusion **6**, 155–169 (2015)
22. Plank, S.: Rapid damage assessment by means of multi-temporal sar - a comprehensive review and outlook to sentinel-1. Remote Sens. **6**, 4870–4906 (2014)
23. Cooner, A., Shao, Y., Campbell, J.: Detection of urban damage using remote sensing and machine learning algorithms: revisiting the 2010 Haiti earthquake. Remote Sens. **8**, 868 (2016)
24. Duarte, D., Nex, F., Kerle, N., Vosselman, G.: Multi-resolution feature fusion for image classification of building damages with convolutional neural networks. Remote Sens. **10**, 1636 (2018)
25. Marinković, D., Stojadinović, Z., Kovačević, M., Stojadinović, B.: 2010 Kraljevo earthquake recovery process metrics derived from recorded reconstruction data. In: Proceedings of the 16th European Conference on Earthquake Engineering, paper ID 10755, June 18–21, Thessaloniki, Greece (2018)

SecondOponionNet: A Novel Neural Network Architecture to Detect Coronary Atherosclerosis in Coronary CT Angiography

Mahmoud Elsayed(✉) and Nenad Filipović

Bioengineering Research and Development Centre (BioIRC), 6 Prvoslava Stojanovića Street, 3400 Kragujevac, Serbia
mahmoud.mohmmed.mustafa@gmail.com, fica@kg.ac.rs

Abstract. Cardiovascular diseases are a family of illnesses that takes millions of lives every year. This has become obvious to researchers and medical practitioners in the recent years. Hence, an increased number of awareness campaigns about their prevention has been noticed as well as a dramatic increase in the research funding of their treatment/diagnosis. There has been a recent tendency in the literature to utilize the advanced machine learning algorithms in the medical field which has been fruitful in terms of demonstrating effective results. In this paper, we propose a novel neural network architecture to detect coronary atherosclerosis in coronary Computed Tomography (CT) angiography. The developed architecture achieved a promising accuracy of 92.63% and AuC of 0.9432. The proposed method is meant to prove a conceptual ideation which is mimicking the clinical practices (such as taking an expert second opinion). This proven concept has the potential to be adopted in many wide medical applications.

Keywords: Neural Networks · Atherosclerosis · CT angiography · Coronary Artery Disease (CAD) · Peripheral Artery Disease (PAD)

1 Introduction

According to the World Health Organization, cardiovascular diseases that includes atherosclerosis are the leading cause of death worldwide, causing 17.9 million death each year [1]. It is furthermore estimated that today more than 200 million people live with Peripheral Artery Disease (PAD) [2]. These high numbers have given a serious alert to the humanity to take strong actions to reduce them as they highly endanger the survival and the wellbeing of the human race. It is noticed that the amount of fund of research on PAD has increased dramatically in the recent years as well as the awareness campaigns raised about their prevision.

Although, in this paper we will be investigating the Coronary Artery Diseases (CAD), the application proposed herewith could be extended to PAD as well. This is because the studies have shown that there is a strong correlation between CAD and PAD. It was found that 46.88% of patients who were diagnosed of PAD also had CAD [3]. In

N. Filipovic (Ed.): AAI 2022, LNNS 659, pp. 186–197, 2023.
https://doi.org/10.1007/978-3-031-29717-5_13

addition to this, the biological process in which they formulate is similar and hence their diagnostics through the medical imaging follow the same procedures.

On the other hand, our technological development and maturity allowed us to use advanced algorithms in the medical field which had a significant positive impact on the accuracy of our diagnosis and the efficiency of our treatment. The literature is rich of examples that perfectly demonstrates this. For example, a manually informed neural networks was used to enhance the performance of heart failure predictors [4]. The neural networks were also used in the accurate prediction and diagnosis of Parkinson's disease [5, 6]. In [7], a technique for artifacts rejection of MEG and EEG signals was designed. In [8], the physical pain was both detected and quantified through the EEG signal analysis. The technological advancement also allows us to generate more medical data that helps us to use more powerful machine learning algorithms [9].

In this paper we propose a novel neural network architecture that was built to detect coronary atherosclerosis in coronary CT angiography. We believe that technologies like the one proposed herewith can help in advancing the field of the automated medical diagnostics and thus help in reducing the horrific number of deaths caused by cardiovascular diseases every year. This will be possible as our network will potentially support medical practitioners to have more efficient diagnosis which will be demonstrated.

2 Materials and Methods

2.1 Dataset

Data Selection

The used data has been chosen based on the following criteria:

1. Has clear full images that represent clear information about the case.
2. Has clear labels or at least has the availability of the labels' information.
3. Has sufficient number of images to facilitate the learning process of the CNN. This creation is based upon the information obtained from the literature.
4. Acquired from a reliable resource like a university or a medical institution

Chosen Dataset

The chosen dataset was collected and published in 2019 by Ohio State University Wexner Medical Center and could be found in [10]. It includes coronary artery image sets for 500 patients which were collected using a CT angiography. Each image represented in a Mosaic Projection View (MPV) which consists of 18 different views of a straightened coronary artery stacked vertically. The data was divided to 5 portions, 4 of which went for the training set (80%) and the rest were divided into two, half for the test set (10%) and half for the validation set (10%), each of which had 50% normal and 50% diseased cases. Figure 1 shows a sample of this dataset demonstrated based on the investigated classes.

As could be seen in the images demonstrated in Fig. 1, it is very hard to distinguish between the negative and positive classes with the bare eye. Hence, this type of images is held to be inconclusive by most medical practitioners. This shows the crucial need of utilizing the engineering tools such image processing and machine learning in such application.

(a) (b)

Fig. 1. A random sample of the investigated classes fetched from the dataset where (a) is negative and (b) is positive.

2.2 Methodology

Augmentation and Classical Image Processing

Although the data seems to be clear from an engineering perspective, their quantity is not that large to be used in a deep learning application which might cause problems like bias or overfitting to avoid this, certain classical image processing techniques have been used as an augmentation strategy to increase the number of images and help the trained neural network to achieve the required generalization. These image processing techniques were mainly in the form of filters such as Gaussian and mean. These two filters will somehow introduce a type of noise to the image which will cause the neural network to be able to generalize and learn to classify even when noise is present.

Mean Filter

The mean filter aims to remove noise from the image by blurring the images and in doing so it is believed that the Signal to Noise Ratio (SNR) is increased. This is done by determining the mean of the pixel values within a n x n kernel. The pixel intensity of the center element is then replaced by the mean. The blur function from the Open-CV [11] library can be used to apply a mean filter to an image. The value of n in the used filter is 3 which makes the kernel size of 3×3. Other values have been tried out during the preprocessing pipeline development, however, this value in particular gave the best results.

Gaussian Filter

The Gaussian filter works on a work principle that is very similar to the mean filter, however, it involves a weighted average of the surrounding pixels and has a parameter called sigma. The kernel represents a discrete approximation of a Gaussian distribution. While the Gaussian filter blurs the edges of an image (like the mean filter), it does a better job of preserving edges than a similarly sized mean filter. The 'Gaussian Blur' function from the Open-CV package was used to implement the filter on the dataset. For consistency we used the same kernel size used in the mean filter (i.e.3×3) and the sigma value was calculated through Eq. 1 where N is obtained from the kernel size which gives us the value of $\sigma = 1/3$.

$$\sigma = \frac{N-1}{6} \tag{1}$$

Other Augmentation Techniques
Other common augmentation techniques have been used in order to help the developed model to achieve the generalization ability. These augmentation techniques include the following:

1. Random rotation
2. Shifts
3. Shear
4. Vertical and horizontal flips

The aforementioned augmentation techniques were implemented with the purpose of increasing the number of images in a way that will facilitate the generalization ability of the trained network.

Classification
It was found that 15% of diagnosis and 37% treatment plans have changed due to taking a second opinion [12]. The clinical impact of a second opinion is proven to be of a significant impact [12–14]. Due to this positive impact, we proposed a neural network that mimic the concept of having a second opinion in the clinical practice by applying the machine learning techniques known as stacking. This is achieved through training two different neural network architecture on the dataset and then taking the features they produce in there last fully connected layer to a diagnosing neural network where the decision or diagnosis will be finalized. Figure 2 shows the overall used methodological neural network architecture.

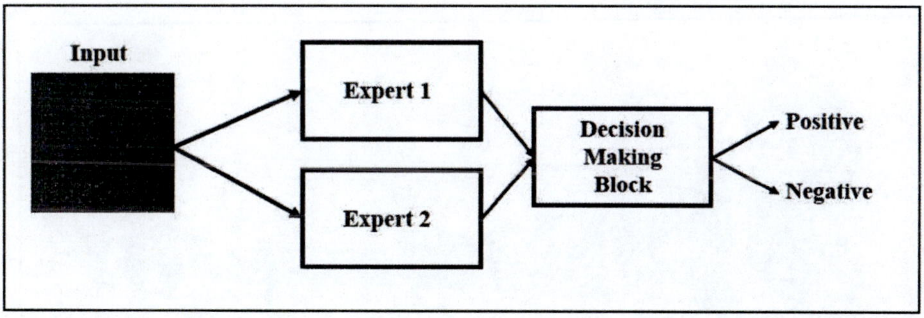

Fig. 2. The General Methodological Architecture of the SecondOponionNet.

Both experts 1 and 2 will investigate the inputted image through different neural networks. The first contains ResNet 18 [15] and the second contains VGG 19 [16]. Figure 3 shows ResNet 18 architecture.

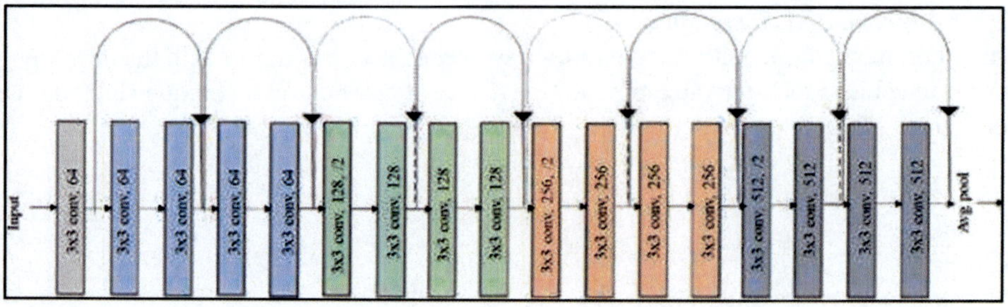

Fig. 3. The Architecture of Expert 1 Building Block [15].

Figure 4 shows the VGG 19 architecture. The decision-making building block contains three fully-connected layers that receive the output of the last fully connected layers of the two experts as an input and output the result of the binary classification (positive or negative) through the SoftMax function shown in Eq. 2 [17],

$$e(\overrightarrow{z})_i = \frac{e^{z_i}}{\sum_{j=1}^{2} e^{z_j}} \tag{2}$$

where $\overrightarrow{z}$ is the input vector, e^{z_i} is the exponential function for the input vector, and e^{z_j} is the exponential function for the output vector. The choice of these two networks in particular have been made after trying out different similar-sized known architectures that did not perform as well as the chosen ones.

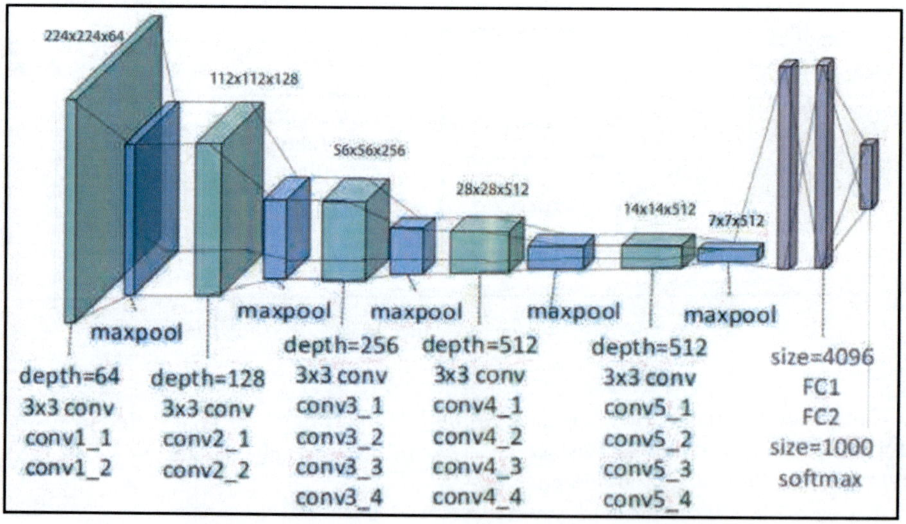

Fig. 4. The Architecture of Expert 2 Building Block [18].

Training Parameters
Although the investigated neural networks architectures have been imported ad pretrained models. They had been trained with no freezing to any of their layers and this pretraining has been considered as weight initialization mechanism. The following hyperparameters have been fine-tunned during the training process:

1. learning rate
2. optimization technique
3. Batch size

Some other hyperparameters have not been tuned due to the fact that they are deeply integrated to the architecture of the neural networks rather than being parameters of training such as the following parameters:

1. activation function
2. loss error
3. Kernel size and depth

According to [19], deep learning and neural networks are complex statistical models that are very hard to converge and optimize, due to the rather high number of variables involved in the process of training. Hence, the best way to get these complex models to learn in the most optimal way is to go through a process of trial and error or in other words, to continuously tune the values of the involved variables while examining the outcomes. Nevertheless, it is important to take into consideration the best practices utilized by the deep learning community. Thus, in order to finalize the training hyperparameters we went through long fine-tuning process which was concluded with the following values:

1. learning rate: 0.0001.
2. optimization technique: ADAM [20].
3. Batch size: 26.

While it may be noticed that the batch size is relatively small, we could not increase it any further due to the computational limitations. For instance, when trying higher batch size such as 27, an out of memory error is produced. The training has been done for 20 epochs with an early stop mechanism that stops the training process if the accuracy of the validation set did not improve for 5 consecutive epochs, to prevent the occurrence of the overfitting problem.

3 Results and Discussion

In order to evaluate the performance of the developed architecture, several evaluation matrices factors had to be calculated. These evaluation matrices include accuracy, precision, recall, and F1 measure which could be calculated from Eqs. 3, 4, 5 and 6 respectively.

$$\boldsymbol{accuracy} = \frac{(number\ of\ correctly\ predicted\ samples)}{(total\ number\ of\ samples)} \tag{3}$$

$$\boldsymbol{precision} = \frac{number\ of\ correctly\ predicted\ samples\ in\ i}{total\ number\ of\ predicted\ samples\ in\ i} x100 \quad (4)$$

$$\boldsymbol{recall} = \frac{number\ of\ correctly\ predicted\ samples\ in\ i}{total\ number\ of\ samples\ in\ i} x100 \quad (5)$$

$$\boldsymbol{F1\ score} = \frac{2.precision.recall}{precision + recall} x100 \quad (6)$$

The developed architecture has been evaluated using the different evaluation matrices as shown in Table 1.

Table 1. The Evaluation Metrics of the Developed Architecture.

Evaluation Matric	Training	Validation	Testing
Accuracy	95.78	92.44	92.83
Sensitivity	91. 14	90. 84	90. 32
Specificity	89.98	89.56	89.12
Recall	92.55	91.42	90.74

When comparing these results with the findings found in similar studies in the literature, an improvement could be noticed as it will be shown later. Furthermore, to further investigate the success of the network a confusion matrix could be produced. Figure 5 and 6 shows the confusion matrix to the training and validation sets respectively where the confusion matrix of the validation resulted from the addition of both test and validation sets. This was done as both gave very similar results as demonstrated in Table 1.

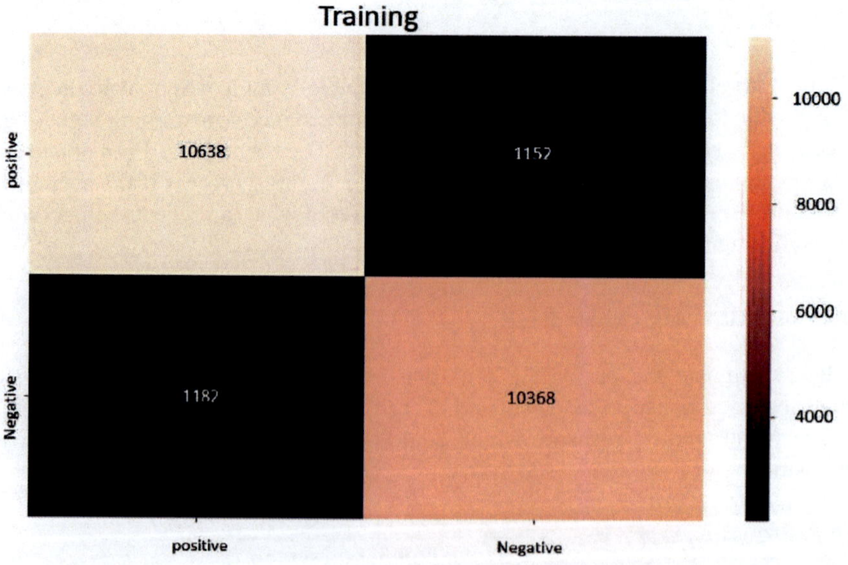

Fig. 5. The Confusion Matrix for the Training Set.

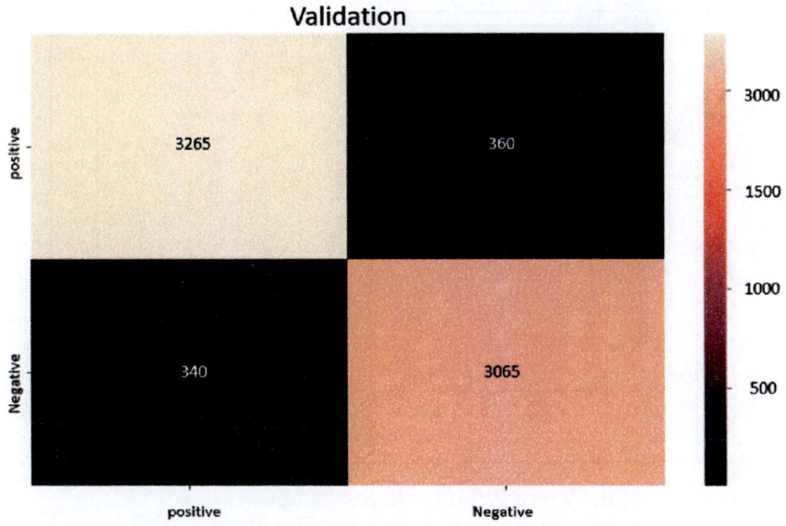

Fig. 6. The Confusion Matrix for the Validation Set.

To further investigate the results, the region of convergence (RoC) has been computed and the Area under the curve (AuC) has been calculated for the testing and validation sets combined by averaging the values of their evaluation matrices. Hereinafter, this combined value will be referred to as validation and will be considered as the main competence indicator to be compared with other studies in the literature as it will be shown later. This is done mainly to come up with a single matrix that could be compared with the other studies in the literature because different studies divide their data differently or even define the validation and testing sets differently. What supported this further is that in our case both testing and validation sets have achieved a very similar results as have been demonstrated in Table 1 and the difference between them could be neglected. Figure 7 shows the RoC for validation set. The AuC is 0.9617 and 0.9432 for the training and validation sets respectively.

Although comparisons between our study and the similar studies found in the literature will not be entirely just and meaningful due to the different experiments set up used in their experiments, a brief comparison between the different evaluations matrices was carried out and it is presented in Table 2.

As could be seen in Table 2, the developed network showed supremacy when compared with the studies found in the literature. This is a very important finding as it confirms a very important philosophical principle which is adoption of the clinical practices in the development of technologies. This principle has many implications that can go far beyond our very specific application. The success of this concept also proves the notion that each well trained neural network is nothing but a synthetic expert, and just like any human expert, the neural network can have its conclusion doubted or confirmed by other expert's opinion.

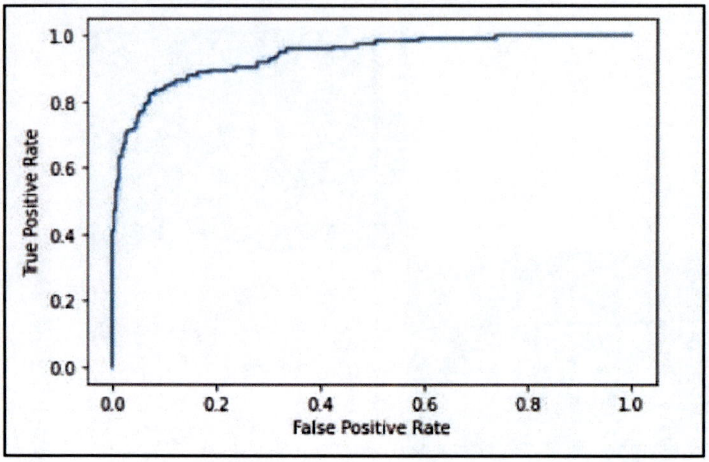

Fig. 7. The Validation Region of Convergence.

Table 2. A Comparison between our method and those found in the literature.

Study	Evaluation matric	
[21]	**Accuracy:**	–
	Sensitivity:	–
	Specificity:	–
	AuC:	0.83
[22]	**Accuracy:**	–
	Sensitivity:	–
	Specificity:	–
	AuC:	0.87
[23]	**Accuracy:**	90.9%
	Sensitivity:	68.9%
	Specificity:	**93.6%**
	AuC:	–
[24]	**Accuracy:**	–
	Sensitivity:	–
	Specificity:	–
	AuC:	0.96
[25]	**Accuracy:**	–
	Sensitivity:	–
	Specificity:	–
	AuC:	0.80
Ours	**Accuracy:**	**92.63%**
	Sensitivity:	**90. 58%**
	Specificity:	89.34%
	AuC:	**0.9432**

To further visualize the inner works of the developed neural networks a saliency maps has been printed to have a look on the regions of interests that network look at to come with a conclusion. The highlighted regions of interests in the saliency maps were then compared with those of radiologists and it turned out that there is a huge overlap between the two regions, which proves that networks conclusions are reliable and in fact well justified. This practice is a very important practice as it helps us make the developed model as an explainable-AI [26, 27, 28]. Figure 8 shows one example of a produced saliency map.

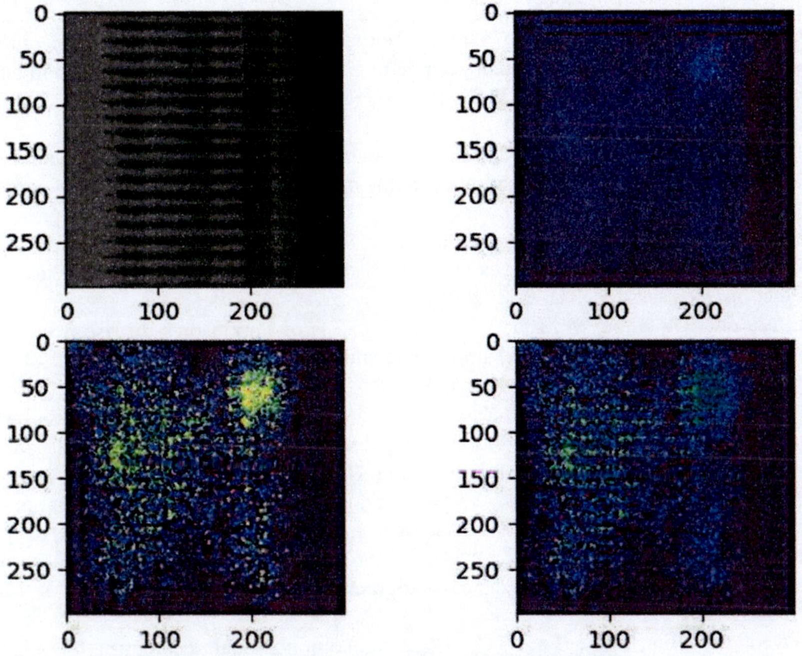

Fig. 8. An Example of Produced Saliency Maps.

4 Conclusion

In this study, we proposed a novel neural network architecture to detect coronary atherosclerosis in coronary CT angiography. This was motivated by the increasing the number of deaths that cardiovascular diseases cause every year. The developed neural network architecture could achieve a promising accuracy of 92.63% and AuC of 0.9432. We believe that our proposed solution will somehow contribute to the growing field of automatic diagnostics which helps in reducing these numbers as our network will potentially support medical practitioners to have more efficient early diagnosis. We also believe that the proposed network has the potential to be adopted in many other medical applications as it proves the theoretical concept that it is possible to adopt the medical practices in guiding the technological advancement of the artificial intelligence algorithms.

Acknowledgment. This research is supported and funded by the European Union's Horizon 2020 research and innovation program under grant agreement No 952603 (SGABU project). This article reflects only the authors' view. The Commission is not responsible for any use that may be made of the information it contains.

References

1. Cardiovascular diseases. https://www.who.int/health-topics/cardiovascular-diseases#tab=tab_1. Accessed 18 Apr 2022
2. Wang, D., Serracino-Inglott, F., Feng, J.: Numerical simulations of patient-specific models with multiple plaques in human peripheral artery: a fluid-structure interaction analysis. Biomech. Model. Mechanobiol. **20**(1), 255–265 (2020). https://doi.org/10.1007/s10237-020-01381-w
3. Sarangi, S., Srikant, B., Rao, D.V., Joshi, L., Usha, G.: Correlation between peripheral arterial disease and coronary artery disease using ankle brachial index-a study in Indian population. Indian Heart J. **64**(1), 2 (2012)
4. Ali, L., Bukhari, S.A.C.: An approach based on mutually informed neural networks to optimize the generalization capabilities of decision support systems developed for heart failure prediction. IRBM **42**(5), 345–352 (2021)
5. Ali, L., Zhu, C., Zhou, M., Liu, Y.: Early diagnosis of Parkinson's disease from multiple voice recordings by simultaneous sample and feature selection. Expert Syst. Appl. **137**, 22–28 (2019)
6. Ali, L., Zhu, C., Zhang, Z., Liu, Y.: Automated detection of Parkinson's disease based on multiple types of sustained phonations using linear discriminant analysis and genetically optimized neural network. IEEE J. Transl. Eng. Heal. Med. **7**, 1-10 (2019)
7. Jas, M., Engemann, D.A., Bekhti, Y., Raimondo, F., Gramfort, A.: Autoreject: automated artifact rejection for MEG and EEG data. Neuroimage **159**, 417–429 (2017)
8. Elsayed, M., Sim, K.S., Tan, S.C.: A novel approach to objectively quantify the subjective perception of pain through electroencephalogram signal analysis. IEEE Access **8**, 199920–199930 (2020)
9. Elsayed, M., Sim, K.S., Tan, S.C.: Effective computational techniques for generating electroencephalogram data. In: 3rd International Conference on Advanced Science and Engineering, ICOASE 2020, pp. 7–11 (2020)
10. Demirer, M., Gupta, V., Bigelow, M., Erdal, B., Prevedello, L., White, R.: Image dataset for a CNN algorithm development to detect coronary atherosclerosis in coronary CT angiography, vol. 1 (2019)
11. OpenCVE
12. Meyer, A.N.D., Singh, H., Graber, M.L.: Evaluation of outcomes from a national patient-initiated second-opinion program. Am. J. Med. **128**(10), 1138.e25-1138.e33 (2015)
13. Epstein Jonathan, W.P.: Clinical and cost impact of second-opinion pathology: review. Am. J. Surg. Pathol. **20**(7), 851–857 (1996)
14. Beishon, M.: A second opinion, because there's no second, GrandRound (2007)
15. He, K., Zhang, X., Ren, S., Sun, J.: Deep residual learning for image recognition. In: Proceedings of the IEEE Computer Society Conference on Computer Vision Pattern Recognition, vol. 2016, pp. 770–778, December 2015
16. Simonyan, K., Zisserman, A.: Very deep convolutional networks for large-scale image recognition. In: 3rd International Conference on Learning Representations, ICLR 2015, ICLR 2015 - Conference Track Proceedings, September 2014

17. Enyinna Nwankpa, C., Ijomah, W., Gachagan, A., Marshall, S.: Activation functions: comparison of trends in practice and research for deep learning (2018)
18. Zheng, Y., Yang, C., Merkulov, A.: Breast cancer screening using convolutional neural network and follow-up digital mammography, p. 4, May 2018
19. Kelleher, J.D.: Deep Learning. MIT Press Essential Knowledge Series, Cambridge (2019)
20. Kingma, D.P., Ba, J.L.: Adam: a method for stochastic optimization. In: 3rd International. Conference Learning Representations, ICLR 2015 - Conference Track Proceedings, December 2014
21. Gupta, V., et al.: Performance of a deep neural network algorithm based on a small medical image dataset: incremental impact of 3D-to-2D reformation combined with novel data augmentation, photometric conversion, or transfer learning. J. Digit. Imaging **33**(2), 431–438 (2019). https://doi.org/10.1007/s10278-019-00267-3
22. Han, D., Liu, J., Sun, Z., Cui, Y., He, Y., Yang, Z.: Deep learning analysis in coronary computed tomographic angiography imaging for the assessment of patients with coronary artery stenosis. Comput. Methods Programs Biomed. **196**, 105651 (2020)
23. Candemir, S., et al.: Automated coronary artery atherosclerosis detection and weakly supervised localization on coronary CT angiography with a deep 3-dimensional convolutional neural network. Comput. Med. Imaging Graph. **83**, 101721 (2020)
24. White, R.D., et al.: Artificial intelligence to assist in exclusion of coronary atherosclerosis during CCTA evaluation of chest pain in the emergency department: preparing an application for real-world use. J. Digit. Imaging **34**(3), 554–571 (2021). https://doi.org/10.1007/s10278-021-00441-6
25. Wu, J.T., et al.: Comparison of chest radiograph interpretations by artificial intelligence algorithm vs radiology residents. JAMA Netw. Open **3**(10), e2022779–e2022779 (2020)
26. Explainable AI: the basics - Policy briefing, November 2019 DES6051. ISBN: 978-1-78252-433-5. The Royal Society. https://royalsociety.org/-/media/policy/projects/explainable-ai/AI-and-interpretability-policy-briefing.pdf
27. Cinà, G., Röber, T., Goedhart, R., Birbil, I.: Why we do need explainable AI for healthcare, June 2022
28. Pawar, U., O'Shea, D., Rea, S., O'Reilly, R.: Explainable AI in healthcare. In: 2020 International Conference Cyber Situational Awareness, Data Analyst Assessment, Cyber SA 2020, June 2020

Ontology-Based Exploratory Text Analysis as a Tool for Identification of Research Trends in Polish Universities of Economics

Edyta Bielińska-Dusza(✉), Monika Hamerska, Magdalena Kotowicz, and Paweł Lula

Krakow University of Economics, Kraków, Poland
{bielinse,monika.hamerska,kotowicm,pawel.lula}@uek.krakow.pl

Abstract. In this article, we present a systematic review of the literature on the current situation in the publication of employees of Polish economic universities in the period 2017–2021. During the research, we used the method of exploratory text analysis supported by domain knowledge defined by ontology (ontology-based exploratory text analysis/ontology-based text). The JEL classification was used as an ontology. We analyzed 4,862 publications from 2017–2021 in the field of economic sciences. Using bipartite graphs (in R package: bipartite), the authors analyzed the connections between topics and individual universities (Krakow University of Economics, University of Economics in Katowice, Poznan University of Economics and Business, Wroclaw University of Economics and Business, Warsaw School of Economics). This review summarizes the current research directions of employees of Polish economic universities in 2017–2021. We found that the primary research efforts centered on Microeconomics, Economic Development, Innovation, Technological Change, and Growth. In the analyzed period, we notice an increase in research in areas: Economic Development, Innovation, Technological Change, and Growth; Agricultural and Natural Resource Economics, Environmental and Ecological Economics; Political Processes: Rent-Seeking, Lobbying, Elections, Legislatures, and Voting Behavior; Economic Systems; Industrial Organization. Moreover, it is possible to point to a relatively uniform research area and the lack of explicit scientific specialization among the analyzed universities.

Keywords: analysis of research publications · research trends · ontology-based exploratory text analysis · ontology-based text mining · bipartite · economic sciences · economic universities · JEL · evaluation

1 Introduction

Bibliometrics [1] and text-mining techniques are actively used in various scientific disciplines as an analysis tools [2, 3, 4, 5]. One such direction is an attempt to identify universities and scientists who have had the most significant impact on the development of a given field, as well as to indicate the factors influencing their impact [1, 6, 7, 8]. Another widely analyzed direction of research is research productivity [9, 10, 11, 12, 13, 14, 15], which uses the central and local database and takes into account a number of different measures (citation, publication, place of publication, and others).

N. Filipovic (Ed.): AAI 2022, LNNS 659, pp. 198–222, 2023.
https://doi.org/10.1007/978-3-031-29717-5_14

Similarly, we can also identify several review papers in the literature dedicated to issues of co-authorship and its development [9, 16, 17, 18, 19, 20, 21, 22, 23, 24, 25, 26]. The dramatic rise in multi-authored papers in economics [18] and scientific collaboration have been widely acknowledged [27]. G. Bukowska et al. point out that the general trend observed in top scientific journals is an increase in the collaborative activities between researchers and institutions, especially with regard to international co-authorship [28]. Research cooperation of Polish scientists is strongly correlated with international cooperation, although its productivity at the national level is lower than at the international level [29]. It seems that scientific cooperation and the implementation of joint research impact the research topics in individual research and teaching centers.

On the other hand, the identification of leading research trends in various fields of science is another important direction for research. Therefore, considering the purpose of our research, studies in the field of economics and techniques based on text-mining are crucial. Similar to the discussed cooperation issues, this direction is often more analyzed in foreign literature [1, 7, 30, 31, 32] than in Polish [28, 33, 34, 35].

Moreover, in the available Polish literature, one can find publications on new trends in economic sciences based on the method of synthesis and summary, supported by a classic review of the literature on the subject, without the use of quantitative methods. These are the works of, among others: [36, 37, 38, 39, 40, 41, 42].

As B. Nogalski rightly points out, each scientific discipline's problems and research areas are critical and monitor the degree and pace of its substantive and methodological development. On the one hand, it allows you to follow the trends of change in science and the empirical world. It also shows the degree of compatibility of scientific work in the field of management with the broadly understood problems occurring in enterprises and the economy [42].

It seems that the number of citations is one of the indicators that can be taken to define the popularity of a given research area. The ranking created by M. Khan et al. [1] identifies the most popular research topics between 1973 and 2020. These are: Business Model, Consumer Engagement, Business Networks, Organizational Innovation, and Relationships.

In addition, the authors show the destination of knowledge outflows and the sources of knowledge inflows in *JBR* Sis used the most cited journals. The list demonstrates that *JBR* (itself) is the most cited source, followed by the *Journal of Marketing*, the *Journal of Consumer Research*, the *Journal of Marketing Research*, and the *Strategic Management Journal* [1].

The results of other research show that development, agricultural and health economics are the most significant topics in Latin America [30].

On the other hand, in the research conducted by A. Rialp et al. [7] the authors included a longitudinal keyword analysis as a proxy to identify the most common research topics or themes as historically revealed in *International Business Review JBR).* The results show the most common themes: Internationalization, Export Performance, International Entrepreneurship, Multinational Enterprises/Firms, Culture, Trust.

Another study in which research trends were determined by indicating the frequency of keywords used was conducted by J. Swacha [43], where the last 31 years in the Polish journal *Przegląd Organizacji* was analyzed. The aforementioned research shows that the most common keywords are: enterprise innovation, enterprise competitiveness, enterprise cooperation, human resource management, knowledge management, and strategic management. On the other hand, the most frequently cited articles represented various sub-disciplines of management sciences: financial management and managerial accounting, strategic management, organizational behavior, business organization management, entrepreneurship, knowledge management, human resource management, and the theory of organization and management. The author also made a compilation of the contributions of economic universities. The two leading positions were taken by the Universities of economics in Wroclaw and Krakow, while the fourth and sixth positions were taken by the Poznan University of Economics and Business and Katowice, respectively [43].

Moreover, according to B. Nogalski, the research areas of Polish management sciences do not differ significantly from the interests of foreign research and development centers [42]. In the research, the author assumed that the promotional activity at the level of the implementation of habilitation proceedings determines the problem structure of the science discipline from the perspective of sources of inspiration and research areas. It identifies two practical and theoretical trends and indicates that strategic management, knowledge and information management, human resources management, marketing management, and organizational behavior are the leading research directions.

It should also be emphasized that due to the legislative changes in Polish higher education, modifying the principles of periodically conducting a comprehensive assessment of the quality of scientific or research and development activities of scientific units, the authors decided to adopt the time frame for the study resulting from the period of the last evaluation (2017–2021).

The literature analysis on the subject shows that there is limited research in this area and, as emphasized by J. Kubiczek et al. [8], a lack of comparisons of scientific achievements is a research gap. In this research [8], the authors consider the number of publications, citations, and the h-index. The research shows that employees of Wroclaw University of Economics and Business have, on average, the most publications, Warsaw School of Economics (WSE) employees were quoted most often, and the highest h-index belongs to one of the employees of Wroclaw University of Economics and Business, while, on average, employees of WSE and Poznan University of Economics and Business have the highest rates [8]. The research of J. Drogosz [44] indicated the most popular journals, publishers, and scientific conferences in the economic disciplines and defined their significance in the next appraisal in 2021.

However, no research would indicate current trends in economic sciences and the evaluation of the research performed in the Polish economic universities in terms of analysis of connections between topics and individual universities.

The research gap identified based on the analysis of the literature on the subject indicates the lack of research not only in the analysis of the links between research areas and individual economic universities in Poland but also in works identifying leading trends in research in the field of economics. Especially if we take into account the analyzed period and the use of exploratory text analysis methods supported by domain knowledge defined by ontology (ontology-based exploratory text analysis/ontology-based text mining).

Considering all of the above, this paper's main objective is to analyze the subject of the publications achievements of employees of Polish economic universities in the period 2017–2021.

The first discusses the essence of the challenges of contemporary publication analysis, the nature of scientific publications, and the evaluation of the scientific activity of Polish universities. The second part focuses on the presentation of the research methodology along with the presentation of the obtained research results. Several detailed methods and techniques were used during the research. Using the methods of exploratory text analysis supported by domain knowledge defined by ontology (ontology-based exploratory text analysis/ontology-based text mining), the analysis of publication descriptions (abstracts + keywords) registered in the Scopus database was performed. The JEL classification was used as an ontology.

On the other hand, the analysis of connections between topics and individual universities (Krakow University of Economics, University of Economics in Katowice, Poznan University of Economics and Business, Wroclaw University of Economics and Business, Warsaw School of Economics) was carried out using bipartite graphs (in R package: bipartite). In addition, to a classic literature review, a narrative review was also used, an additional tool and a starting point for further analysis and critique of the literature, as well as structural and causal analysis. All calculations were made in program R; and the last part contains conclusions and indications for future research directions.

2 The Theoretical Background

2.1 Challenges of Contemporary Publication Analysis

The analysis of research trends based on text mining is carried out in many fields of science [2, 7, 45, 46] and takes into account various research directions.

The research B. Frączek [47] confirms the topicality of research issues in various science areas. The author focuses on analyzing articles published in the Scopus and Web of Science databases, which include bibliometric methods. The number of works in Social Sciences published in Scopus by 2014 amounted to 1,442 records (ranking second). It should be emphasized that social sciences are a group of empirical sciences dealing with society and an individual's activities, and economic sciences are included in it, in addition to several others.

Hence, the number of Social Sciences records does not equal the number of economic science records. In turn, works in Business, Management, and Accounting only occupy the tenth place with 242 records [47].

On the other hand, in the Web of Science database, works on Business Economics are ranked third (192 records). The most popular sciences are Medicine (4609 records) and Information Science (853 records). The high Medicine score should not be surprising as systematic reviews play a significant role in evidence-based Medicine [48].

Articles affiliated with Polish universities were not among the top 10 countries with the highest number of records [47]. This indicates a relatively low use percentage of bibliometric analysis among Polish scientists, especially in economic sciences.

Furthermore, research [46] shows that in the Web of Science database from 1980–2019, among the top ten most popular scientific disciplines which use text-mining as an analysis tool, there are no economic sciences. However, since the papers related to text mining are being widely applied to various academic studies, and the quantitative trend is increasing, this could indicate the potential increase of their use in this discipline.

Depending on a given scientific field, journals devoted to a given research topic are dominated by various types of publications [49] as well as various research issues. It can be presented, using the JEL classification, which is the standard method for classifying articles, dissertations, books, book reviews, and economical working papers [50]. The JEL classification and others are becoming more and more common among researchers and journal editors.

In turn, the issues of bibliometry occupy an essential place among the undertaken research topics [47]. The analysis of the thematic areas of publications in which bibliometric methods were used shows the growing popularity [1, 2], dependent on, inter alia, the scientific discipline, country, or affiliation of the researcher.

According to D. Antons [3], the text mining techniques for reviewing academic literature and informing domain-specific case studies started in 2007.

In addition, the increasing number of scientific and research papers [51, 52, 53], sample sizes threatening to exceed the cognitive limits of human processing capacities [54], as well as the growing complexity of research problems contributes to the rapid increase in the importance of using text mining techniques [46] to semi-automate the steps in the analysis process [55]. They are also the most scientific and sophisticated literature review methods [45].

Moreover, as indicated by [54], the progress of knowledge in this field is evolving and is particularly promising. It is worth emphasizing that these issues can be considered in two ways, as a topic or a research method.

Researchers note [54, 56, 57, 58] that the use of this type of tools allows for the reduction of costs and manual labor involved, increase in the practicality of the entire process, shorten the time required to produce products, screening, and data extraction, as well as provide new insights. Although, as emphasized by [54], the results obtained through machine learning algorithms require human interpretation and insightful syntheses, novel explanations, and theory building because the best algorithm remains unclear [56].

Importantly, bibliometric reviews are based on the analysis of large data sets in the form of bibliometric analysis and, as noted by Tranfield et al. [59], are an important part of any research project and are the most commonly used method for synthesizing and summarizing a field in the scientific body of knowledge [45].

In addition, S. Nerur et al. [60] suggest that bibliometrics has an advantage over qualitative studies as it objectively presents statistical results from a selected scientific database with lesser room for subjective bias [1]. Additionally, the results of any bibliometric study may also provide clear research directions to be developed in the future based on the current mainstreams of a given research field or journal [7].

2.2 Characteristics of Research Publication

Conducting scientific activity, including scientific research, development work, and artistic creation, is one of the basic tasks that Polish legislation places on national universities and treats as their mission (Articles 2, 4, and 11 Sect. 1,3) [61]. Conducting scientific activity is one of the basic duties of university research and research - teaching staff and scientific publications are one of the types of scientific achievements resulting from this activity. The basic scope of responsibilities of research workers can therefore be compared to the key tasks of universities to direct their activity [62].

Looking through the prism of Polish legislation, scientific activity, including publishing, is not only of key importance for the scientific and professional development of a university employee but above all, its effects are significant from the point of view of the evaluation of scientific activity, which is obligatory for academic universities.

Evaluation is a comprehensive system for assessing the quality of scientific activity (Article 265 et seq.) [61, 63], covering the achievements of all employees conducting scientific activity at a university and carried out within scientific disciplines [64] and covers achievements arising in connection with employment or education at a university. The Evaluation and Science Committee carry out the evaluation, and its effect is the award of the university by the Ministry of Education and Science to the appropriate scientific category A +, A, B +, B, or C, in individual disciplines subject to evaluation. Category A + is the highest category in this list, and category C - the lowest. Having specific categories is important, among others, for the possibility of defining the status of a university as an academic university, the possibility of a university to create courses of study on its own, or to award academic degrees in individual disciplines.

The evaluation is carried out every 4 years and covers the period of 4 years preceding the year of its conduct. The first evaluation of scientific activity began on January 1, 2022, and covers the years 2017–2021 (Article 324 Sect. 1) [65].

Since the applicable legal regulations have thoroughly reformed the model of institutional evaluation of science [66], therefore the research conducted for the purposes of this article includes scientific publications created during the evaluated period.

The concept of scientific publication has been defined to assess the level of scientific activity conducted in a given discipline. Scientific publications are scientific achievements that are substantively related to scientific research conducted in the evaluated entity within a specific discipline. These achievements are: scientific articles published in scientific journals, reviewed materials from international scientific conferences and monographs, scientific editorials of monographs and chapters in such monographs, both included in the list of journals and materials and in the list of publications - scored based on the measure of their reputation, as well as scientific articles and monographs not included in these lists (§8, §11 Sect. 2) [67, 68] (Article 324) [65].

The concepts of a scientific article and a scientific monograph are defined as follows: a scientific article is a peer-reviewed article published in a scientific journal or in peer-reviewed materials from an international scientific conference: 1) presenting a specific scientific issue in an original and creative, problem or cross-sectional manner; 2) provided with footnotes, bibliography or other scientific apparatus appropriate for a given scientific discipline. An academic article is also a review article published in a scientific journal included in the journal list. It was indicated that the scientific article is not an editorial, abstract, extended abstract, letter, errata, or an editorial note; a scientific monograph is a peer-reviewed book publication: 1) presenting a specific scientific issue originally and creatively; 2) provided with footnotes, bibliography or other scientific apparatus appropriate for a given scientific discipline. A scientific monograph is also: 1) reviewed and annotated, bibliography or other scientific apparatus appropriate for a given scientific discipline translation: a) a work essential for science or culture into Polish, b) a work essential for science or culture, published in Polish 2) scientific edition of source texts into another modern language (§ 8–10) [68]. It should be mentioned that scientific monographs and articles are included in the evaluation if information about them is entered into a database accessible via the researcher's electronic identifier.

Scientific publications can be divided according to their type or structure. Polish Scientific Bibliography distinguishes the following types of scientific publications: original-article, review article, case-study/case-report, gloss or legal commentary (short-communication/commentary-on-the-law), scientific-review, guidelines, popular-science-article, others-non citable, others-citable, scientific monograph (scholarly-monograph). Depending on the scientific field, various types of publications will dominate in journals devoted to a given research topic [49].

Scientific publications also differ in terms of topics related to specific fields of science and scientific disciplines. In Polish legislation, the fields of science have been classified into eight categories: humanities, engineering and technical sciences, medical and health sciences, agricultural, social, exact and natural sciences, theology, and arts. However, the fields of science have been divided into specific scientific/artistic disciplines. Broadly understood economic sciences are included in the field of social sciences, including such disciplines as economics and finance, socio-economic geography and spatial management, political and administrative sciences, management and quality sciences, legal sciences, sociological sciences, pedagogy, canon law, psychology [64].

3 Methodology

3.1 Research Methodology

Research Scope and Goals

The article's main purpose is to analyze the subject of the publication achievements of employees of Polish economic universities in the period 2017–2021. The analyzed period of four years is the period of the last parametric evaluation of Polish research units carried out by the Ministry of Education and Science. The result of the evaluation of scientific units depends on the scientific achievements of scientists of a given unit. On the basis of parametric evaluation, universities receive the appropriate scientific category.

The data used in the analysis come from the Scopus database and concern five Polish Universities of Economics:

1. Krakow University of Economics
2. University of Economics in Katowice
3. Poznan University of Economics and Business
4. Wroclaw University of Economics and Business
5. Warsaw School of Economics

The study included 4862 publications in which at least one of the authors indicates an affiliation with one of the economic universities listed above.

3.2 Ontology-Based Analysis of Publication Abstracts

The analysis of publications' contents was performed using an ontology-based approach with the use of the JEL classification system originated with the Journal of Economic Literature. It is a standard method of classifying scholarly literature in the field of economics. JEL classification, presented in Table 1, has 20 primary JEL categories, and each JEL primary category has secondary and tertiary subcategories [33].

Table 1. JEL classification

JEL Symbol	Primary category
A	General Economics and Teaching
B	History of Economic Thought, Methodology, and Heterodox Approaches
C	Mathematical and Quantitative Methods
D	Microeconomics
E	Macroeconomics and Monetary Economics
F	International Economics
G	Financial Economics
H	Public Economics
I	Health, Education, and Welfare
J	Labor and Demographic Economics
K	Law and Economics
L	Industrial Organization
M	Business Administration and Business Economics, Marketing, Accounting, Personnel Economics
N	Economic History
O	Economic Development, Innovation, Technological Change, and Growth
P	Economic Systems
Q	Agricultural and Natural Resource Economics, Environmental and Ecological Economics
R	Urban, Rural, Regional, Real Estate, and Transportation Economics
Y	Miscellaneous Categories
Z	Other Special Topics

Source: own elaboration based on [69]

For every category, a set of patterns was defined. Every pattern allowed us to find phrases related to a given category in the text. For every sequence of tokens recognized in the text, a similarity measure based on the Jaccard coefficient to a pattern was calculated, and – if this measure was above a threshold value – a considered JEL category was assigned to a text. The process presented here was conducted twice – for the original text and for a text after the lemmatization process (conducted with the dictionary-based approach). The details of the identification process were presented in [70].

The program realizing the process of JEL concept identification, was prepared in R language.

The process of recognition of JEL categories for every paper and embraced paper's title, abstract and keywords delivered by authors. The results formed the matrix **Q**:

$$\mathbf{Q} = \begin{array}{c} \\ p_1 \\ \dots \\ p_N \end{array} \begin{array}{c} c_1 \ \dots \ c_M \\ \begin{bmatrix} q_{11} & \cdots & q_{1M} \\ \dots & \dots & \dots \\ q_{N1} & \cdots & q_{NM} \end{bmatrix} \end{array} \quad (1)$$

where p_i represents papers, c_j categories from the JEL classification, and q_{ij} indicates if c_j category appeared in the p_i paper ($q_{ij} = 1$) or not ($q_{ij} = 0$).

3.3 Graph-Based Model of Publication Activity

Undirected Graph Model for the Description of JEL Categories Co-Occurrences

Let us assume that the graph G shows JEL categories co-occurrences in the same document. Nodes represent JEL categories, and edges between two nodes appear if they occurred in the same document. Weights assigned to edges inform how many times given two nodes appeared together and were calculated as:

$$\mathbf{W} = \mathbf{Q}^T\mathbf{Q} \quad (2)$$

The structure of the graph can be characterized by:

- number of nodes and number of edges,
- density – calculated as a number of edges divided by the maximum number of edges that this graph can have,
- number of compartments (connected subgraphs),
- diameter expressed as the length of the longest geodesic (the shortest path between two nodes),
- degree (number of edges incident to a given node) distribution.

The importance of nodes in the G graph can be expressed by:

- the number of occurrences of concepts represented by a given node,
- node strength calculated as the sum of weights assigned to edges incident to a given node,
- weighted betweenness centrality for the k-th node is defined as:

$$c_k^{WB} = \sum_{i \neq k} \sum_{\substack{j \neq k \\ j \neq i}} \frac{f_{ikj}}{f_{ij}} \quad (3)$$

where $i, j \in \{1, \ldots, N\}$, N – the number of all nodes, f_{ij} – the number of all shortest paths between i-th and j-th node, f_{ikj} – the number of all shortest paths between i-th and j-th node which pass through the k-th node. The length of an edge existing between the p-th and r-th node was calculated as:

$$l_{pr} = \max_{i,j}(w_{ij}) + 1 - w_{pr} \tag{4}$$

Weighted betweenness centrality measure can also be used for evaluation of the importance of edges. Then the significance of the e-th edge can be defined as:

$$c_e^{WB} = \sum_i \sum_{j \neq i} \frac{f_{iej}}{f_{ij}} \tag{5}$$

where f_{iej} is the number of all shortest paths between i-th and j-th node which pass through the e-th node.

Bipartite Graph Model for Description of Relationships Between JEL Categories and Authors' Affiliations

Let us assume that matrix **V** defined as:

$$\mathbf{V} = \begin{matrix} & \begin{matrix} u_1 & \ldots & u_T \end{matrix} \\ \begin{matrix} p_1 \\ \ldots \\ p_N \end{matrix} & \begin{bmatrix} v_{11} & \ldots & v_{1T} \\ \ldots & \ldots & \ldots \\ v_{N1} & \ldots & v_{NT} \end{bmatrix} \end{matrix} \tag{6}$$

where p_i represents papers, u_j Polish universities of economics, and v_{ij} indicates if at least one researcher from the u_j university belongs to the authors' team prepared the p_i paper ($v_{ij} = 1$) or not ($v_{ij} = 0$).

Relationships between *JEL* categories and universities can be represented by the bipartite graph B connecting nodes representing *JEL* categories ($c_1, \ldots, c_M$) and nodes representing universities ($u_1, \ldots, u_T$). Interactions between categories and universities by weights forming matrix **B** calculated as:

$$\mathbf{B} = \mathbf{Q}^T \mathbf{V} \tag{7}$$

For every node in a bipartite graph many different statistics can be calculated. It seems that for productivity analysis, the most important are the following:

- node degree – the number of nodes from the opposite set (partner nodes) which are connected to a given node,
- node strength – the sum of normalized weights assigned to edges that are incident to a given node,
- specificity index – coefficient of variation calculated for weights assigned to incident edges normalized to the range [0; 1].

Also, many statistics are defined for the whole bipartite graph model. Particularly useful for publication achievements analysis may be:

- number of nodes,
- number of edges,
- number of compartments (connected subgraphs),
- connectance expressed by the density coefficient,
- mean number of shared partners (nodes from the opposite set connected to a given node),
- niche overlap – defined for two nodes from the same set and calculated as a similarity measure between sets of their partners,
- H_2' - specificity index based on entropy measure and normalized to the range [0; 1]. Low values indicate low level of specificity. The H_2' index reaches the maximum value (equal to 1) for a node which is connected only to one partner from the opposite set.

4 The Analysis of the Publication Activity of Researchers from Polish Universities of Economics

4.1 Publication Activity of Researchers from Polish Universities of Economics

In the first stage, the number of publications was analyzed and presented, considering the analyzed economic universities and the individual years covered by the analysis. The summary of publications is presented in Table 2.

Table 2. Publications of Polish economic universities in 2017–2021 based on the Scopus database

Units	Number of publications				
	2017	2018	2019	2020	2021
Krakow University of Economics	120	118	166	225	262
University of Economics in Katowice	99	110	122	127	167
Poznan University of Economics and Business	145	154	184	213	300
Wroclaw University of Economics and Business	148	184	194	244	260
Warsaw School of Economics	188	199	209	257	351

Source: own elaboration based on data from the Scopus database

4.2 Analysis of JEL Category Importance

The process of analysis allowed us to identify 60053 keywords and key phrases related to JEL categories. The number of occurrences of references by main categories of the JEL classification is presented in Fig. 1.

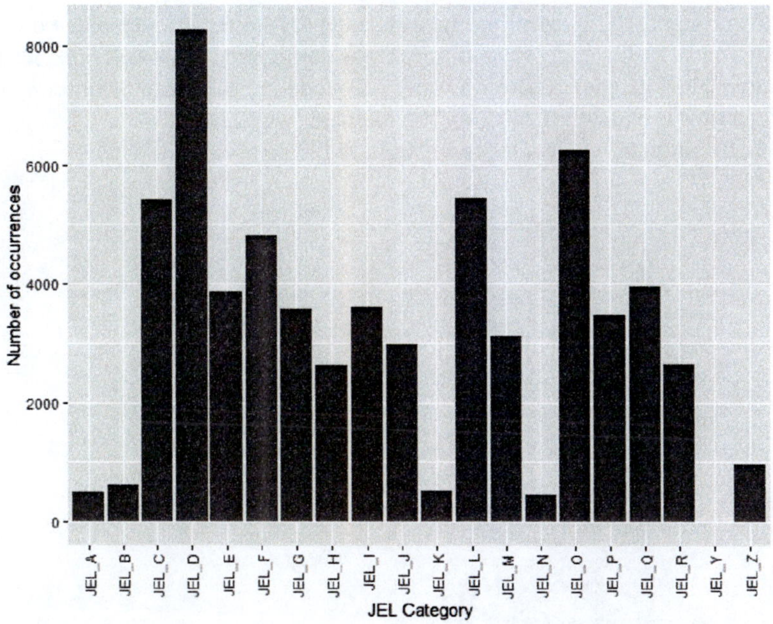

Fig. 1. The number of references to concepts defined in the JEL classification Source: own elaboration.

In Fig. 2, the variability in years for the number of occurrences of every JEL category was presented.

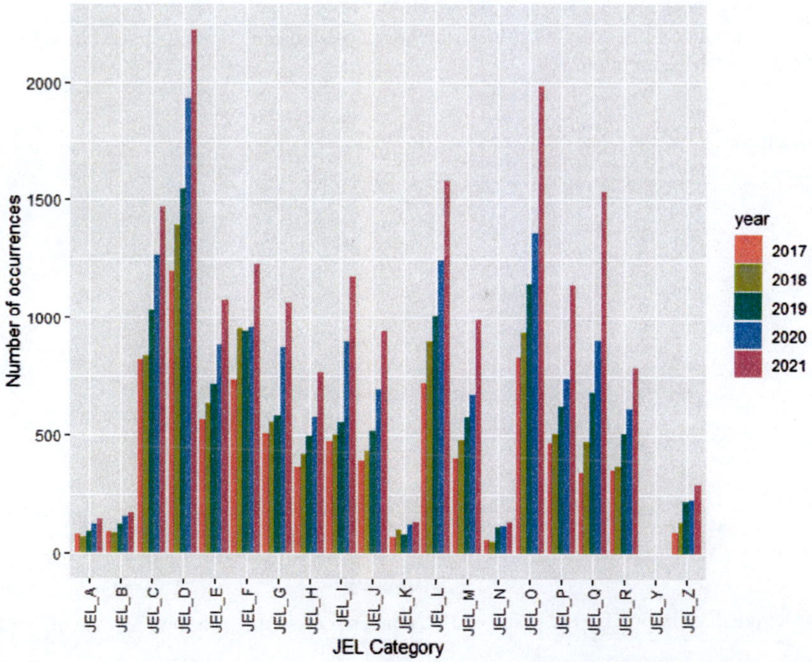

Fig. 2. The variability of the number of references to concepts defined in the JEL classification in years. Source: own elaboration.

Next, the graph G representing the JEL category co-occurrence was built. It had 1008 nodes and 121877 edges. The G graph was composed of 238 components. One of them contained 771 nodes. There was only one node in each of the other components. Graph density was equal to 0.2401, and its diameter was 4.

The distribution of weights is presented in Fig. 3.

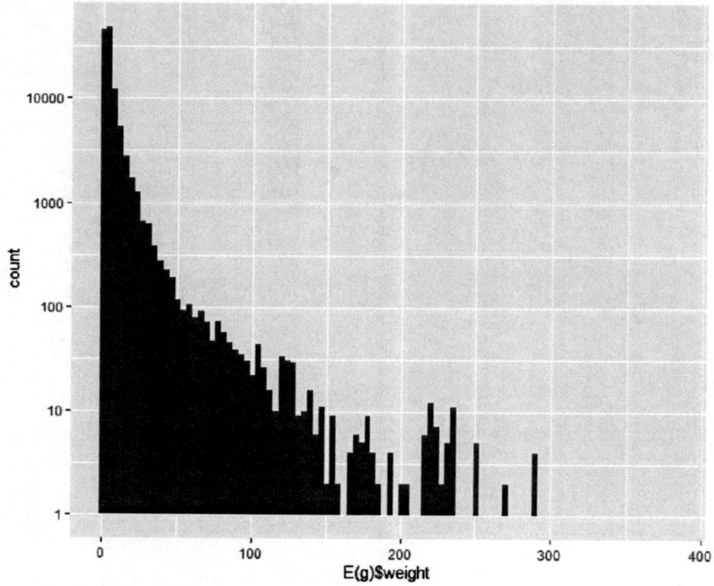

Fig. 3. The distribution of weights in JEL category co-occurrence graph (y axis in logarithmic scale) Source: own elaboration.

The distribution of node degrees is presented in Fig. 4.

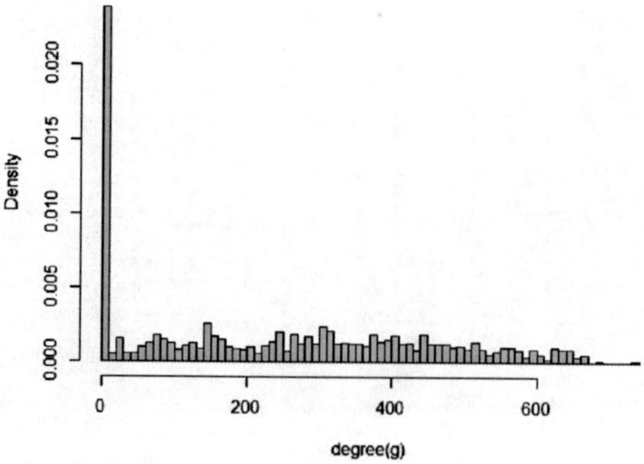

Fig. 4. Degree distribution for the graph of JEL concept co-occurrence Source: own elaboration.

Degree values are described by the following statistics:

- minimum: 0.00,
- 1st quarter: 26.75,
- median: 234.50,
- mean: 241.82,
- 3rd quarter: 405.00,
- maximum: 731.00.

The importance of JEL categories was calculated using the information about the number of occurrences of every category and selected centrality measures.

Assuming that the importance of JEL categories depends on the number of references to them, the results presented in Table 3 were obtained.

Table 3. Category importance measured by the number of references

Category	Category description	Number of references
JEL_D.7.2	Political Processes: Rent-Seeking, Lobbying, Elections, Legislatures, and Voting Behavior	1010
JEL_M	Business Administration and Business Economics, Marketing, Accounting, Personnel Economics	665
JEL_O.3	Innovation, Research and Development, Technological Change, Intellectual Property Rights	621
JEL_O.1.5	Human Resources, Human Development, Income Distribution, Migration	596
JEL_O.1.3	Agriculture, Natural Resources, Energy, Environment, Other Primary Products	561
JEL_I.1.5	Health and Economic Development	484
JEL_C.8	Data Collection and Data Estimation Methodology, Computer Programs	469
JEL_L.8.2	Entertainment, Media	431
JEL_P.4.8	Legal Institutions. Property Rights. Natural Resources. Energy. Environment. Regional Studies	420
JEL_C.3.8	Classification Methods, Cluster Analysis, Principal Components, Factor Models	411

Source: own elaboration

Using node strength as a measure of category importance, the results shown in Table 4 were obtained.

Table 4. Category importance measured by node strength

Category	Category description	Strength
JEL_D.7.2	Political Processes: Rent-Seeking, Lobbying, Elections, Legislatures, and Voting Behavior	15432
JEL_O.1.5	Human Resources, Human Development, Income Distribution, Migration	12114
JEL_I.1.5	Health and Economic Development	11521
JEL_O.1.3	Agriculture, Natural Resources, Energy, Environment, Other Primary Products	11286
JEL_O.3	Innovation; Research and Development; Technological Change; Intellectual Property Rights	10529
JEL_P.4.8	Legal Institutions. Property Rights. Natural Resources. Energy. Environment. Regional Studies	10019
JEL_M	Business Administration and Business Economics, Marketing, Accounting, Personnel Economics	9433
JEL_R.5.8	Regional Development Planning and Policy	8919
JEL_O	Economic Development, Innovation, Technological Change, and Growth	8896
JEL_E.5.2	Monetary Policy	8838

Source: own elaboration

Table 5 shows category importance calculated with the use of weighted betweenness centrality measure.

Table 5. Category importance measured by weighted betweenness centrality

Category	Category description	Weighted betweenness centrality
JEL_D.7.2	Political Processes: Rent-Seeking, Lobbying, Elections, Legislatures, and Voting Behavior	37247.828
JEL_O.1.5	Human Resources, Human Development, Income Distribution, Migration	10005.915
JEL_O.1.3	Agriculture, Natural Resources, Energy, Environment, Other Primary Products	8458.341
JEL_M	Business Administration and Business Economics, Marketing, Accounting, Personnel Economics	6968.871
JEL_O.3	Research and Development; Technological Change; Intellectual Property Rights	5412.699

(*continued*)

Table 5. (*continued*)

Category	Category description	Weighted betweenness centrality
JEL_O.1.6	Financial Markets; Saving and Capital Investment; Corporate Finance and Governance	4712.765
JEL_E.5.2	Monetary Policy	3418.365
JEL_I.1.5	Health and Economic Development	3407.616
JEL_O.1.8	Urban, Rural, Regional, and Transportation Analysis; Housing; Infrastructure	3405.901
JEL_P.4.8	Legal Institutions. Property Rights. Natural Resources. Energy. Environment. Regional Studies	3192.608

Source: own elaboration

Table 6. Relationships between categories importance measured by weighted betweenness centrality

Relationship	Relationship description	Weighted betweenness centrality
JEL_N.5 - JEL_D.7.2	Agriculture, Natural Resources, Environment, and Extractive Industries Political Processes: Rent-Seeking, Lobbying, Elections, Legislatures, and Voting Behavior	713.3167
JEL_D.7.2 - JEL_K.4.1	Political Processes: Rent-Seeking, Lobbying, Elections, Legislatures, and Voting Behavior - Litigation Process	629.0333
JEL_B.4 - JEL_D.7.2	Economic Methodology - Political Processes: Rent-Seeking, Lobbying, Elections, Legislatures, and Voting Behavior	624.3079
JEL_O.3 - JEL_D.2.9	Innovation, Research and Development, Technological Change, Intellectual Property Rights - Production and Organizations: Other	591.2000

(*continued*)

Table 6. (*continued*)

Relationship	Relationship description	Weighted betweenness centrality
JEL_N.9 - JEL_D.7.2	Regional and Urban History - Political Processes: Rent-Seeking, Lobbying, Elections, Legislatures, and Voting Behavior	560.6667
JEL_D.7.2 - JEL_K.1.5	Political Processes: Rent-Seeking, Lobbying, Elections, Legislatures, and Voting Behavior - Civil Law, Common Law	553.3333
JEL_D.7.2 - JEL_K.1.6	Political Processes: Rent-Seeking, Lobbying, Elections, Legislatures, and Voting Behavior - Election Law	493.8333
JEL_D.7.2 - JEL_F.3.7	Political Processes: Rent-Seeking, Lobbying, Elections, Legislatures, and Voting Behavior - International Finance Forecasting and Simulation: Models and Applications	463.9826
JEL_D.7.2 - JEL_K.1	Political Processes: Rent-Seeking, Lobbying, Elections, Legislatures, and Voting Behavior - Basic Areas of Law	463.0000
JEL_C.8.0 - JEL_K.3.6	Data Collection and Data Estimation Methodology, Computer Programs: General - Family and Personal Law	458.8667

Source: own elaboration

Also, the importance of relationships between JEL categories was evaluated. In Table 6, the ten most important links between categories were listed.

4.3 Analysis of Connections Between JEL Categories and Universities

To analyze relationships between JEL categories mentioned in research papers and authors' affiliation the bipartite graph *B* was built (Table 7).

Table 7. Main characteristics of the bipartite model presenting publication activity of researchers from Polish universities of economics

Statistics	Value
Number of universities	5
Number of categories	771
Number of compartments	1
Connectance	0.9012
Mean number of shared partners (universities)	647.6
Mean number of shared partners (categories)	4.07
Niche overlap (universities)	0.9036
Niche overlap (categories)	0.7656
$H'2$ index	0.0435

Source: own elaboration

Table 8. Characteristic of Polish universities of economics

	University of Economics in Katowice	Krakow University of Economics	Poznan University of Economics and Business	Wroclaw University of Economics and Business	Warsaw School of Economics
Degree	653	709	692	681	739
Strength	90.15	166.79	144.97	128.28	240.80
Specificity index	0.0531	0.0471	0.0481	0.0515	0.0449

Source: own elaboration

The *B* graph is connected. The connectance measure representing network density is very high (0.9012). Niche overlap index for universities (0.9036) confirms the similarity of topics mentioned in publications. $H_2^{'}$ index informs about a very low degree of specialization.

Statistics for universities are presented in Table 8.

The strength measure allows to evaluate the level of publication activity for every university. The specificity index informs about the relatively low variation of interactions between nodes representing each university and JEL categories.

Statistics for JEL categories are shown in Table 9.

The average specificity is equal to 0.313, a relatively low value, but some categories have very high specificity (it means they are assigned only to one university).

However, Fig. 5 indicates that categories with high specificity index have low strength.

Table 9. Characteristics of JEL categories

	Min	1st Quartile	Median	Mean	3rd Quartile	Max
Degree	1.000	4.000	5.000	4.506	5.000	5.000
Strength	0.000	0.001	0.003	0.006	0.008	0.078
Specificity index	0.000	0.182	0.268	0.313	0.386	1.000

Source: own elaboration

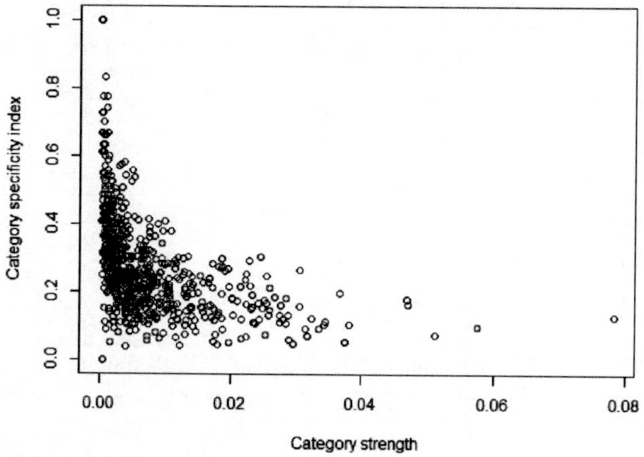

Fig. 5. Relationship between category strength and category specificity. Source: own elaboration.

5 Conclusions

The article's main purpose was to analyze the subject of the publication achievements of employees of Polish economic universities in the period 2017–2021. The study covered 5 Polish Economic Universities and 4862 publications in which at least one of the authors indicates an affiliation with one of the analyzed economic universities.

Based on the analysis of 60053 keywords and key phrases related to the first level of the JEL categories hierarchy, it was identified that the most significant number of occurrences relates to JEL_D – Microeconomics, and JEL_O – Economic Development, Innovation, Technological Change, and Growth. The following JEL categories, in terms of the number of occurrences, were: JEL_C – Mathematical and Quantitative Methods and JEL_L – Industrial Organization. On the other hand, the smallest number of occurrences appeared for the following JEL categories:

- JEL_A – General Economics and Teaching
- JEL_B – History of Economic Thought, Methodology, and Heterodox Approaches
- JEL_K – Law and Economics
- JEL_N – Economic History
- JEL_Y – Miscellaneous Categories
- JEL_Z – Other Special Topics

When analyzing the lower levels of the JEL classification, it should be noted that the most frequently discussed scientific issues concern: Political Processes: Rent-Seeking, Lobbying, Elections, Legislatures, and Voting Behavior (JEL_D.7.2) with 1010 references.

Other popular research areas raised by universities of economics were, among others: Innovation, Research and Development, Technological Change, Intellectual Property Rights (JEL_O.3) and Human Resources, Human Development, Income Distribution, Migration (JEL_O.1.5). Also, research results [1, 42, 43] identify similar research topics.

It is also worth emphasizing in the analyzed period, the interest in research in areas such as:

- JEL_O – Economic Development, Innovation, Technological Change, and Growth,
- JEL_Q – Agricultural and Natural Resource Economics, Environmental and Ecological Economic
- JEL_D – Political Processes: Rent-Seeking, Lobbying, Elections, Legislatures, and Voting Behavior
- JEL_P – Economic Systems
- JEL_L – Industrial Organization.

Based on the analysis of JEL categories mentioned in research papers and authors' affiliation, it should be noted that the topics discussed within individual economic universities are very similar, which means a lack of scientific specialization among analyzed units.

In addition, visualization of changes taking place in the subject of research problems seems to influence the perception of dependencies both in the sphere of researchers' scientific activity and their awareness in the field of scientific reality.

The broadest range of research areas concerns the Warsaw School of Economics (Degree: 739), whereas the narrowest scope is characteristic of the University of Economics in Katowice (Degree: 653). Despite these differences, it should be noted that the results for individual universities do not indicate significant differences in the scope of the topics covered. More considerable differences than in the scope of the issues covered can be indicated by analyzing the publication activity in which the Warsaw School of Economics stands out (Strength: 240.80). The next position was taken by the Krakow University of Economics (strength: 166.79) and the lowest by the University of Economics in Katowice (90.15), which may be the result of, for example, different sizes of the analyzed economic universities.

In August 2022, the Evaluation and Science Committee published the results of the parametric evaluation of scientific units. All analyzed economic universities were awarded the B + scientific category regarding their implemented scientific disciplines. Considering the similarity of universities in terms of conducted research, it should be noted that the level of research is also very similar, from the point of view of a research unit.

Acknowledgements. The research has been carried out as part of a research initiative financed by the Ministry of Science and Higher Education within "Regional Initiative of Excellence" Program for 2019–2022. Project no.: 021/RID/2018/19. Total financing: 11,897,131.40 PLN.

References

1. Khan, M.A., Pattnaik, D., Ashraf, R., Ali, I., Kumar, S., Donthu, N.: Value of special issues in the journal of business research: a bibliometric analysis. J. Bus. Res. **125**, 295–313 (2021). https://doi.org/10.1016/J.JBUSRES.2020.12.015
2. Donthu, N., Kumar, S., Mukherjee, D., Pandey, N., Lim, W.M.: How to conduct a bibliometric analysis: an overview and guidelines. J. Bus. Res. **133**, 285–296 (2021). https://doi.org/10.1016/J.JBUSRES.2021.04.070
3. Antons, D., Grünwald, E., Cichy, P., Salge, T.O.: The application of text mining methods in innovation research: current state, evolution patterns, and development priorities. R&D Manag. **50**(3), 329–351 (2020). https://doi.org/10.1111/RADM.12408
4. Rey-Martí, A., Ribeiro-Soriano, D., Palacios-Marqués, D.: A bibliometric analysis of social entrepreneurship. J. Bus. Res. **69**(5), 1651–1655 (2016). https://doi.org/10.1016/J.JBUSRES.2015.10.033
5. Guerras-Martín, L.Á., Madhok, A., Montoro-Sánchez, Á.: The evolution of strategic management research: recent trends and current directions. BRQ Bus. Res. Q. **17**(2), 69–76 (2014). https://doi.org/10.1016/J.BRQ.2014.03.001
6. Podsakoff, P.M., MacKenzie, S.B., Podsakoff, N.P., Bachrach, D.G.: Scholarly Influence in the field of management: a bibliometric analysis of the determinants of university and author impact in the management literature in the past quarter century. J. Manag. **34**(4), 641–720 (2008). https://doi.org/10.1177/0149206308319533
7. Rialp, A., Merigó, J.M., Cancino, C.A., Urbano, D.: Twenty-five years (1992–2016) of the international business review: a bibliometric overview. Int. Bus. Rev. **28**(6), 101587 (2019). https://doi.org/10.1016/J.IBUSREV.2019.101587
8. Kubiczek, J., Derej, W., Kantor, A.: Scientific achievements of economic academic workers in Poland: bibliometric analysis. EKONOMISTA **1**, 67–92 (2022). https://doi.org/10.52335/DVQIGJYKFF49
9. Lula, P., Cieraszewska, U., Hamerska, M.: Scientific cooperation in the field of economics in selected European countries. Adv. Intell. Syst. Comput. AISC **1197**, 1683–1692 (2021). https://doi.org/10.1007/978-3-030-51156-2_196/COVER/
10. Jedlikowska, D.: Analiza tematyczna wybranych artykułów naukoznawczych. Zagadnienia Naukoznawstwa **55**(2), 51–78 (2021). https://doi.org/10.12775/ZN.2019.014
11. Kowalska, M., Osińska, V.: Bazy danych i wizualizatory jako narzędzia oceny produktywności naukowej. NAUKA **2**, 93–114 (2018)
12. Ong, D., Chan, H.F., Torgler, B., Yang, Y.A.: Collaboration incentives: endogenous selection into single and coauthorships by surname initial in economics and management. J. Econ. Behav. Organ. **147**, pp. 41–57 (2018). https://doi.org/10.1016/J.JEBO.2018.01.001
13. Abramo, G., Andrea D'angelo, C., di Costa, F.: Research productivity: are higher academic ranks more productive than lower ones? Scientometrics **88**(3), 915–928 (2011). https://doi.org/10.1007/s11192-011-0426-6
14. Osiewalska, A.: Analiza cytowań z wybranych polskojęzycznych czasopism ekonomicznych. In: Pietruch-Reizes, D. (ed.) Zarządzanie informacją w nauce, Wydawnictwo Uniwersytetu Śląskiego, pp. 244–257 (2008)
15. Acedo, F.J., Barroso, C., Casanueva, C., Galán, J.L.: Co-authorship in management and organizational atudies: an empirical and network analysis. J. Manag. Stud. **43**(5), 957–983 (2006). https://doi.org/10.1111/J.1467-6486.2006.00625.X
16. Kwiek, M., Roszka, W.: Are female scientists less inclined to publish alone? The gender solo research gap. Scientometrics **127**(4), 1697–1735 (2022). https://doi.org/10.1007/s11192-022-04308-7

17. Henriksen, D.: The rise in co-authorship in the social sciences (1980–2013). Scientometrics **107**(2), 455–476 (2016). https://doi.org/10.1007/S11192-016-1849-X/TABLES/2
18. Kuld, L., O'Hagan, J.: Rise of multi-authored papers in economics: Demise of the 'lone star' and why? Scientometrics **114**(3), 1207–1225 (2018). https://doi.org/10.1007/S11192-017-2588-3
19. Cainelli, G., Maggioni, M.A., Uberti, T.E., de Felice, A.: The strength of strong ties: how co-authorship affect productivity of academic economists? Scientometrics **102**(1), 673–699 (2014). https://doi.org/10.1007/s11192-014-1421-5
20. Marušić, A., Bošnjak, L., Jerončić, A.: A systematic review of research on the meaning, ethics and practices of authorship across scholarly disciplines. Getting to Good: Res. Integrity Biomed. Sci. 191–207 (2018). https://doi.org/10.1371/JOURNAL.PONE.0023477
21. Abramo, G., D'Angelo, C.A., di Costa, F.: The collaboration behavior of top scientists. Scientometrics **118**(1), 215–232 (2019). https://doi.org/10.1007/S11192-018-2970-9
22. Bukowska, G., Łopaciuk-Gonczaryk, B.: Publishing patterns of polish authors in domestic and foreign economic journals. Ekonomista **4**, 442–466 (2018)
23. Kosch, O., Szarucki, M.: Transatlantic affiliations of scientific collaboration in strategic management: a quarter-century of bibliometric evidence. J. Bus. Econ. Manage. **21**(3), 627–646 (2020). https://doi.org/10.3846/JBEM.2020.12395
24. Kozak, M., Bornmann, L., Leydesdorff, L.: How have the eastern European countries of the former warsaw pact developed since 1990? a bibliometric study. Scientometrics **102**(2), 1101–1117 (2015). https://doi.org/10.1007/S11192-014-1439-8/TABLES/3
25. Teodorescu, D., Andrei, T.: The growth of international collaboration in East European scholarly communities: a bibliometric analysis of journal articles published between 1989 and 2009. Scientometrics **89**(2), 711–722 (2011). https://doi.org/10.1007/S11192-011-0466-Y
26. Kulczycki, E., et al.: Publication patterns in the social sciences and humanities: evidence from eight European countries. Scientometrics **116**, 463–486 (2018). https://doi.org/10.1007/s11192-018-2711-0
27. Hudson, J.: Trends in multi-authored papers in economics. J. Econ. Perspect. **10**(3), 153–158 (1996). https://doi.org/10.1257/JEP.10.3.153
28. G. Bukowska, J. Fałkowski & B. Łopaciuk-Gonczaryk, Teaming up or writing alone-authorship strategies in leading Polish economic journals (146.), 2014
29. Kwiek, M.: The internationalization of the polish academic profession a comparative European approach. CPP RPS **80**, 681–695 (2014)
30. Bonilla, C.A., Merigó, J.M., Torres-Abad, C.: Economics in Latin America: a bibliometric analysis. Scientometrics **105**(2), 1239–1252 (2015). https://doi.org/10.1007/s11192-015-1747-7
31. Furrer, O., Thomas, H., Goussevskaia, A.: The structure and evolution of the strategic management field: a content analysis of 26 years of strategic management research. Int. J. Manage. Rev. **10**(1), 123 (2008). https://doi.org/10.1111/j.1468-2370.2007.00217.x
32. Cummings, S., Daellenbach, U.: A guide to the future of strategy? The history of long range planning. long range planning **42**(2), 234–263 (2009). https://doi.org/10.1016/J.LRP.2008.12.005
33. Cieraszewska, U., Hamerska, M., Lula, P., Zembura, M.: The Significance of Medical Science Issues in Research Papers Published in the Field of Economics. In: Jajuga, K., Najman, K., Walesiak, M. (eds.) SKAD 2020. SCDAKO, pp. 133–152. Springer, Cham (2021). https://doi.org/10.1007/978-3-030-75190-6_9
34. Polowczyk, J.: Wpływ ekonomii behawioralnej na zarządzanie strategiczne w świetle badan bibliometrycznych. Przegląd Organizacji **6**, 3–9 (2012)
35. Kos, B.: Ekonomia jako obszar badań naukowych - trendy, perspektywy rozwoju: praca zbiorowa. Wydawnictwo Uniwersytetu Ekonomicznego (2010)

36. Dyduch, W.: Przyszłość zarządzania czyli jak działać po koronakryzysie. In: Bajor, E. (ed.), Przyszłość zarządzania. Wyzwania w dobie postglobalizacji, Wyd. Dom Organizatora, pp. 125–132 (2020)
37. Cyfert, S.: Przyszłość zarządzania w dobie postkoronakryzysu. In: Bojar, E. (ed.), Przyszłość zarządzania. Wyzwania w dobie postglobalizacji, Wyd. Dom Organizatora, pp. 69–76 (2020)
38. Bajor, R.: Wyzwania i nowe trendy w zarządzaniu w świetle badań. Studia Lubuskie, Państwowa Wyższa Szkoła Zawodowa Sulechowie, XII, pp. 83–98 (2016)
39. Sopińska, E., Mierzejewska, W.: Ewolucja zarządzania strategicznego w świetle badań polskich i zagranicznych. In R. Krupski (Ed.), Zarządzanie Strategiczne. Rozwój koncepcji i metod, Wydawnictwo Wałbrzyskiej Wyższej Szkoły Zarządzania i Przedsiębiorczości w Wałbrzychu, pp. 31–47 (2017)
40. Bratnicki, M., Dyduch, M. Dokonania i przyszłe kierunki badań nad przedsiębiorczością organizacyjną. Prace Naukowe Uniwersytetu Ekonomicznego w Katowicach. Nauki o Zarządzaniu: Dokonania, Trendy, Wyzwania, pp. 34–35, 2017, Available: https://www.researchgate.net/publication/328703993
41. Sułkowski, R.: R. Współczesne tendencje rozwoju nauk o zarządzaniu. In: Kożuch, B., Sułkowski, Ł. (eds.), Instrumentarium zarządzania publicznego, pp. 13–22 (2015)
42. Nogalski, B.: Inspiracje, problemy i obszary badawcze w naukach o zarządzaniu – spojrzenie retrospektywne. Edukacja Ekonomistów I Menedżerów **4**(38), 11–27 (2015)
43. Swacha, J.: Ewolucja tematyki badań z zakresu nauk o zarządzaniu w Polsce w latach 1990–2021 na podstawie publikacji Przeglądu Organizacji. Organizacja i Kierowanie, **2**(191), 13–31 (2022). www.sgh.waw.pl/oik/
44. Drogosz, J.: Dorobek piśmienniczy jednostek naukowych z grupy nauk ekonomicznych w świetle oceny parametrycznej z roku 2017. Studia Prawno-Ekonomiczne, CXV, pp. 203–226 (2020)
45. Kumar, S., et al.: What do we know about business strategy and environmental research? insights from business strategy and the environment. Bus. Strategy Environ. **30**(8), 3454–3469 (2021). https://doi.org/10.1002/BSE.2813
46. Jung, H., Lee, B.G.: Research trends in text mining: Semantic network and main path analysis of selected journals. Expert Syst. Appl. **162**, 113851 (2020). https://doi.org/10.1016/J.ESWA.2020.113851
47. Frączek, R.: Upowszechnianie wyników badań naukowych w międzynarodowych bazach danych. Analiza biometryczna na przykładzie nauk technicznych, ze szczególnym uwzględnieniem elektrotechniki. Wydawnictwo Uniwersytetu Śląskiego (2017)
48. Sauerland, S., Seiler, C.M.: Role of systematic reviews and meta-analysis in evidence-based medicine. World J. Surgery **29**(5), 582–587 (2005). https://doi.org/10.1007/S00268-005-7917-7
49. Zdonek, I., Hysa, B.: Analiza publikacji z obszaru nauk o zarządzaniu pod względem stosowanych metod badawczych. Zeszyty Naukowe Politechniki Śląskiej, Seria: Organizacja I Zarządzanie **102**(1975), 391–406 (2017)
50. JEL Classification System (2022). https://www.aeaweb.org/econlit/jelCodes.php?view=jel
51. Schryen, G., Wagner, G., Benlian, A., Paré, G.: A knowledge development perspective on literature reviews: validation of a new typology in the is field communications of the association for information systems. Commun. Assoc. Inf. Syst. **46**, 134–168 (2020). https://doi.org/10.17705/1CAIS.04607 Accessed 21 Aug 2022
52. Larsen, K.R., West, J.D.: Understanding the elephant: a discourse approach to corpus identification for theory review articles. J. Assoc. Inf. Syst. **20**(7), 887–927 (2019). https://www.academia.edu/37713592/Understanding_the_Elephant_A_Discourse_Approach_to_Corpus_Identification_for_Theory_Review_Articles

53. Morrison, A.J., Inkpen, A.C.: An analysis of significant contributions to the international business literature. J. Int. Bus. Stud. **22**(1), 143–153 (1991). https://doi.org/10.1057/PALGRAVE.JIBS.8490297
54. Wagner, G., Lukyanenko, R., Paré, G.: Artificial intelligence and the conduct of literature reviews. J. Inf. Technol. 1–18 (2021). https://doi.org/10.1177/02683962211048201
55. van den Bulk, L.M., Bouzembrak, Y., Gavai, A., Liu, N., van den Heuvel, L.J., Marvin, H.J.P.: Automatic classification of literature in systematic reviews on food safety using machine learning. Curr. Res. Food Sci. **5**, 84–95 (2022) https://doi.org/10.1016/J.CRFS.2021.12.010
56. Popoff, E., Besada, M., Jansen, J.P., Cope, S., Kanters, S.: Aligning text mining and machine learning algorithms with best practices for study selection in systematic literature reviews. Syst. Rev. 9(1), 1–12 (2020). https://doi.org/10.1186/S13643-020-01520-5/FIGURES/5
57. Marshall, I.J., Wallace, B.C.: Toward systematic review automation: a practical guide to using machine learning tools in research synthesis. Syst. Rev. **8**(1), 163 (2019). https://doi.org/10.1186/s13643-019-1074-9
58. Queiros, L., Mearns, E., Ademisoye, E., McCarvil, M., Alarcão, J., Garcia, M., Abogunrin, S.: Is artificial intelligence replacing humans in systematic literature reviews? a systematic literature review. Value Health **25**(7), S522 (2022). https://doi.org/10.1016/J.JVAL.2022.04.1229
59. Tranfield, D., Denyer, D., Smart, P.: Towards a methodology for developing evidence-informed management knowledge by means of systematic review. Br. J. Manag. **14**(3) 207–222 (2003). https://doi.org/10.1111/1467-8551.00375
60. Nerur, S.P., Rasheed, A.A., Natarajan, V.: The intellectual structure of the strategic management field: an author co-citation analysis. Strategic Manag. J. **29**(3), 319–336 (2008). https://doi.org/10.1002/SMJ.659
61. Ustawa, consolidated text Pub. L. No. Dz. U. 2022 r., poz. 574, z późń. zm., Ustawa z dnia 20 lipca 2018 r. Prawo o szkolnictwie wyższym i nauce (2018). https://isap.sejm.gov.pl/isap.nsf/DocDetails.xsp?id=WDU20220000574
62. Bocheńska, A.: Komentarz do wybranych przepisów ustawy - Prawo o szkolnictwie wyższym i nauce . In Akademickie prawo zatrudnienia. Komentarz, red. K. W. Baran (No. 115), (2020). https://sip.lex.pl/komentarze-i-publikacje/komentarze/komentarz-do-wybranych-przepisow-ustawy-prawo-o-szkolnictwie-587808893
63. Dańda, A., Szkup, B., Banaszak, B., Wewiór, P., Wawer, Ł., Rojek, M.: (2 C.E.). Ewaluacja jakości działalności naukowej przewodnik. https://www.gov.pl/attachment/c28d4c75-a14e-46c5-bf41-912ea28cda5b
64. Rozporządzenie w sprawie dziedzin i dyscyplin, Pub. L. No Dz.U. z 2018 r. poz. 1818, Rozporządzenie Ministra Nauki i Szkolnictwa Wyższego z dnia 20 września 2018 r. w sprawie dziedzin nauki i dyscyplin naukowych oraz dyscyplin artystycznych (2018). https://isap.sejm.gov.pl/isap.nsf/DocDetails.xsp?id=WDU20180001818
65. Ustawa wprowadzająca, Pub. L. No. Dz.U. 2018 r. poz. 1669 z późn. zm., Ustawa z dnia 3 lipca 2018 r. Przepisy wprowadzające ustawę – Prawo o szkolnictwie wyższym i nauce, 2018, Available: https://isap.sejm.gov.pl/isap.nsf/DocDetails.xsp?id=WDU20180001669
66. Prawo o szkolnictwie wyższym i nauce. Komentarz, WKP, 2019, Available: https://sip.lex.pl/komentarze-i-publikacje/komentarze/prawo-o-szkolnictwie-wyzszym-i-nauce-komentarz-587806848
67. Rozporządzenie w sprawie wykazów, consolidated text Pub. L. No. Dz.U.2020. poz.349, Rozporządzenie Ministra Nauki i Szkolnictwa Wyższego z dn. 7.11. 2018 r. w sprawie sporządzania wykazów wydawnictw monografii naukowych oraz czasopism naukowych i recenzowanych materiałów z konfer. Międz (2018). https://isap.sejm.gov.pl/isap.nsf/DocDetails.xsp?id=WDU20200000349

68. Rozporządzenie w sprawie ewaluacji, consolidated text Pub. L. No. Dz. U. z 2022 r., poz. 661, Rozporządzenie Ministra Nauki i Szkolnictwa Wyższego z dnia 22 lutego 2019 r. w sprawie ewaluacji jakości działalności naukowej (2019). https://isap.sejm.gov.pl/isap.nsf/DocDetails.xsp?id=WDU20190000392
69. American Economic Association., JEL Classification System (2022). https://www.aeaweb.org/econlit/jelCodes.php
70. Lula, P., Oczkowska, R., Wiśniewska, S., Wójcik, K.: Ontology-based system for automatic analysis of job offers. In: Minist, J., Tvrdíková, M. (eds.). pp. 205–2013 (2018). http://www.cssi-morava.cz/new/doc/IT2018/sbornik.pdf

Improved Three-Dimensional Reconstruction of Patient-Specific Carotid Bifurcation Using Deep Learning Based Segmentation of Ultrasound Images

Milos Anić[1,2(✉)] and Tijana Đukić[2,3]

[1] Faculty of Engineering, University of Kragujevac, 6 Sestre Janjić Street, 34000 Kragujevac, Serbia
anic.milos@kg.ac.rs

[2] Bioengineering Research and Development Center, BioIRC, 6 Prvoslava Stojanovića Street, 34000 Kragujevac, Serbia
tijana@kg.ac.rs

[3] Institute for Information Technologies, University of Kragujevac, Jovana Cvijića Bb, 34000 Kragujevac, Serbia

Abstract. Clinical examination is crucial during diagnostics of many diseases, including carotid artery disease. One of the most commonly used imaging techniques is the ultrasound (US) examination. However, the main drawback of US examination is that only two-dimensional (2D) cross-sectional images are obtained. For a more detailed analysis of the state of the patient's carotid bifurcation it would be very useful to analyze a three-dimensional (3D) model. Within this study, an improved methodology for the 3D reconstruction is proposed. US images were segmented by using deep convolutional neural networks, and lumen and arterial wall regions are extracted. Instead of using a generic model of the carotid artery as the basis that is further adapted to the particular patient with individual US cross-sectional images, in the presented approach the longitudinal cross-sectional US image of the whole carotid bifurcation is used to extract the shape of the whole geometry, which ensures more realistic 3D model. Computer AI-based 3D reconstruction of patient-specific geometry could ensure more complete view of the carotid bifurcation, but also this geometry could be further used within numerical simulations such as blood flow simulation or simulation of plaque progression, that could provide additional quantitative information useful for clinical diagnostics and treatment planning.

Keywords: deep learning · image segmentation · 3D reconstruction · finite element mesh

1 Introduction

Computer-aided systems for automated detection and classification based on artificial intelligence have been presented in literature [1, 2]. Machine learning and deep learning

N. Filipovic (Ed.): AAI 2022, LNNS 659, pp. 223–248, 2023.
https://doi.org/10.1007/978-3-031-29717-5_15

techniques have been applied for segmentation of medical images [3, 4]. In recent years, research has shown a great potential of deep convolutional neural networks (CNN) in dealing with complex computer vision problems. From classification problems [5, 6], segmentation and object detection problems, CNNs often outperformed human accuracy [7]. As requirements for the tasks that have to be solved using CNNs are growing because of increasingly higher complexity of the problems, increase in performance of CNNs can be achieved as soon as better hardware becomes available [5, 8]. In medical image segmentation, CNNs have shown a great potential. From brain [8] and left heart ventricle [9], to even smallest organs like pulmonary acinus of mice lungs [10], CNNs have been used to segment regions and classify them.

As datasets in medicine are increasingly larger day by day, the task of segmentation is becoming harder. Here, automatic and semi-automatic segmentation tasks are done using CNNs, where main task is to extract features from complex data [11, 12]. Thanks to this, CNNs became the golden standard for a variety of problems like classification, interpretation of medical image data and image segmentation.

Clinical examination is crucial during diagnostics of many diseases, including carotid artery disease. Within this particular examination, several imaging techniques are applied in order to analyze the state of patient's arteries and to detect possible atherosclerotic lesions. Imaging techniques include computed tomography (CT), magnetic resonance imaging (MRI) and ultrasound (US) examination. US imaging is fast, noninvasive and inexpensive and is therefore most commonly the first method applied in diagnostics.

One of the primary advantages of deep learning within US image segmentation is the efficiency of the usage of given data and the improved and accurate prediction of desired information which is accomplished by employing different processing layers (convolution, pooling, fully connected layers, etc..) that are complex and more data-specific than generic imaging features analyzed by traditional ultrasound computer-aided diagnosis (CAD) systems [13]. Generic images generated by traditional US CAD systems, additionally to being less patient specific, have reduced image quality which is manifested with blurring and artifacts which can often obscure diagnostically significant features. Even with these features of US images and other, difficult to work with types of images, deep learning algorithms proved to be able to extract desired features from them, segment them and analyze them which is why the method is widely used in medical image applications [3, 4]. Because of the mentioned great potential that CNN possesses, in this Chapter the CNNs are used to accurately segment US images of carotid artery walls and lumens as a prestep for 3D reconstruction of carotid arteries.

Another drawback of US examinations is that only two-dimensional (2D) cross-sectional images are obtained. In order to perform a more detailed analysis of the state of the patient's carotid bifurcation it would be very useful if a 3D model were available. For this purpose, the 3D reconstruction was performed in literature, by using the available 2D cross-sections [14]. The accuracy of this approach was validated against clinically measured parameters [15]. But, the main drawback of the presented method is that it uses a generic carotid bifurcation model as the basis and then the available segmented data is attached onto this model to adapt it to the specific patient. This method is further expanded within this study, by segmenting US images that contain whole carotid bifurcation and also by including more US transversal cross-sections. This way, a more detailed 3D geometry of the patient-specific carotid bifurcation is obtained and this geometry can be further used for both simple visual analysis but also for quantitative analysis of blood flow parameters obtained in numerical simulations.

Numerical simulations have a great potential to help elucidate many diverse phenomena in biomedicine, more specifically in modeling processes related to the cardiovascular system, including simulations of stent implantation [16, 17], in silico studies for validation of stent mechanical behavior [18], analysis of blood flow through coronary arteries [19, 20], atherosclerotic plaque formation and progression [21–23], analysis of treatment outcomes for stenotic carotid arteries [24]. With appropriate patient-specific reconstructed model, it is possible to analyze in detail the state of hemodynamic parameters, such as velocity, wall shear stress (WSS) and pressure. The 3D reconstructed model enables easier visual analysis of current plaques within the carotid bifurcation, but it also gives the possibility to analyze potential locations for further progression of atherosclerotic plaque. In this Chapter, clinical data from US examination for one patient will be used to perform 3D reconstruction using the methodology previously presented in literature [14] and the improved model presented in this Chapter. Consequently, both models will be used to perform unsteady simulations of blood flow to illustrate the benefits of the improvements of the reconstruction methodology.

The Chapter is organized as follows. Details of the applied improvements of the reconstruction methodology are discussed in Sect. 2. The numerical model used to perform blood flow simulations is also discussed in Sect. 2. Results of the 3D reconstruction of carotid bifurcation for a particular patient are presented in Sect. 3, together with the results of the blood flow simulations. Section 4 concludes the paper.

2 Materials and Methods

The methodology for 3D reconstruction using US images segmented with deep learning techniques is described in Sects. 2.1 and 2.2. Deep learning techniques are used to segment the US images and extract the lumen and wall of the carotid bifurcation. This part of the methodology is discussed in Sect. 2.1. The data from transversal US images is used to define the cross-sections of the appropriate branches of the carotid artery (internal carotid artery (ICA), external carotid artery (ECA) and common carotid artery (CCA)). The data from longitudinal US images is used to define the shape of the arterial branches. Segmented data are preprocessed and further used to generate the 3D finite element mesh of the reconstructed geometry. Details about this part of the methodology are discussed in Sect. 2.2. I order to analyze the benefits of the presented improved methodology, blood flow simulations are performed and the applied numerical model is presented in Sect. 2.3.

2.1 Deep Learning Segmentation

A major challenge in medicine nowadays is the ability to accurately analyze increasingly large datasets of both images and data inside repositories. In the recent years, a great promise in the field of extraction of features from complex data was shown by deep learning-based algorithms [11, 12], thus, they became the golden standard for a variety of problems including classification, interpretation of medical image data and image segmentation.

Conventional US systems emphasize feature selection and extraction [25] while traditional US systems focus on feature and image classification. Texture feature is one of the most common features in ultrasound imaging which can represent the nature of the lesion surface.

As already mentioned, in this Section deep learning algorithms were used to detect carotid artery wall and lumen from ultrasound images in order to achieve 3D reconstruction. Before imaging gathering procedure began, unique protocol of US examination was created and distributed among clinical partners, which led to harmonization of US datasets acquired from them. In order to develop deep learning module, acquired data had to be processed. This dataset included 214 patients where each patient had captured common carotid artery and its branches in longitudinal and transversal projections taken in B-mode. Total number of images was 1861 which resulted in average 8.7 images per patient. All images were anonymized as per international regulations.

Preprocessing included manual annotation of both transversal and longitudinal images where different datasets had to be prepared in order to detect lumen and wall on both types of images, resizing/cropping of images to the size of 512 × 512 pixels, classification of longitudinal and transversal US images. This resulted in 4 different datasets, two datasets per image section type, one for lumen and one for wall for both transversal and longitudinal images. An example of the dataset is shown on Fig. 1.

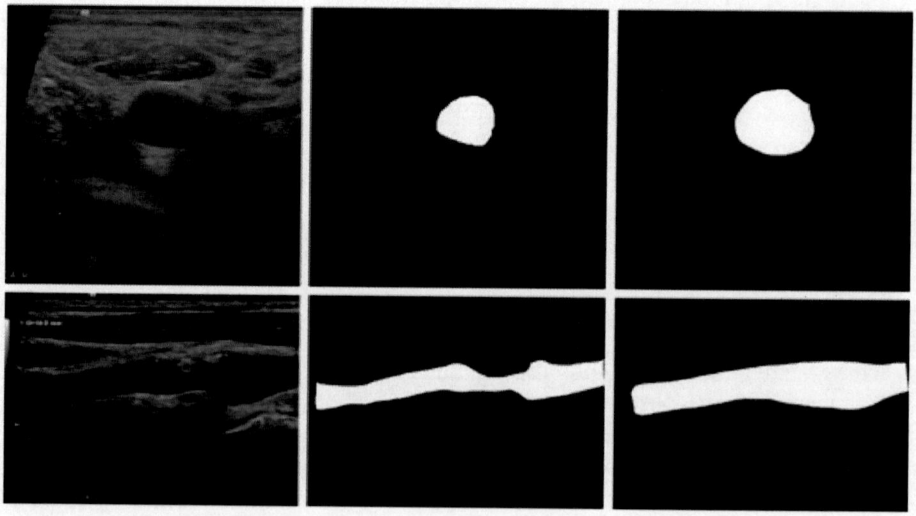

Fig. 1. Example of US dataset preprocessed images. The first column represents original images, second represents lumen masks and third represents wall masks.

Images that were not in B-mode or showed blood flow were not used, example of which can be found on Fig. 2.

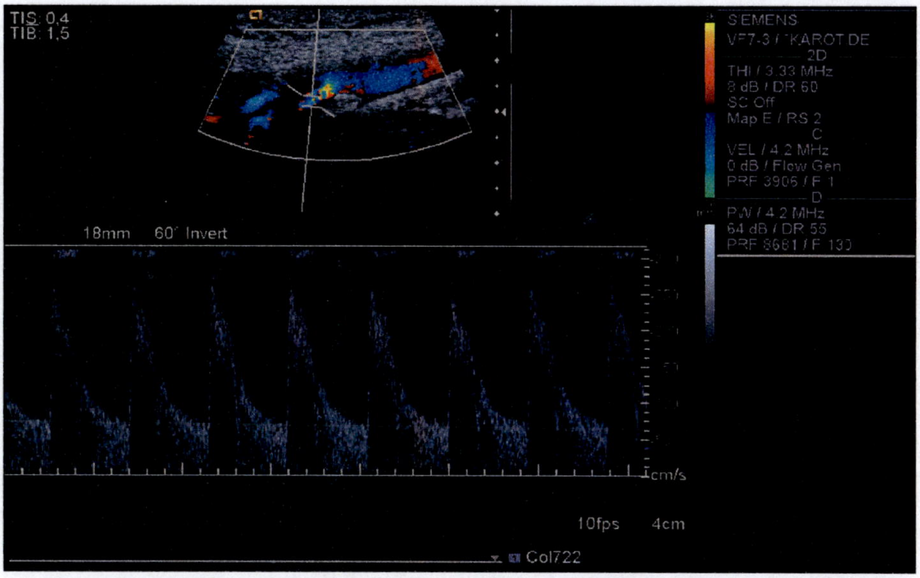

Fig. 2. Example of images that were not used in dataset.

Semantic image segmentation was performed to extract significant features further used for classification task. Automatic segmentation of carotid artery was done using FCN-8s [26, 27], SegNet [28] and U-Net [29] based deep convolutional neural networks. Modified versions of U-Net and SegNet networks, from the aspect of depth, were used beside original architectures. Introduced modifications are discussed in next paragraph. The results produced using U-Net modified architecture were substantially better despite SegNet [28] architecture having almost twice less trainable parameters which was observed in overall details of segmented images. This was a crucial part in determination of the most reliable network which could produce data for trustworthy 3D models.

The most significant application of U-Net deep convolutional neural network is segmentation of medical images. This convolutional network is founded in encoder-decoder model where encoder consists of max-pooling and convolutional layers which gradually reduce the size of image and increase channel count while decoder part, symmetrically to encoder part, consists of upconvolution and convolution operations. Unlike encoder, decoder doubles dimensions of features and reduces channel count. Additionally, skip connections are utilized to enhance the quality of decoder features. Unlike the original network, U-Net convolutional network used in this Chapter was enhanced from the aspect of depth which was introduced in [10].

Compared to the original model of U-Net [30], additional convolutional layer was added and batch normalization is used after each convolutional layer which proved to work better on our dataset then original model. It should be also mentioned that these layers are padded which results in image size being the same in both input and output. Each of mentioned layers is followed by ReLU activation and the model is trained with combination of binary cross-entropy and soft dice coefficient as loss function expressed as:

$$Loss = binary\_crossentropy(y_{true}, y_{pred}) + 1 - dice\_coeff(y_{true}, y_{pred}) \quad (1)$$

where y_{true} and y_{pred} express flattened ground truth and flattened predicted probabilities of image. Aforementioned modifications are also applied on SegNet model.

At the end of image acquisition and preprocessing, dataset that consisted of 214 patients was taken into consideration and the images were randomly distributed into training, validation and testing sets by a ratio of 8:1:1, where training set contained 1500 images.

In case of this study, binary classification task for image segmentation was taken into consideration. Our method was compared to SegNet [28] and FCN-8s [26, 27] model with VGG16 as backbone classifier. The VGG16 is used for this comparison since it is a usual choice in similar research in literature. As it was already mentioned, SegNet architecture has twice less trainable parameters then U-Net. On the other hand, FCN has twice as many parameters then U-Net. Both SegNet and FCN networks led to overall lower quality of results, thus, modified U-Net architecture [10] was used for the task of image segmentation, the results of which are shown in Table 1.

Table 1. Modified U-Net results on test dataset for lumen and wall in transversal and longitudinal projections

Segmentation	Projection	F1-score
LUMEN	Transversal	0.92
	Longitudinal	0.96
WALL	Transversal	0.84
	Longitudinal	0.81

In Fig. 3, graphical results of modified U-Net architecture can be observed. Figure 3A represents original image from US, Fig. 3B represents prediction for lumen and Fig. 3C represents prediction for wall region for the particular transversal US image respectively.

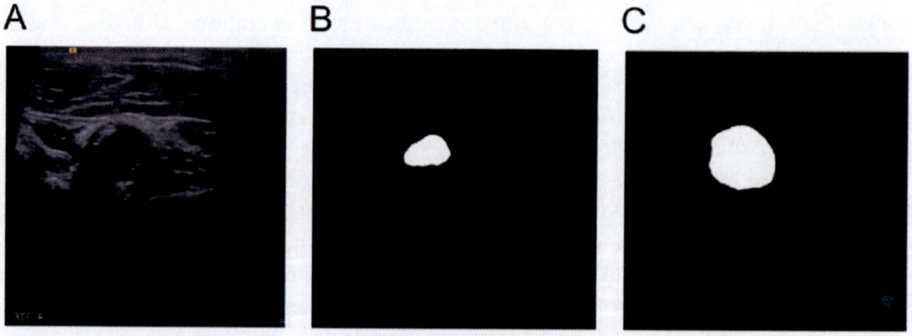

Fig. 3. Results of segmentation of transversal US image of a carotid artery for one patient. A - original image; B - predicted lumen region; C – predicted wall region.

2.2 Improved 3D Reconstruction

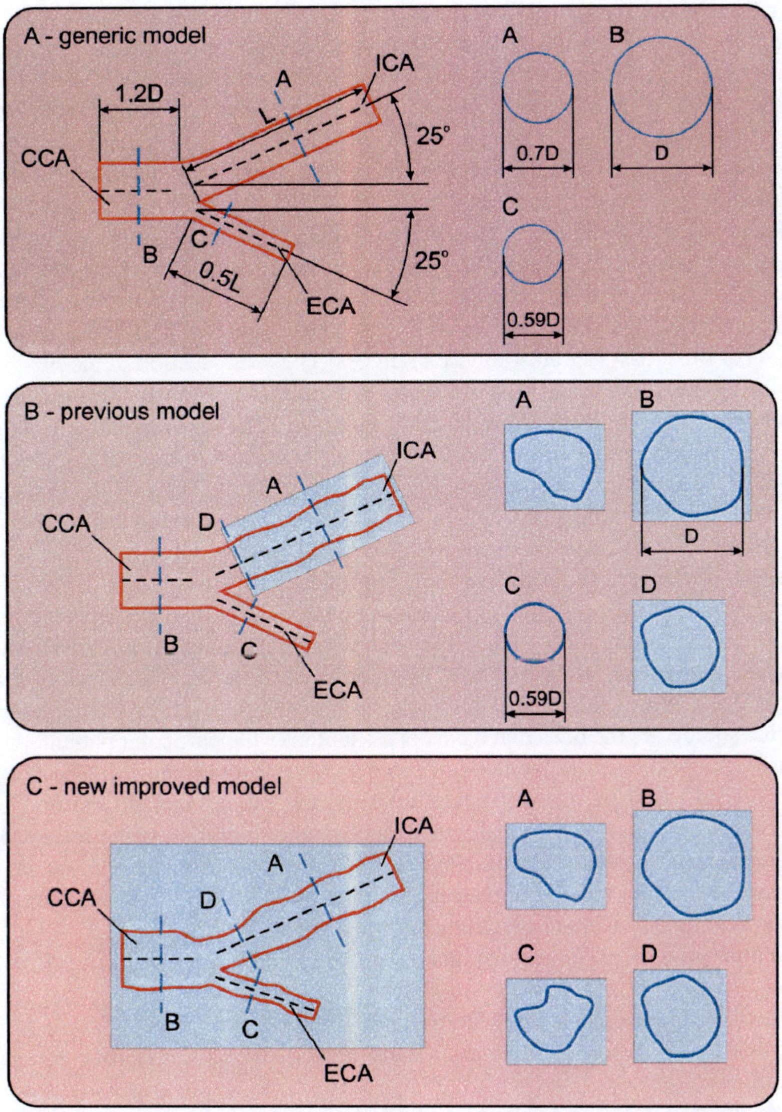

Fig. 4. Illustration of the evolution of the proposed 3D reconstruction methodology; A – generalized model proposed in literature [31, 32]; B –previous approach used for the adaptation of the generalized model to patient-specific data; C – new improved approach used for creation of patient-specific 3D model.

The 3D reconstruction of patient-specific carotid bifurcation is performed using the available clinical imaging data for the particular patient. The generalized model of the carotid artery is adapted to the particular patient in literature [14, 15], by using the longitudinal US image only for the ICA branch. Within this previously developed methodology, the longitudinal cuts from CCA and ECA branches were missing within the available dataset. Also, another problem with the clinical data set is the limited number of 2D transversal cuts that were available. In order to overcome the problem with the missing cuts, the generalized model proposed in literature [31, 32] was used as the basis. This basis is then adapted to the specific patient, by including the available data in the geometry. The generalized model is illustrated in Fig. 4A. The generalized model was defined according to data presented in literature [31, 32]. The adapted model presented in [14, 15] is illustrated in Fig. 4B. Parts of the model marked with a blue square are the ones that have been adapted to the particular patient. The transversal cuts of the CCA and ICA branches (annotated by the A, B and D lines in Fig. 4B) are used to define the shapes of the cross-sections of these branches. The cross-section of the ECA branch (annotated by the C line in Fig. 4B) is defined as circular, since mostly the transversal cut from this branch was missing from the clinical dataset. The longitudinal cut of the ICA is used to extract the centerline of the ICA and the diameters in this segment, while the ECA and CCA branches are considered to be straight. The lengths of the branches are still defined according to the generic model presented in literature [31, 32]. The arterial wall is also reconstructed using the available clinical data, in combination with generic data presented in literature, using the same approach that is used for lumen. Within the improved methodology presented in this Chapter, the longitudinal US images contained the whole carotid bifurcation. Hence the segmented data included lines of lumen and wall for all three branches. These lines are then used to define the shapes of all three branches, like it is illustrated in Fig. 4C. Like in Fig. 4B, the parts of the model marked with a blue square are the ones that have been adapted to the particular patient. As it can be observed, now the whole model is adapted to the particular patient. The lengths of the branches and their positions in space are also now patient-specific and not generic. Instead of using a single transversal US image for the definition of cross-sections for the whole branch, which was the case in the previous model, now the transversal cross-sections are inserted in the appropriate place along the carotid bifurcation, by matching the values of diameter from longitudinal and transversal US images.

The entire process of 3D reconstruction can be divided in overall 6 steps:

1. Extraction of vessel centerlines from longitudinal US images
2. Definition of location of individual transversal cross-sections along the vessels' centerlines
3. Placement of patches and definition of NURBS surfaces
4. Generation of Meshes for Each Vessel
5. Merging the Meshes into a Unique Mesh Representing the Entire Carotid Bifurcation

Each of these steps will explained in detail in the sequel.

2.2.1 Extraction of Vessel Centerlines

The deep learning module processes the US images and provides the resulting lines defining the boundaries of lumen area and wall area, independently. These lines are then used to extract the centerline of each branch of the carotid bifurcation, by performing intersections with both lines at a predefined distance from the start of the vessel. This distance was approximately 1.5 mm, to ensure accurate calculation. The result of this step is shown in Fig. 5. The lines defining the boundary of the lumen domain are shown as red lines, the extracted centerline is shown as a yellow line and blue lines represent the measured diameters of the vessel along the centerline.

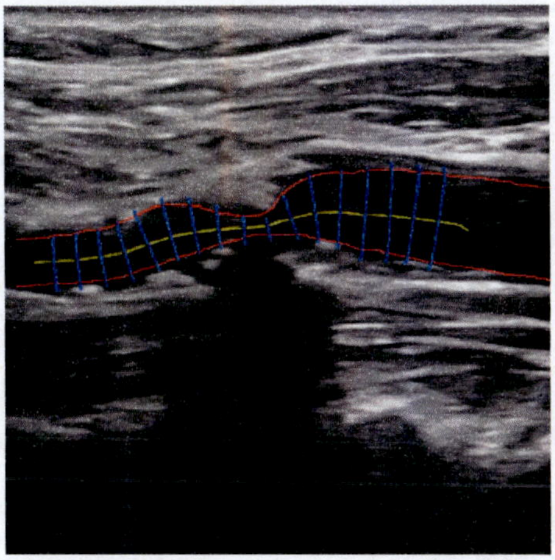

Fig. 5. Illustration of the extraction of centerline (yellow line) and diameters (blue lines) from one longitudinal US image with contours (red lines) extracted using deep learning module.

The centerlines are then converted to nonuniform B-spline curves, such that its line is parameterized and defined by:

$$\vec{c}(t) = \sum_{i=1}^{q} \mathbf{x}_i N_{i,k}(t) \tag{2}$$

where $0 \leq t \leq 1$ and $2 \leq k \leq i + 1$. In Eq. (2) the control points of the centerline are defined with $\mathbf{x}_i$, and $N_{i,k}$ denotes the k-th order basis functions that are defined within the Cox–de Boor recursive algorithm [87] and calculated as:

$$\begin{aligned} N_{i,1}(t) &= \begin{cases} 1 \; if \;\; s_i \leq t < s_{i+1} \\ 0 \quad otherwise \end{cases} \\ N_{i,k}(t) &= \frac{(t - s_i)N_{i,k-1}(t)}{s_{i+k-1} - s_i} + \frac{(s_{i+k} - t)N_{i+1,k-1}(t)}{s_{i+k} - s_{i+1}} \end{aligned} \tag{3}$$

In Eq. (3), **s** represents the knot vector that is calculated using the Chord length algorithm.

2.2.2 Definition of Location of Individual Transversal Cross-Sections

Like already discussed, within the deep learning module both longitudinal and transversal US images were segmented. However, the exact position of these transversal cross-sections in space is not known. Hence, it is necessary to position the transversal contours obtained within segmentation along the vessels' centerlines. Within the previous version of the 3D reconstruction methodology, data from transversal US images is used to define all cross-sections of the appropriate branches of the carotid artery. This was done so that along the entire length of each branch, the same cross-section is used, only the contour was scaled according to the measured diameter.

In the new improved methodology presented in this Chapter, instead of using a single transversal US image for the definition of cross-sections for the whole branch, several available transversal cross-sections are inserted in the appropriate place along the carotid bifurcation. Namely, in the previous step (discussed in Sect. 2.2.1), the diameters along the centerline were calculated for each branch of the carotid bifurcation. The change of diameter for one case is shown in Fig. 6. Now the diameters of contours extracted from transversal US images are matched with these measured values and the cross-sections are placed accordingly.

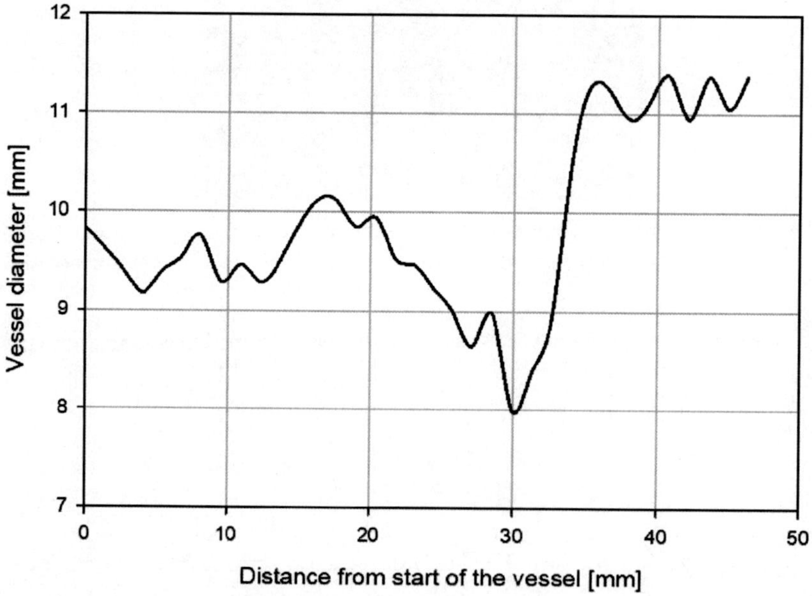

Fig. 6. Change of diameter along the centerline of the vessel, measured from longitudinal US images.

In order to accurately position the extracted contours in space, it was necessary to calculate the trihedron for all points of the parameterized centerlines defined in Eq. (1). The trihedron is calculated using the Frenet–Serret formulas [33] and consists of three vectors - tangent $\vec{T}(t)$, normal $\vec{N}(t)$ and binormal $\vec{B}(t)$ of the considered curve. The points of all cross-sections are projected onto the trihedron normal-binormal plane in each point of the centerline to obtain the patches of the surface. This is performed for all three branches of the carotid bifurcation, for both lumen and wall. The vectors defining the trihedron for one branch are illustrated in Fig. 7. The tangent vectors are shown as yellow arrows, normal vectors are shown as red arrows and binormal vectors are shown as green arrows. The contours defining the lumen areas are colored in blue and contours defining the wall areas are colored in black. During the positioning of the contours, it was necessary to prevent twisting of the geometry and to preserve the continuity and equal distribution of points in individual contours. For this reason, all contours are first converted to non-uniform B-spline curves and then all points of each curve are converted to polar coordinates in normal-binormal plane and sorted in circular direction.

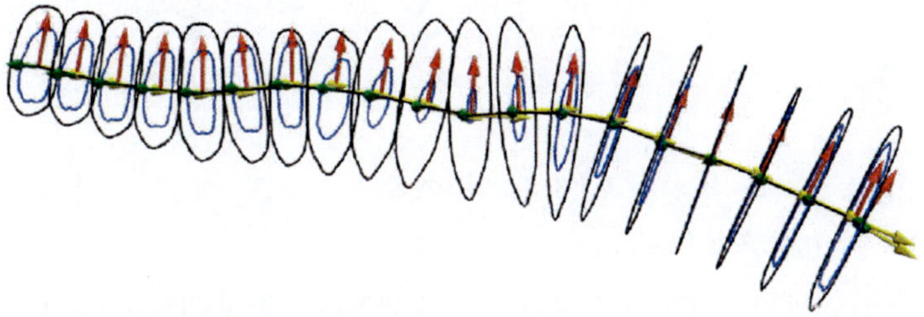

Fig. 7. Illustration of positioning of lumen and wall cross-sections along the parameterized centerline, with displayed trihedron vectors.

2.2.3 Definition of NURBS Surfaces

In order to be able to generate a 3D mesh of the reconstructed elements mentioned so far, it is necessary to define the NURBS surface representation. The patches of the surface obtained in the previous step discussed in Sect. 2.2.2 represent the control net polygon denoted by $\mathbf{B}_{i,j}$ and the surface is defined as:

$$\vec{S}^i(u, v) = \sum_{i=1}^{q} \sum_{j=1}^{w} \mathbf{B}_{i,j} N_{i,k}(u) M_{j,l}(v) \tag{4}$$

where u and v have values in the interval [0, 1] and $N_{i,k}$ and $M_{j,l}$ represent the k-th and l-th order basis functions, respectively. These basis functions are calculated using the already defined Eq. (3).

The control points used for the creation of the NURBS surface for one case are shown in Fig. 8.

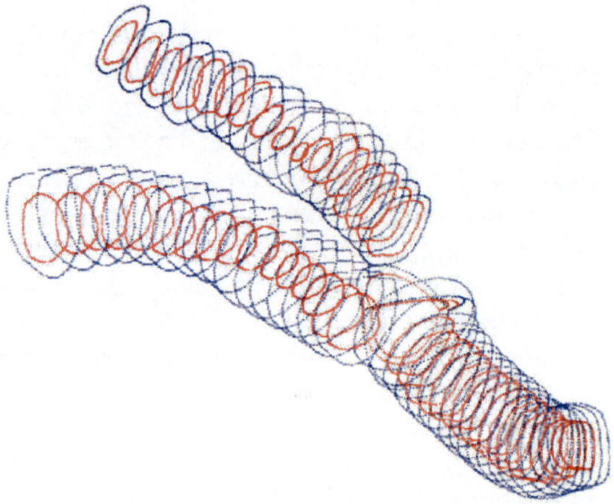

Fig. 8. The control net polygon used to generate the NURBS surface.

2.2.4 Generation of Meshes for Each Vessel

The parameterized centerline and NURBS surface are used to discretize the reconstructed branches of the carotid bifurcation and generate finite element mesh. The finite element mesh generated within the presented methodology consists of hexahedral elements. One generated mesh for a branch of the carotid bifurcation is illustrated in Fig. 9 and a single hexahedral finite element is separately shown, colored in red. During the mesh generation, two parameters are defined – the number of nodes in longitudinal and radial direction. The meshing procedure was described in literature [34]. It should be noted that during generation, the elements of the lumen and wall are numbered independently, in order to be able to distinguish them during subsequent simulations. In Fig. 9, elements of the wall are shown transparent, while the elements of the lumen are shown colored in blue.

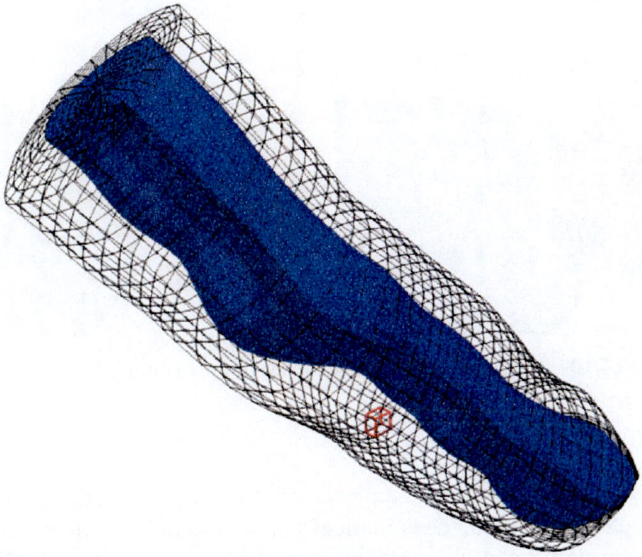

Fig. 9. Generated hexahedral finite element mesh for one branch of the carotid bifurcation. Elements of the lumen are colored in blue, a single hexahedral finite element is shown in red.

2.2.5 Merging the Meshes into a Unique Mesh Representing the Entire Carotid Bifurcation

The reconstruction of individual branches is however not enough to reconstruct a complete carotid bifurcation. It is necessary to merge the three branches together. In order to do that, a specific procedure was performed, that was discussed also in literature [35, 36]. It is necessary to adapt the orientations of the trihedrons of ending points of the centerline curves. Also, the final patches for all branches have to be trimmed, to ensure more appropriate fitting. The trimming process actually means that the NURBS control points of the ending patches are replaced by a new shared contour for each pair (CCA-ICA, CCA-ECA and ICA-ECA). The shared contour was defined by averaging the projections of removed ending patches of two connected branches onto the trim-planes. This procedure ensures a continuity of the elements between connected branches, like it is illustrated in Fig. 10. In Fig. 10A the whole carotid bifurcation is shown, with CCA branch colored in red, ICA branch colored in blue and ECA branch colored in green. In Fig. 10B the CCA branch is transparent, to illustrate the mentioned continuity of the elements. After the meshing procedure, an additional smoothing of connecting nodes is performed, to obtain a smooth transition between branches.

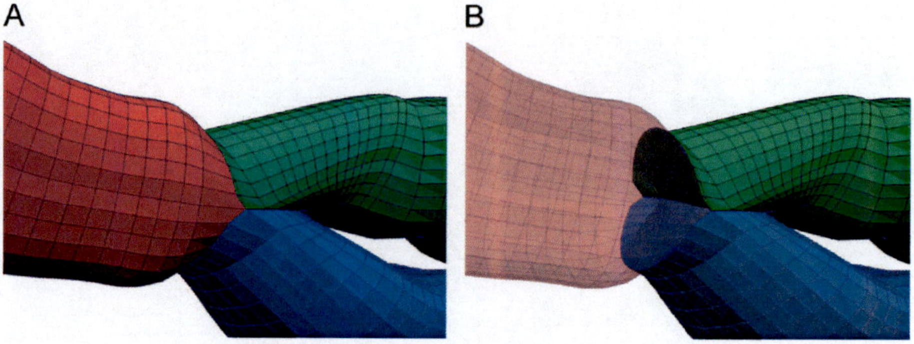

Fig. 10. Merging of the three branches of the carotid bifurcation.

2.3 Fluid Flow Simulations

In order to analyze the benefits of the presented improved methodology for 3D reconstruction of carotid bifurcation, the numerical simulations of blood flow were performed. Simulations were performed using an in-house developed software PakF [37], that is based on the finite element method [38]. This software was primarily developed and applied for biomedical simulations of 3D fluid flow in human arteries and it can also be used for simulations of plaque formation and progression [21–23]. It was validated in literature [23, 37, 38] and experimentally during several international projects.

Within numerical simulations blood is considered as a viscous incompressible Newtonian fluid. The flow is governed by the Navier-Stokes and continuity equations that are given by:

$$\rho\left(\frac{\partial v_i}{\partial t} + v_j\frac{\partial v_i}{\partial x_j}\right) = -\frac{\partial p}{\partial x_i} + \mu\left(\frac{\partial^2 v_i}{\partial x_j \partial x_j} + \frac{\partial^2 v_j}{\partial x_j \partial x_i}\right) \tag{5}$$

$$\frac{\partial v_i}{\partial x_i} = 0 \tag{6}$$

In Eqs. (5) and (6) v_i represents blood velocity in direction x_i, ρ represents fluid density, p is blood pressure, μ is the dynamic viscosity; and summation is assumed on the repeated (dummy) indices, $i, j = 1, 2, 3$.

Equations (6) and (7) are transformed into an incremental-iterative form, that is given by:

$$\begin{bmatrix} \frac{1}{\Delta t}\mathbf{M_v} + {}^{t+\Delta t}\mathbf{K}_{\mathbf{vv}}^{(i-1)} + {}^{t+\Delta t}\mathbf{K}_{\mu\mathbf{v}}^{(i-1)} + {}^{t+\Delta t}\mathbf{J}_{\mathbf{vv}}^{(i-1)} & \mathbf{K_{vp}} \\ \mathbf{K_{vp}^T} & \mathbf{0} \end{bmatrix} \begin{Bmatrix} \Delta\mathbf{v}^{(i)} \\ \Delta\mathbf{p}^{(i)} \end{Bmatrix} = \begin{Bmatrix} {}^{t+\Delta t}\mathbf{F}_{\mathbf{v}}^{(i-1)} \\ {}^{t+\Delta t}\mathbf{F}_{\mathbf{p}}^{(i-1)} \end{Bmatrix} \tag{7}$$

where i represents the equilibrium iteration, and Δt represents the time step. The left upper index "t + Δt" denotes that the quantities are evaluated at the end of time step. In Eq. (7) the matrices are defined as follows: $\mathbf{M_v}$ is the mass matrix, $\mathbf{K_{vv}}$ and $\mathbf{J_{vv}}$ are convective matrices, $\mathbf{K_{\mu v}}$ is the viscous matrix, $\mathbf{K_{vp}}$ is the pressure matrix, and $\mathbf{F_v}$ and $\mathbf{F_p}$ are forcing vectors.

Pressure is eliminated at the element level, because in this software the penalty formulation is used [37]. For this formulation, the incompressibility constraint is defined as:

$$div\,\mathbf{v} + \frac{p}{\lambda} = 0 \tag{8}$$

where λ is a relatively large positive scalar. This ensures that p/λ is a small number, practically close to zero.

Hence, the final incremental-iterative form of the equilibrium equation is given by:

$$\left(\frac{1}{\Delta t}\mathbf{M_v} + {}^{t+\Delta t}\mathbf{K}_{\mathbf{vv}}^{(i-1)} + {}^{t+\Delta t}\mathbf{K}_{\mu\mathbf{v}}^{(i-1)} + {}^{t+\Delta t}\hat{\mathbf{K}}_{\mu\mathbf{v}}^{(i-1)} + {}^{t+\Delta t}\mathbf{J}_{\mathbf{vv}}^{(i-1)} + \mathbf{K}_{\lambda\mathbf{v}}\right)\Delta\mathbf{v}^{(i)} = {}^{t+\Delta t}\hat{\mathbf{F}}_{\mathbf{v}}^{(i-1)} \tag{9}$$

where the matrices and vectors are calculated as:

$${}^{t+\Delta t}\hat{\mathbf{K}}_{\mu\mathbf{v}}^{(i-1)} = \int_V \mu\mathbf{H}^T\mathbf{H}dV$$

$$\mathbf{K}_{\lambda\mathbf{v}} = \lambda\int_V \mathbf{H}^T\mathbf{H}dV$$

$${}^{t+\Delta t}\hat{\mathbf{F}}_{\mathbf{v}}^{(i-1)} = {}^{t+\Delta t}\mathbf{R_B} + {}^{t+\Delta t}\hat{\mathbf{R}}_{\mathbf{S}}^{(i-1)} - \left({}^{t+\Delta t}\mathbf{K}_{\mathbf{vv}}^{(i-1)} + {}^{t+\Delta t}\mathbf{K}_{\mu\mathbf{v}}^{(i-1)} + {}^{t+\Delta t}\hat{\mathbf{K}}_{\mu\mathbf{v}}^{(i-1)} + \mathbf{K}_{\lambda\mathbf{v}}\right){}^{t+\Delta t}\mathbf{v}^{(i-1)}$$

$${}^{t+\Delta t}\left(\hat{\mathbf{R}}_{\mathbf{S}}\right)_{i\alpha}^{(i-1)} = \int_S \mathbf{H}^T\left[\lambda^{t+\Delta t}\nabla\mathbf{v}^{(i-1)}\cdot\mathbf{n} + \left({}^{t+\Delta t}\nabla\mathbf{v}^{(i-1)} + {}^{t+\Delta t}\nabla^T\mathbf{v}^{(i-1)}\right)\cdot\mathbf{n}\right]dS$$

$$\mathbf{M_v} = \rho\int_V \mathbf{H}^T\mathbf{H}dV$$

$${}^{t+\Delta t}\mathbf{K}_{\mathbf{vv}}^{(i-1)} = \rho\int_V \mathbf{H}^{\mathbf{T}}(\mathbf{H}^{t+\Delta t}\mathbf{v}^{(i-1)})\nabla^T\mathbf{H}dV$$

$${}^{t+\Delta t}\mathbf{K}_{\mu\mathbf{v}}^{(i-1)} = \int_V \mu\nabla\mathbf{H}^T\nabla^T\mathbf{H}dV$$

$${}^{t+\Delta t}\mathbf{J}_{\mathbf{vv}}^{(i-1)} = \rho\int_V \mathbf{H}^{\mathbf{T}}(\nabla\mathbf{H}^{t+\Delta t}\mathbf{v}^{(i-1)})\mathbf{H}dV$$

$${}^{t+\Delta t}\mathbf{R_B} = \int_V \mathbf{H}^{T\,t+\Delta t}\mathbf{f}^{\mathbf{B}}dV$$

$${}^{t+\Delta t}\mathbf{R}_{\mathbf{S}}^{(i-1)} = \int_S \mathbf{H}^T\left(-{}^{t+\Delta t}p^{(i-1)}\mathbf{n} + \nabla^{t+\Delta t}\mathbf{v}^{(i-1)}\cdot\mathbf{n}\right)dS \tag{10}$$

In simulations of blood flow through carotid bifurcation, higher values of Reynolds number (with value of several hundreds, to thousand) have to be considered. For this reason, the standard Petrov-Galerkin upwind stabilization technique [37] is applied in simulations.

Within simulations the convergence criterion is defined such that the convergence is reached when the maximum absolute change in the non-dimensional velocity between two adjacent time steps is less than 10^{-3}.

The initial velocity profile is prescribed at the inlet of the CCA branch, by assuming a parabolic velocity profile at the inlet. The change of flow rate over time measured for a specific patient, for one cardiac cycle is shown in Fig. 11. This data is used to calculate and prescribe the values of inlet velocity for unsteady flow simulations. For all nodes belonging to the wall of the artery the velocity is defined to be equal to zero. The outflow zero pressure boundary condition is applied at the outlets of the carotid bifurcation.

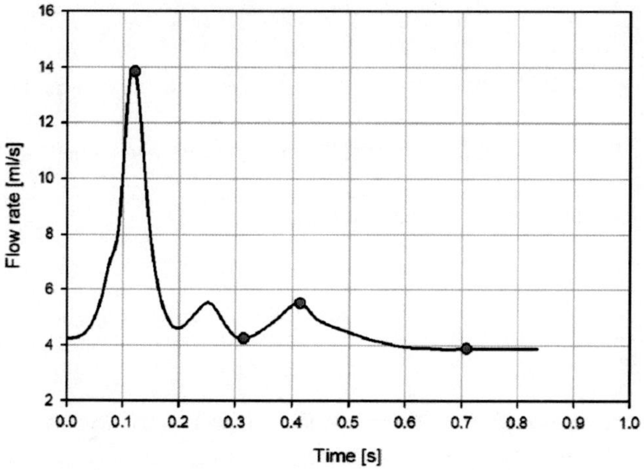

Fig. 11. Change of flow rate over time during one cardiac cycle, with denoted specific moments in time that are used for comparison of results.

3 Results

The reconstruction process for one particular patient is shown in Fig. 12. The set of clinical US images acquired for this patient are shown in Fig. 12A, while the segments obtained using the deep learning module are shown in Fig. 12B, colored in red. The obtained reconstructed 3D finite element mesh for the particular patient is shown in Fig. 12C. The cross-sections of the reconstructed geometry are shown together with extracted plaque components. These plaque components are extracted like it was already described in literature [24]. In this Chapter the main focus is on the improvements of the methodology and since within this improved methodology the extraction of plaque components was not changed, plaque reconstruction was not additionally addressed.

A comparison of a patient-specific reconstructed mesh using the previously proposed approach [14, 15] and the approach presented in this Chapter is shown in Fig. 13. Both models are generated using the same patient-specific data. Both models are shown with transparent arterial wall, in order to illustrate both the shapes of lumen and arterial wall. As it can be observed, the shape of the ECA and CCA branches is no longer generic and straight, but adapted to the particular patient. Also, the exact orientation, i.e. the angle between the branches, is now patient-specific instead of generic like in the previous model. It is obvious that the changes introduced within the methodology provide a more detailed and realistic model of the carotid bifurcation. For this specific

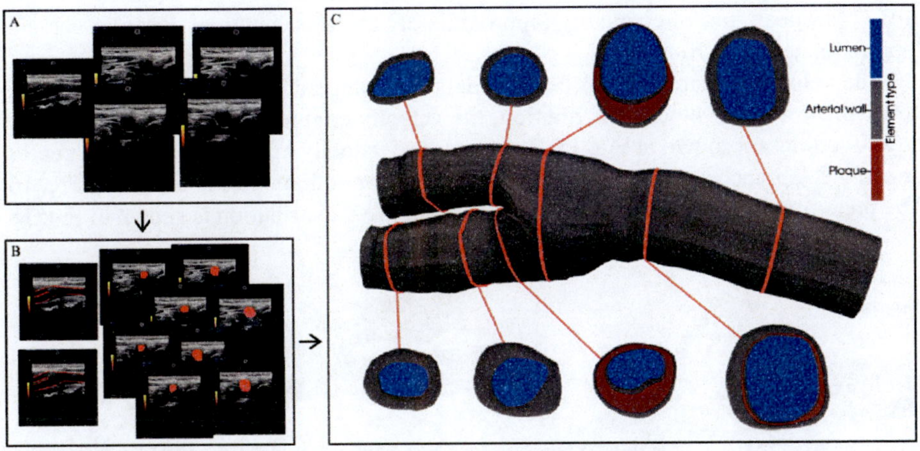

Fig. 12. Reconstruction process of the proposed methodology for 3D reconstruction; A - Original US images; B - Segmented data using deep learning module; C - Reconstructed model.

patient, the atherosclerotic segment is located between the CCA and ICA branches, in the bifurcation itself. With the previous version of the reconstruction methodology, this part of the artery is totally ignored and only the segment of ICA branch alone is reconstructed using US images. This caused almost complete overlook of the largest part of the stenosis. This is different with the improved methodology that was able to capture this stenosis much better. The effect of this discrepancy becomes even more obvious in numerical simulations, like it will be shown in the sequel of this Section.

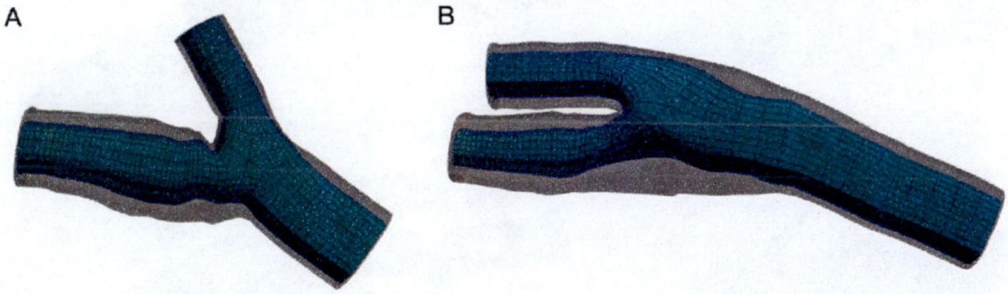

Fig. 13. Comparison of two reconstructed carotid bifurcations for the same patient; A – reconstructed geometry using the generic model as the basis, like presented in [14]; B –reconstructed geometry using the improved methodology presented in this study.

In order to show the benefits of the presented improved reconstruction methodology, unsteady blood flow simulations during one cardiac cycle are performed using both geometries. Four specific moments during cardiac cycle were chosen to analyze the obtained results and these moments in time are illustrated with red dots in Fig. 8. These moments in time are related to the velocity values and represent the peak systolic velocity (approximately at 0.125s), incisura between systole and diastole (approximately at 0.314s), peak diastolic velocity (approximately at 0.41s) and end-diastolic velocity (approximately at 0.708s).

The velocity streamlines for both models are comparatively shown in Fig. 14. To better illustrate the velocity distribution, the velocity contours at several characteristic cross-sections are shown in Fig. 15. The change of velocity magnitude along the centerline of CCA branch and subsequently ICA branch for both models is plotted in Fig. 16. The pressure distribution is shown in Fig. 17 and WSS distribution is shown in Fig. 18.

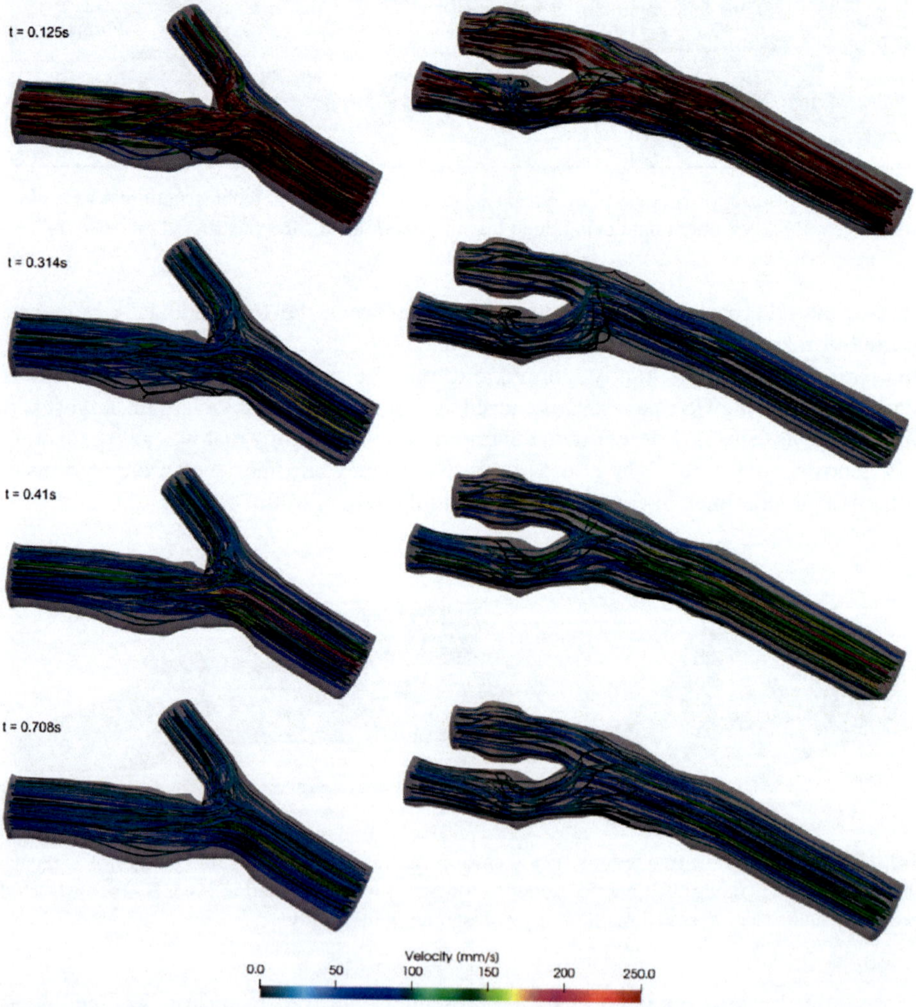

Fig. 14. Results of blood flow simulations - Velocity streamlines in 4 considered moments in time, marked in Fig. 11. Model based on generic geometry is shown on the left, improved model is shown on the right.

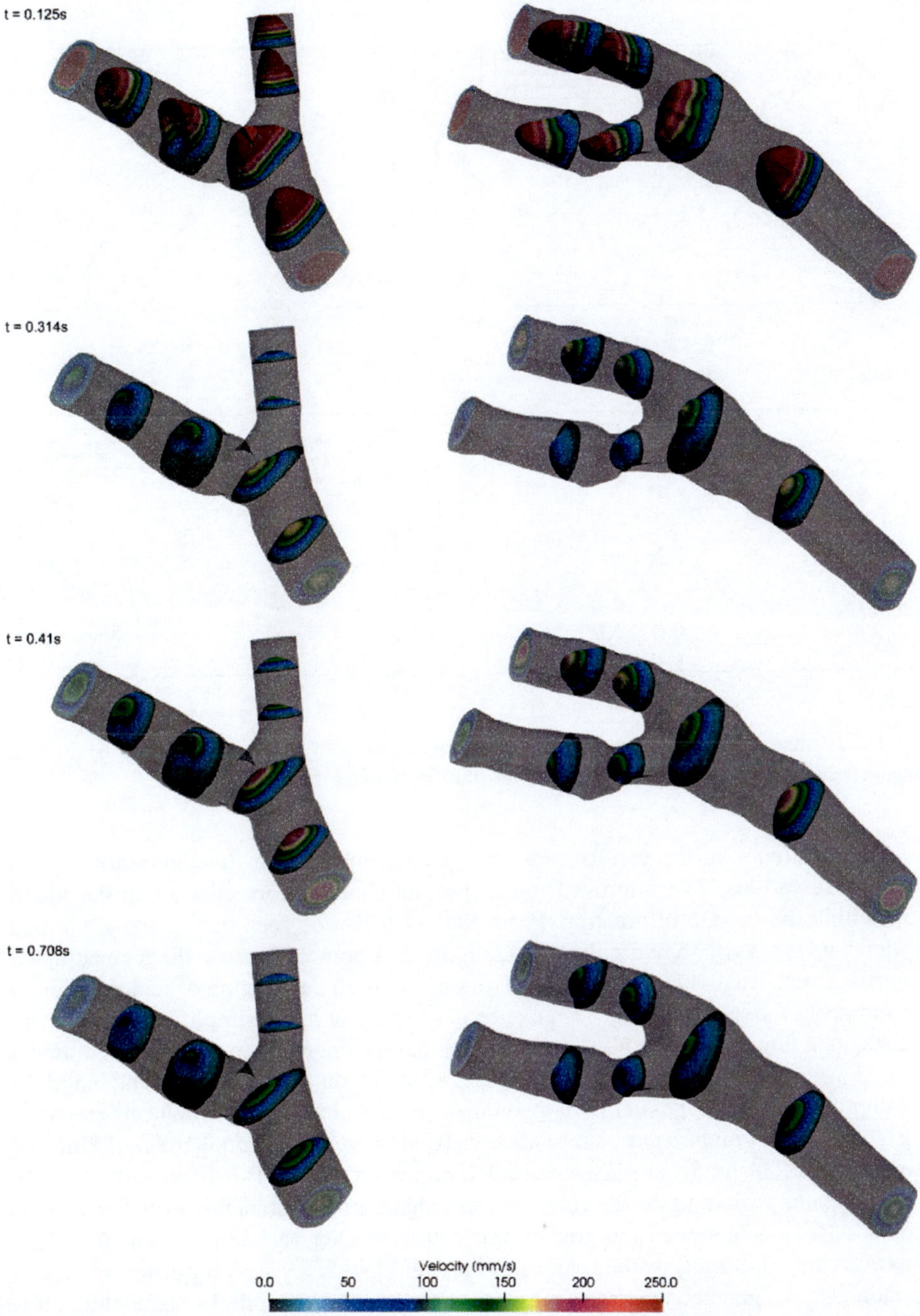

Fig. 15. Results of blood flow simulations - Velocity contours in several cross-sections, for 4 considered moments in time, marked in Fig. 11. Model based on generic geometry is shown on the left, improved model is shown on the right.

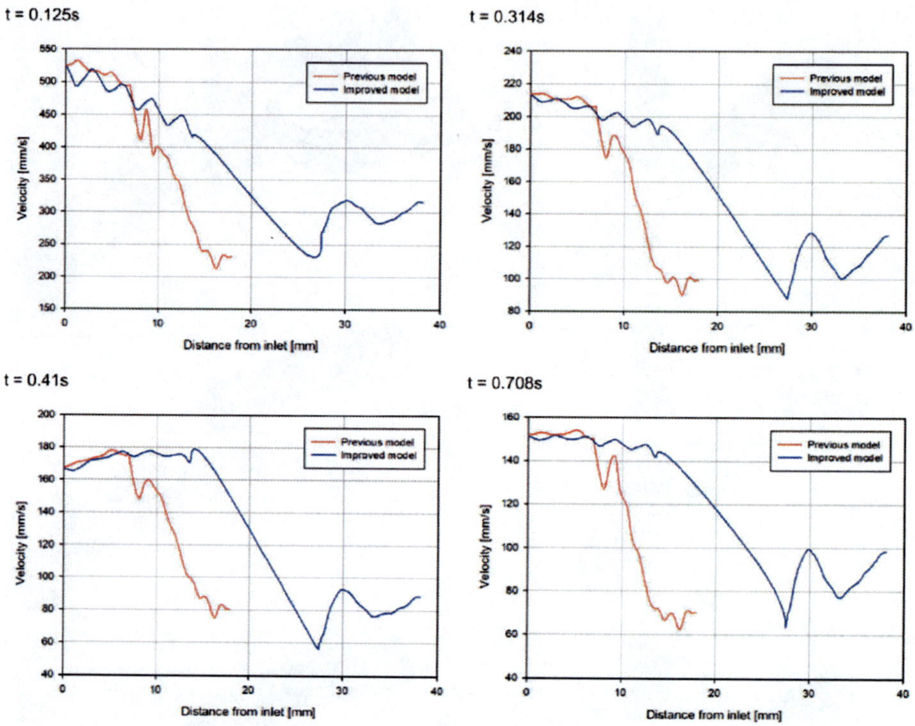

Fig. 16. Change of velocity magnitude along the centerlines of CCA and ICA branches for both considered models, for 4 considered moments in time, marked in Fig. 11.

The difference in the results additionally confirms the conclusions reached with visual observations. The improved model provides much more data about the blood flow within the carotid bifurcation. From Fig. 14 it can be seen that in the simplified model there is almost no recirculation within the ICA branch, because the reconstructed stenosis is very small. On the other hand, there is a significant region of recirculation in the improved model. From Fig. 15 it can be observed that in the simplified model, there is a stagnant flow within the ICA branch on the outer side, while in the improved model there is a smaller stagnant flow within the ICA branch on the inner side. The length of the improved model is greater than the simpler one, which is also evident in graphs in Fig. 16. On these graphs it can also be observed that the velocities along the centerlines of the CCA and ICA branches vary more in the improved model. This is caused primarily by the greater variations in the diameter and shape of the branches in the improved model. Also, as the stenosis is greater, this causes greater variation of velocity in the improved model. Similarly, the regions of high and low WSS are very different, due to variations in the geometry. These differences only confirm the initial statement that it is very important to reconstruct the patient-specific geometry as accurately as possible, to ensure more realistic results of numerical simulations.

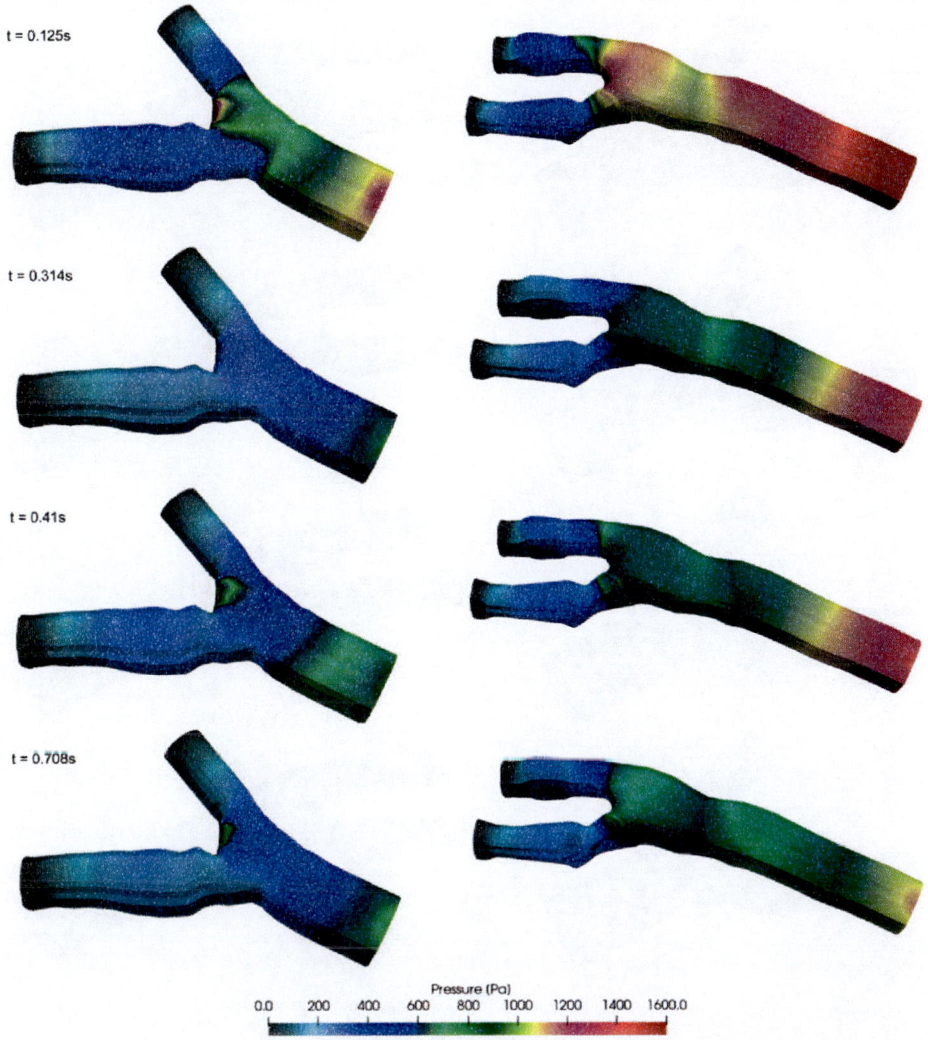

Fig. 17. Results of blood flow simulations – Pressure distribution in 4 considered moments in time, marked in Fig. 11. Model based on generic geometry is shown on the left, improved model is shown on the right.

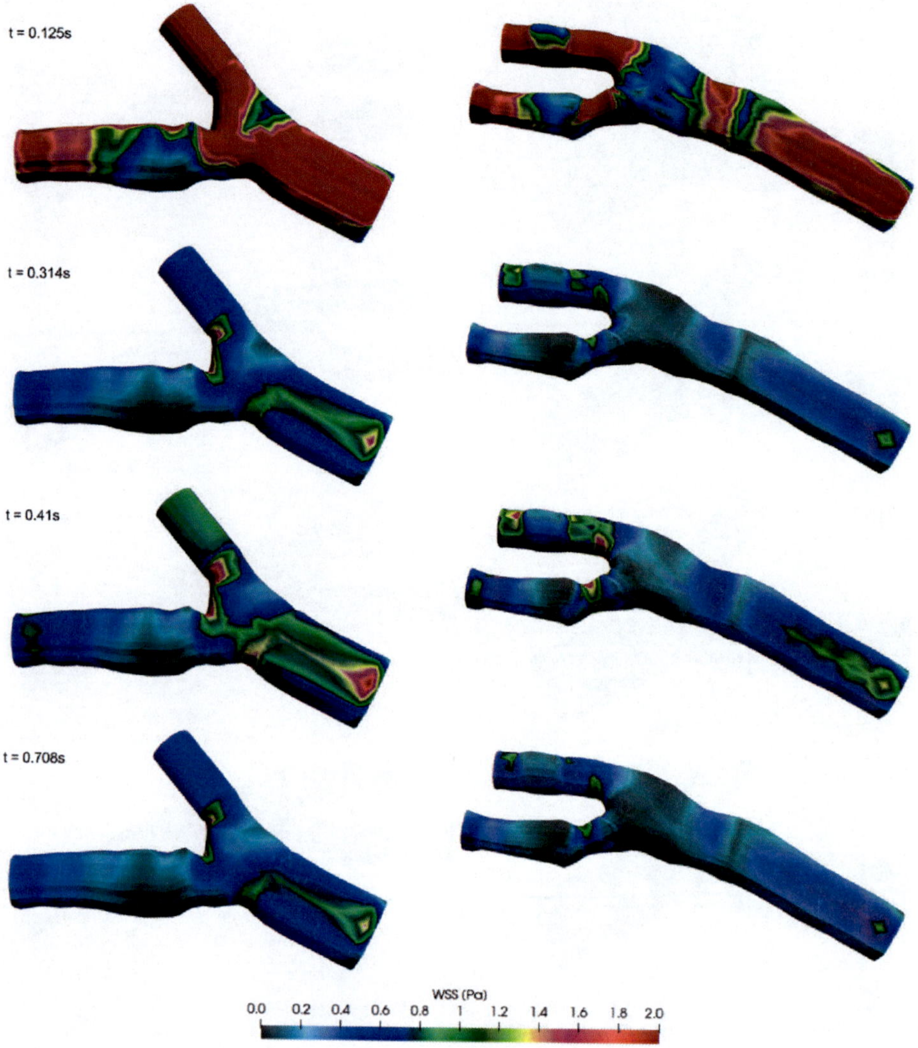

Fig. 18. Results of blood flow simulations – Distribution of WSS in 4 considered moments in time, marked in Fig. 11. Model based on generic geometry is shown on the left, improved model is shown on the right.

4 Conclusions

A detailed analysis of the state of patients' arteries is a crucial step in the diagnostics. Computer AI-based models could be a very useful tool that could help determine and plan the most appropriate treatment. The 3D reconstruction of patient-specific geometry could ensure more complete overview of the carotid bifurcation. Another application of 3D reconstruction methodology is in numerical simulations, which could provide additional relevant quantitative information about the state of arteries. Numerical simulations can provide robust and reliable results without clinically invasive or expensive measurements. This was proven through the validation of numerical simulations that was performed in literature. Velocity within carotid bifurcation phantom was measured using MRI technique in [39] and compared with results of numerical simulations. Another study reported the good agreement between results of numerical simulations and MRI measurements [40]. Due to the high incidence of stenosis within this vessel and the simplicity of clinical imaging, carotid bifurcation and its hemodynamics were also examined in other studies in literature [41–44]. Lopes et al. [43] analyzed the influence of rigid and flexible walls on the results of blood flow simulations. They compared the velocity and WSS distributions and concluded that the flexible walls provide better results. Markl et al. [44] analyzed the distribution of WSS using in vivo measurements that were acquired with flow-sensitive 4D MRI. Results presented in cited studies in literature are comparable with the ones presented in this Chapter. Since the simulations in this Chapter considered only rigid arterial walls and the reconstruction provides details about the arterial wall that were not used within presented simulations, the further improvement of the methodology will be concentrated on including deformable arterial wall within blood flow simulations. Nevertheless, the improved reconstruction methodology presented within this study is a step towards a more broad application of AI-based computer models in clinical practice.

Acknowledgments. The research presented in this study was part of the project that has received funding from the European Union's Horizon 2020 research and innovation programme under grant agreement No. 755320–2 - TAXINOMISIS. This article reflects only the author's view. The Commission is not responsible for any use that may be made of the information it contains.

References

1. Sun, T., et al.: Comparative evaluation of support vector machines for computer aided diagnosis of lung cancer in CT based on a multi-dimensional data set. Comput. Method Programs Biomed. **111**(2), 519–524 (2013)
2. Shi, J., Su, Q., Zhang, C., Huang, G., Zhu, Y.: An intelligent decision support algorithm for diagnosis of colorectal cancer through serum tumor markers. Comput. Method Programs Biomed. **100**(2), 97–107 (2010)
3. Shen, D., Wu, G., Suk, H.I.: Deep learning in medical image analysis. Annu. Rev. Biomed. Eng. **19**, 221–248 (2017)
4. Ravì, D., et al.: Deep learning for health informatics. IEEE J. Biomed. Health Inform. **21**(1), 4–21 (2016)
5. Krizhevsky, A., Sutskever, I., Hinton, G.E.: Imagenet classification with deep convolutional networks. In: Advances in Neural Information Processing Systems, pp. 1097–1105 (2012)
6. Szegedy, C., et al.: Going deeper with convolution. In: IEEE Conference Computer Vision and Pattern Recognition (2015)
7. He, K., Zhang, X., Ren, S., Sun, J.: Delving deep into rectifiers: Surpassing human-level performance on imagenet classification. In: Computer Vision and Pattern Recognition (2015)
8. Milletari, F., et al.: Hough-CNN: Deep learning for segmentation of deep brain regions in MRI and ultrasound. Comput. Vis. Image Unrestanding **164**, 92–102 (2017)

9. Sustersic, T., Anic, M., Filipovic, N.: Heart left ventricle segmentation in ultrasound images using deep learning. In: 20th IEEE Mediterranean Electrotechnical Conference, MELECON 2020 - Proceedings, pp. 321–324 (2020)
10. Arsic, B., Obrenovic, M., Anic, M., Tsuda, A., Filipovic, N.: Image segmentation of the pulmonary acinus imaged by synchrotron x-ray tomography. In: Proceedings - 2019 IEEE 19th International Conference on Bioinformatics and Bioengineering (2019)
11. Goodfelow, I.C.A., Bengio, Y.: Deep Learning. In: Goodfellow, Y., Bengio, Y., Aaron, C. (eds.) Google Books, MIT Press, Cambridge (2016)
12. Lanza, G., Giannandrea, D., Lanza, J., Ricci, S., Gensini, G.F.: Personalized-medicine on carotid endarterectomy and stenting. Ann. Transl. Med. **8**(19), 1274 (2020)
13. Alzubaidi, L., et al.: Review of deep learning: concepts, CNN architectures, challenges, applications, future directions. J. Big Data **8**(1), 1–74 (2021). https://doi.org/10.1186/s40537-021-00444-8
14. Djukic, T., Arsic, B., Koncar, I., Filipovic, N.: 3D Reconstruction of patient-specific carotid artery geometry using clinical ultrasound imaging. In: Miller, K., Wittek, A., Nash, M., Nielsen, P.M.F. (eds.) Computational Biomechanics for Medicine. pp. 73–83. Springer, Cham (2021). https://doi.org/10.1007/978-3-030-70123-9_6
15. Đukić, T., Arsić, B., Đorović, S., Končar, I., Filipović, N.: Validation of the machine learning approach for 3D reconstruction of carotid artery from ultrasound imaging. In: IEEE 20th International Conference on Bioinformatics and Bioengineering (BIBE) (2020)
16. Đukić, T., Saveljić, I., Pelosi, G., Parodi, O., Filipović, N.: Numerical simulation of stent deployment within patient-specific artery and its validation against clinical data. Comput. Methods Programs Biomed. **175**, 121–127 (2019)
17. Đukić, T., Saveljić, I., Pelosi, G., Parodi, O., Filipović, N.: A study on the accuracy and efficiency of the improved numerical model for stent implantation using clinical data. Comput. Methods Programs Biomed. **207**, 106196 (2021)
18. Milošević, M., Anić, M., Nikolić, D., Milićević, B., Kojić, M., Filipović, N.: InSilc computational tool for in silico optimization of drug-eluting bioresorbable vascular scaffolds. Comput. Math. Methods Med. **2022**, 5311208 (2022)
19. Đukić, T., Filipović, N.: Simulating fluid flow within coronary arteries using parallelized sparse lattice Boltzmann method. In: 8th International Congress of Serbian Society of Mechanics, Kragujevac, Serbia (2021)
20. Đukić, T., Topalović, M., Filipović, N.: Validation of lattice boltzmann based software for blood flow simulations in complex patient-specific arteries against traditional CFD methods. Math. Comput. Simul. **203**, 957–976 (2022)
21. Filipović, N., Teng, Z., Radović, M., Saveljić, I., Fotiadis, D., Parodi, O.: Computer simulation of three-dimensional plaque formation and progression in the carotid artery. Med. Biol. Eng. Comput. **51**, 607–616 (2013)
22. Filipović, N., et al.: Three-dimensional numerical simulation of plaque formation and development in the arteries. IEEE Trans. Inf. Technol. Biomed. **16**(2), 272–278 (2012)
23. Parodi, O., et al.: Patient-specific prediction of coronary plaque growth from CTA angiography: a multiscale model for plaque formation and progression. IEEE Trans. Inf. Technol. Biomed. **16**(5), 952–956 (2012)
24. Đukić, T., Filipović, N.: Simulation of carotid artery plaque development and treatment. In: Cardiovascular and Respiratory Bioengineering, Elsevier, pp. 101–133 (2022)
25. Ravindraiah, R., Tejaswini, K.: A survey of image segmentation algorithms based on fuzzy clustering. Int. J. Comput. Sci. Mob. Comput. **2**(7), 200–206 (2013)
26. Long, J., Shelhamer, E., Darrell, T.: Fully convolutional networks for semantic segmentation. In: Proceedings of the IEEE Computer Society Conference on Computer Vision and Pattern Recognition (2015)

27. Sinonyan, K., Zisserman, A.: Very deep convolutional networks for large-scale image recognition. In: 3rd International Conference on Learning Representations, ICLR 2015 - Conference Track Proceedings (2015)
28. Badrinarayanan, V., Kendall, A., Cipolla, R.: SegNet: a deep convolutional encoder-decoder architecture for image segmentation. IEEE Trans. Pattern Anal. Mach. Intell. **39**(12), 2481–2495 (2017)
29. Ronneberger, O., Fischer, P., Brox, T.: U-Net: convolutional networks for biomedical image segmentation. In: Navab, N., Hornegger, J., Wells, W.M., Frangi, A.F. (eds.) MICCAI 2015. LNCS, vol. 9351, pp. 234–241. Springer, Cham (2015). https://doi.org/10.1007/978-3-319-24574-4_28
30. Zhou, X.Y., Yang, G.Z.: Normalization in training U-Net for 2-D biomedical semantic segmentation. IEEE Robot. Autom. Lett. **4**(2), 1792–1799 (2019)
31. Perktold, K., Peter, R.O., Resch, M., Langs, G.: Pulsatile non-newtonian blood flow in three-dimensional carotid bifurcation models: a numerical study of flow phenomena under different bifurcation angles. J. Biomed. Eng. **13**(6), 507–515 (1991)
32. Perktold, K., Resch, M., Peter, R.O.: Three-dimensional numerical analysis of pul-satile flow and wall shear stress in the carotid artery bifurcation. J. Biomech. **24**, 409–420 (1991)
33. Vukicevic, A.M., Stepanovic, N.M., Jovicic, G.R., Apostolovic, S.R., Filipovic, N.D.: Computer methods for follow-up study of hemodynamic and disease progression in the stented coronary artery by fusing IVUS and X-ray angiography. Med. Biol. Eng. Comput. **52**(6), 539–556 (2014). https://doi.org/10.1007/s11517-014-1155-9
34. Vukicevic, A., Çimen, S., Jagic, N., Jovicic, G., Frangi, A.F., Filipovic, N.: Three-dimensional reconstruction and NURBS-based structured meshing of coronary arteries from the conventional X-ray angi-ography projection images. Sci. Rep. **8**, 1711 (2018)
35. Antiga, L., Steinman, D.: Robust and objective decomposition and mapping of bifurcating vessels. IEEE Trans. Med. Imaging **23**(6), 704–713 (2004)
36. Zhang, Y., Bazilevs, Y., Goswami, S., Bajaj, C., Hughes, T.: Patient-specific vascular NURBS modeling for isogeometric analysis of blood flow. Comput. Methods. Appl. Mech. Eng. **196**(29–30), 2943–2959 (2007)
37. Filipovic, N., Mijailovic, S., Tsuda, A., Kojic, M.: An implicit algorithm within the arbitrary Lagrangian-Eulerian formulation for solving incompressible fluid flow with large boundary motions. Comp. Meth. Appl. Mech. Engrg. **195**, 6347–6361 (2006)
38. Kojić, M., Filipović, N., Stojanović, B., Kojić, N.: Computer modeling in bioengineering: Theoretical Background, Examples and Software. Wiley, Chichester (2008)
39. Long, Q., Xu, X., Köhler, U., Robertson, M.B., Marshall, I., Hoskins, P.: Quantitative comparison of CFD predicted and MRI measured velocity fields in a carotid bifurcation phantom. Biorheology **39**, 467–474 (2002)
40. Cibis, M., Potters, W., Selwaness, M., Gijsen, F., Franco, O., Arias Lorza, A.E.A.: Relation between wall shear stress and carotid artery wall thickening MRI versus CFD. J Biomech. **49**(5), 735–741 (2016)
41. Gharahi, H., Zambrano, B.Z.D., DeMarco, K., Seungik, B.: Computational fluid dynamic simulation of human carotid artery bifurcation based on anatomy and volumetric blood flow rate measured with magnetic resonance imaging. Int. J. Adv. Eng. Sci. Appl. Math. (2016). https://doi.org/10.1007/s12572-016-0161-6
42. Rispoli, V., Nielsen, J., Nayak, K., Carvalho, J.: Computational fluid dynamics simulations of blood flow regularized by 3D phase contrast MRI. Biomed. Eng. (2015). https://doi.org/10.1186/s12938-015-0104-7
43. Lopes, D., Puga, H., Teixeira, J., Teixeira, S.: Influence of arterial mechanical properties on carotid blood flow: comparison of CFD and FSI studies. Int. J. Mech. Sci. **160**, 209–218 (2019)

44. Markl, M., et al.: In vivo wall shear stress distribution in the carotid artery: effect of bifurcation geometry, internal carotid artery stenosis, and recanalization therapy. Circ. Cardiovasc. Imaging 647–55 (2010) https://doi.org/10.1161/CIRCIMAGING.110.958504

Seat-to-Head Transfer Functions Prediction Using Artificial Neural Networks

Slavica Mačužić Saveljić(✉)

Faculty of Engineering, University of Kragujevac, 6 Sestre Janjic Street, 34000 Kragujevac, Serbia
s.macuzic@kg.ac.rs

Abstract. Drivers in vehicles are exposed daily to vibrations that are transmitted from the seat to the driver's body. Vibrations from vehicles have the greatest impact on comfort, regardless of their intensity and shape. The transmission response function (STHT) between the seat and the head represents a strong connection between the vibration-induced head motion responses transmitted by the seat/head interface. The aim of this work was to form a model of an artificial neural network (ANN) based on experimental measurements. Twenty healthy male subjects participated in the experiment. They are exposed to vertical vibrations. Based on the obtained values of transfer functions, an artificial neural network was trained. The results showed that the developed model has the ability to predict the values of transfer functions in the range of trained values when changing the input parameters.

Keywords: Artificial neural network · Experimental measurements · Vertical body vibration

1 Introduction

The human body is a complex biomechanical system. Its sensitivity to vibrations depends on several factors such as vibration frequency, amplitude, the direction of action, body position, muscle tension, and others. The direct effects of vibrations acting on the human body during long periods of exposure can seriously and permanently damage some body organs, so vibrations are not only an engineering problem, but also a health problem. Exposure to vibrations has different effects on the human body, starting with the appearance of slight discomfort and ending with a decrease in work performance and damage to health. Today, there are guidelines for defining the tolerance of the human body exposed to vibrations of the whole body in the international standard ISO 2631-1:1997 [1]. Vibrations absorbed by the human body lead to muscle contractions that can cause muscle fatigue, especially at resonant frequencies. For example, vertical vibrations in the range of 5–10 Hz cause resonance in the thoracic-abdominal system, 20–30 Hz in the head region, and 4–8 Hz in the spine [2].

Assessment of the oscillatory comfort of the human body can be achieved by subjective and objective methods. Subjective methods are implemented according to the

N. Filipovic (Ed.): AAI 2022, LNNS 659, pp. 249–270, 2023.
https://doi.org/10.1007/978-3-031-29717-5_16

principle of interval scales that the subject uses to assess his current discomfort, while objective methods use systems within which the subject's discomfort is determined. In addition to these, there are also biodynamic studies of the behavior of the human body, which play a very important role in the analysis and prediction of oscillatory comfort in the vehicle. Measurements of the dynamic response of the driver's body exposed to vibration can be represented by frequency response functions that can be classified based on measurements at different points of the body and on the basis of measurements at the same point. Seat-To-Head Transmissibility function STHT in the seat-driver system analyzes the ratio of head acceleration to seat acceleration in the frequency domain, which is a measure of transmitted vibrations through the subject's body. Researching the impact of vibrations on the head is very important because if, as a vital part of the body, it is exposed to discomfort, the effects of vibrations, in addition to fatigue and pain, directly affect the field of vision [3].

For many years, the effect of vibrations on the driver's body has been investigated both in operational and laboratory conditions. The advantage of the laboratory research is that stable parameters of the micro-environment and reproducibility of the results can be ensured. Within this chapter, the STHT functions of twenty male subjects will be experimentally determined, and based on the obtained experimental results, a model will be formed using artificial neural network (ANN) methods that will be able to predict the response of the STHT function of a man exposed to vibrations. The scheme of the methodology is shown in Fig. 1.

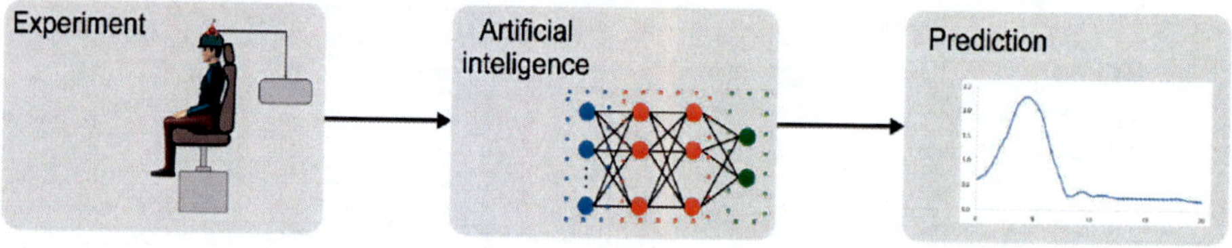

Fig. 1. The scheme of the methodology

A review of the literature found that a significant number of studies dealt with the impact of vibrations, both on individual parts of the body and on the entire human body, in different environments. As part of the experimental research, the researchers classified the influence of vibrations according to the type of excitation, the place of examination, the direction of action of the excitation, the applied method, the position of the body, etc.

1.1 Literature According to the Type of Vibration Excitation

Many studies on vibration-induced discomfort have used harmonic excitations, which represent the simplest form of vibration. The changing characteristics of harmonic vibrations change in time according to the harmonic law, sine/cosine. Numerous processes in nature take place in a harmonic way, so the research of this form of vibration is of great importance. Research on the behavior of the human body to a series of vertical sinusoidal vibrations was investigated in the paper [4]. The results were similar for sinusoidal vibrations and random broadband vibrations. The frequency response of the human body is nonlinear for vertical vibrations, while it has different directions for horizontal vibrations. Vibrations that occur on a person's body in passenger vehicles refer to the fact that the person is exposed to vibrations over some surface; in this case, it's a car seat. The authors [6] also studied how the effect of vibration changes as a function of its duration. Three experiments were conducted when subjects were exposed to vertical sinusoidal vibration of different durations. The findings revealed that subjective effects, expressed as log m/s^2, increase linearly with logarithmic exposure duration up to 4 s. After that time, the effect continues to increase, but much more slowly. In addition to the research dealing with harmonic excitations, the movements experienced in real situations are never sinusoidal and contain a wider range of frequencies. Therefore, it is important to know how the results obtained for single-frequency movements can be applied to different types of movements, i.e., closer to the stimuli that occur in the real transport environment.

For a stationary random vibration, the average sample over a sufficiently long period is independent of the time period over which the sample is taken. Some experimental studies with sinusoidal and stationary random vibration have investigated whether the response of the human body exposed to random vibration can be predicted based on the knowledge of the response to the harmonic sinusoidal motion. The knowledge of the outcome of such studies is necessary to apply the results of sinusoidal studies to random vibration environments. Vibrations occurring throughout the body are a risk factor for low back disorders [6]. The aim of this research was to measure the effect of vibrations on the stability of the spine in a sitting position. Twenty healthy subjects were exposed to random vibrations with a frequency range of 2–20 Hz and an acceleration value of 1.15 m/s^2 r.m.s. The results showed the stiffness of the human body and the delay of body reflexes on the seat without a backrest. The stiffness of the human body and the slow reflexes of the body are significantly reduced when testing vibrations on a seat with a backrest. Research in the paper [7] shows that the resonances of the human body exposed to random vibrations range from 4–6 Hz, including the lower part of the spine and pelvis, while for the upper part of the body with the spinal column, the resonances range from 10–14 Hz. Also, the paper [8] investigated the biodynamic responses of the human body exposed to multiaxial vibrations that were transmitted from the seat to the subject's head, arms, and back. The response characteristics exposed to longitudinal, vertical, and lateral vibrations reveal significant dynamic interactions between the upper body and the backrest, except for those parts of the body located on the seat. The results also showed that the energy transmitted to the body is influenced by the backrest, the magnitude of the excitation and the mass, while the effect of the seat height is almost negligible.

1.2 Literature According to the Place of Vibration Examination

Users of passenger vehicles are exposed to random vibrations in the fore-and-aft, lateral, and vertical directions in real conditions of exploitation. This is actually the main reason for conducting two types of tests according to the location of the vibration test- laboratory and road.

Laboratory tests of the oscillatory properties of the human body exposed to vibrations are extremely important during the design of motor vehicles. These studies of whole-body vibration discomfort established a relationship between the amplitude, duration, frequency content, and waveform of the signal. Higher amplitudes are more unpleasant than lower ones [9]. The paper [10] investigated the behavior of the human body in a sitting position based on the occurrence of random back-and-forth vibrations in the case of 30 male subjects. The excitation amplitudes were 1.75 m/s^2 and 2.25 m/s^2, while the seat was with a backrest and without a backrest. After testing, the results showed that the human body behaved like a non-linear system and its responses depended on the spatial position. The analysis of the results proved that there is an influence of the size of the frequency and the seat back on the obtained results. The paper [11] also studied the effect of prolonged sitting on the occurrence of lower back pain. Five subjects' head movement and discomfort were observed for five-seat backs, three of which were commercially available. The results showed that there was significantly less head movement and discomfort when using lumbar back support than with a simple flat backrest. It is very important to point out that the frequency weighting curves, which are the basis of all standards related to the vibrations of the human body, were developed on the basis of laboratory research. Road tests are carried out in order to check the comfort of passengers in several positions in the vehicle, taking into account the operation of the vehicle's suspension system, the speed of the vehicle, the characteristics of the road, etc. Research has shown that vibrations are transmitted to passengers at all points of contact between the passenger (driver) and the seat, and for this reason, vibrations are considered the main factor of discomfort in the vehicle. The contribution of the study [12] is the characterization of the role of vibrations in passenger sleepiness. The results of this study clearly show that exposure to low-frequency random vibration between 1 Hz and 15 Hz has a significant effect on subjective levels of sleepiness, and more importantly, on human psychomotor performance and attention.

1.3 Literature According to the Direction of Action of the Vibration Excitation

In order to fully understand the response of the human body to vibrations, experimental research studies vibrations according to the direction of action of the excitation. In real conditions, the body of the driver is exposed to the effects of vibrations in several directions. The driver's dominant load is in the vertical direction and was caused as a result of shocks from bumps, shocks from the operation of the power unit, etc. The suspension system of the vehicle should also be mentioned because it is one of the important components of the vehicle that directly affects safety, performance and vibration level. The main purpose of vehicle suspension vibration control is to reduce the acceleration of the supported mass to improve ride comfort and to maintain proper wheel kinematics to ensure road-holding ability [13–15].

In the process of accelerating or braking, the significant vibration load of the driver is in the fore-and-aft direction. Due to the action of the centrifugal force, loads in the lateral direction cannot be ignored. The complex effects of multicomponent vibrations are difficult to predict, so the effects of uniaxial vibrations are examined to identify the coupling. There is a limitation in laboratory capabilities to investigate the simultaneous effects of vibrations in all three directions. For this reason, the human body is exposed to vibrations both in the direction of one axis and in the direction of several axes at the same time. Exposure of the human body to the action of excitation in the direction of one axis is carried out in the horizontal and vertical directions. Uniaxial vibrations along the vertical and horizontal axis showed significant movements in the sagittal plane (vertically, fore-and-aft and laterally) of the upper body, suggesting strong coupling effects [16, 17]. Body resonances in horizontal back-and-forth vibrations were identified from measured biodynamic responses. In the research [18] and [19] the authors observed a primary resonance around 0.7 Hz, while the secondary one was around 2.8 Hz and 4.75 Hz. The primary resonance is attributed to the movement and oscillations of the upper body under the effect of horizontal excitation, while the secondary resonance originates from the horizontal movements of the musculoskeletal structure. Vertical vibrations, over a longer period of exposure, lead to pain in the lower back [20–22]. The cause of low back pain is not well understood, but it is clear that solving this problem requires a better understanding of the movement of the spine in vehicles under the influence of vertical vibrations. Analyses of the effects of vertical vibrations are usually performed in the frequency domain.

In the case of multiaxial vibrations, the authors [9] showed that the peak of the vertical apparent mass (AM) under the effect of biaxial vibrations along the y and z axis is lower for all frequency values below 6 Hz compared to vibrations in the z-direction. This effect was also observed for triaxial vibration (xyz) responses. The observed differences could be partially attributed to the larger effective amplitudes of biaxial and triaxial vibration used in these studies compared to uniaxial vibration [19]. A similar effect was also observed in the vibration responses in the x and z directions reported by Qui and Griffin [23], which showed a decreasing peak AM value and corresponding frequency as the excitation value in the x direction increased. AM responses to biaxial and triaxial vibrations were primarily studied in a seated position with no back support and arms on the lap.

1.4 Literature According to Human Body Position

The seat geometries, backrests, and seating positions considered in the reported studies vary widely and are not fully representative of the vehicle environment. Considering the influence of these parameters on biodynamic responses, the identification of appropriate and representative seat geometry and seating positions for certain classes of vehicles (commercial vehicles and cars) is of primary importance [24]. Body position can have a major impact on the level of vibration transmitted to the seated person and can determine the degree of adverse effects. The influence of the seat back on the reactions of the body in the sitting position exposed to vibrations in the vertical and horizontal directions showed a significant influence. A change in body position that changes contact with a vibrating surface, such as a backrest, also modifies the effects of vibration [14, 25–27]. The authors [28] showed that under random vertical vibrations, the frequency response functions of the STHT increased with the presence of a backrest (reclined at 6°) in the frequency range of 0.25–20 Hz. In the synthesis of works presented in the research of the group of authors [29], it was shown that under the influence of fore-and-aft vibrations, the presence of the seat back causes significantly higher values of STHT in the fore-and-aft direction, as well as in the vertical direction in the entire frequency range, while the

STHT function in the fore-and-aft direction was significantly lower without the backrest present. The effect of the seat back on the STHT values in the case of vertical vibrations was lower, but it caused more head movement.

2 Seat-to-Head Transmissibility Function

The human body exposed to whole-body vibrations is evaluated by biodynamic response functions. Mechanical Impedance (MI), apparent mass, and seat-to-head transmissibility function are three biodynamic response functions. In this chapter, the STHT function was determined experimentally, and for that reason, in the further part of the chapter, the emphasis was on the STHT function. The STHT function represents the ratio of head acceleration to seat acceleration in the frequency domain [30]:

$$H(f) = \frac{Gxy}{|Gxx|} \tag{1}$$

where Gxy is the cross-spectrum between input and output, Gxx is the autospectrum of the input signal.

The STHT function is a biodynamic function "through the body" that represents the level of transmitted incoming vibrations to parts of the human body, in this case to the head. Although the influence of various factors on MI and AM has been extensively studied, the complexity associated with measuring the response to head acceleration has resulted in relatively few studies investigating their influence on seat-to-head vibration transmission [31–35].

2.1 Influencing Factors on STHT Function

The human body is not a rigid body but a complex mechanical system that has elastic properties. The response of the human body to whole-body vibrations is not only related to the magnitude and duration but also depends on the frequency of the vibrations. In addition, the biodynamic response of the human body exposed to vibrations is influenced by a considerable number of factors.

Body mass influences the biodynamic responses of the body in a sitting position. Greater body mass causes a greater contact area and uniform contact force between the thighs and the seat, which could significantly alter the biodynamic responses "towards the body" [36]. The author [37] suggested that variations in body mass and age produce only small variations in the MI responses of subjects exposed to vertical vibrations. The vast majority, however, showed a large scatter in the strength of the apparent mass response, particularly in the low-frequency range, attributed to variations in body mass.

Several studies have investigated the influence of gender on the biodynamic responses of sedentary subjects. Some studies conclude that the influence of gender is mostly insignificant [17, 38, 39], while others suggest otherwise. A research study [40] found greater amplitudes of STHT functions in females compared to males at frequencies above 5 Hz, and the opposite trend was observed at frequencies up to 4 Hz. Next, the differences were found to be significant, and the amplitudes of the STHT functions in females were almost twice as large compared to males at some higher frequencies. A group of authors

[41] observed a significant effect of gender on AM responses and STHT functions of 31 male and 27 female subjects, which was significantly associated with various anthropometric parameters. Females showed a lower frequency of primary resonance compared to male subjects. A clear effect of gender was found when considering data for male and female subjects of comparable body masses. While the peak amplitudes for both sexes of similar body masses were comparable, female subjects showed a higher amplitude near the secondary frequency resonance, which is attributed to the relatively greater pelvic and visceral mass of women.

The influence of hand position on the body's biodynamic response was also a topic of research. Only a few studies have measured the responses of subjects in a seated position with their hands on the steering wheel [17, 42–44, 46, 48]. Some of them also investigated the relative effects of hand position. The steering wheel represents another source of vibrations on the subject's body. In addition, hands on the steering wheel reduce the proportion of body weight on the seat. Also, this hand position can cause body stiffness and change the orientation of the pelvis, which has been shown to affect biodynamic responses [49]. The authors [45] reported that a reclined seat in combination with hands on the steering wheel produced a more pronounced secondary peak around 10 Hz, while the effect above 12 Hz was not too significant. However, a relatively smaller effect of position on vibration transmission was observed by the authors [47].

The next influencing factor is the seat back. The optimal position and angle of inclination of the seat back are different according to the anthropometric characteristics and personal preferences of the driver. Many studies have aimed to determine the optimal seatback angle by analyzing joint angles [50] with some success, but individual differences between drivers make the task difficult. The group of authors [25] also conducted experimental research on 12 adult male subjects who were exposed to random whole-body vibrations in the frequency range of 0.5–15 Hz in the case of fore-and-aft and vertical vibrations. Three seating conditions were considered: without a backrest, with a vertical backrest, and with an inclined backrest. The results showed a great influence of the seat back on the amplitude of the STHT function for both directions of vibration action. The results also revealed a non-linear response of the body in the sitting position with respect to the magnitude of the stimulus, while the hand position was judged to be insignificant. The authors [47] reported that the backrest has a significant effect on the transmission of vertical vibrations through most of the vertebrae of the entire spinal column and the head, while the effect of the seat backrest on horizontal vibrations was measured in the lower parts of the body, namely the thoracic and lumbar regions, and was observed only in the lower frequency range.

The effect of excitation magnitude on the transmission of vibration from the seat to the head has been studied by many researchers trying to determine whether the human body behaves linearly or non-linearly. They also studied whether the vibration strength has an effect on the STHT function. Several studies have concluded that the dynamic response of the body in a seated position varies depending on the magnitude of the vibration stimulus [31, 32]. The effect of stimulus strength on the STHT function is small compared to the proportion of within-subject variation. Variations in seat back angle and head position cause greater effects on the STHT function than those caused by differences in vibration magnitude. Vibration excitation levels can cause significant changes in the frequency response function, but it is difficult to distinguish whether the change is caused by the nonlinearity of the human body or by the subject's spontaneous control of the body's dynamic response to reduce the discomfort that appears under high vibration levels. Research [35] and [34] has shown that the primary and secondary resonance frequencies decrease with increasing vibration strength. The STHT function responses show shifts in resonance toward lower frequencies with increasing vertical vibration.

3 Neural Networks

Observing humanity over a long period of time, it can be concluded that the idea of developing intelligent machines, which would independently perform certain types of work, has existed for many years. The term "artificial intelligence" itself refers to an inanimate system that exhibits the ability to navigate new situations. It is based on the behavior of human beings in order to fully replicate human behavior [51].

Today, intelligent systems that offer artificial intelligence capabilities often rely on machine learning. Machine learning describes the capacity of systems to learn from problem-specific training data to automate the process of analytical model building and solve associated tasks. Deep learning is a machine learning concept based on artificial neural networks. In this chapter, the machine learning algorithms used in the prediction of the STHT function will be presented. Neural networks can be described as a relatively new concept used in data analysis. Their wide application is reflected in social and technical sciences, economics, and many other fields. Research and development of artificial neural networks are based on existing knowledge about the way the human brain functions. Neural networks were created as a result of the fusion of several different directions of research: signal processing, neurobiology, and physics. They represent a typical example of an interdisciplinary field. There are two categories of neural networks: artificial and biological.

Artificial neural networks are similar to biological neural networks in terms of structure, function, and information processing, but they are artificial creations. The ability to learn from a limited set of examples is one of the most important properties of neural networks. Artificial neural networks are data-driven networks, so the quality of the model depends on the amount of data. Artificial neurons have a simple structure and have similar functions to biological neurons. The body of a neuron is called a node or unit. Figure 2 shows a model of an artificial neuron.

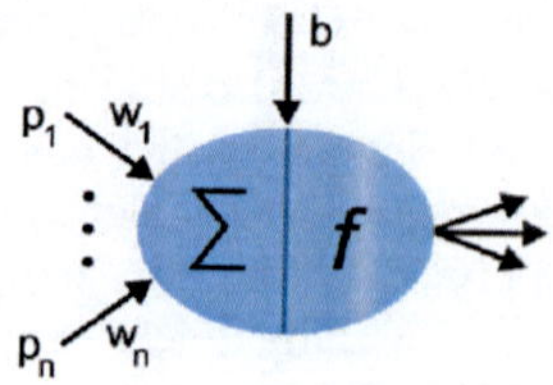

Fig. 2. A model of an artificial neuron

The output from the neuron is:

$$y = f\left(\sum_{i=1}^{n} p_i w_i + b\right) \tag{2}$$

where, *pi*, *p2*,…*pn* are input signals, *wi*, *w2*,…,*wn* are weighting coefficients (gains per synapse), *b* is the activation threshold, *f* is the activation function.

Artificial neural networks have been widely used in a wide range of applications [51–54] because they can model complex nonlinear systems by learning from input and system output signals. Some properties of ANNs are typical of the functioning of the human brain. ANN can learn from examples, and adaptability is one of the main properties [55].

In the paper [56], the authors showed that ANN has acceptable accuracy for biodynamic modeling. Also, the complexity of this model is low, and it is suitable for modeling and predicting acceleration and force in the time and frequency domains. Despite other biodynamic models such as Wan-Schimmel [57], Muksian and Nash [58], and Allen [59], this ANN model has a better accuracy of almost 100%. Research [60] and [61] used lumped parameter models, which have certain constraints, such as fixed body weight. The paper presents a new ANN model that can predict spinal acceleration excitation signals, human body mass, and height. The accuracy of the model is 96%, making it useful in real-time and offline analysis.

3.1 A Multi-layer Perceptron

A multilayer perceptron, in addition to the input and output, also has at least one hidden neuron layer. In practice, networks with a maximum of 3 hidden layers are used. Figure 3 shows the general model of a multilayer perceptron.

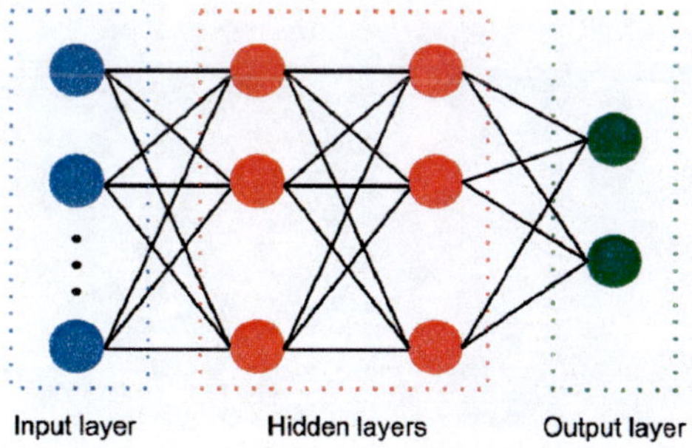

Fig. 3. A multi-layer perceptron.

Multilayer perceptrons are successfully applied to solve some difficult and diverse problems. It has three characteristics:

1. The model of each neuron in the output and hidden layers of the network includes an activation function.

2. The network contains one or more layers of hidden neurons that do not belong to either the input or output part of the network. These hidden neurons enable the network to learn complex tasks.
3. The network has a high degree of connectivity, as determined by the network's weight coefficients (synapses).

3.2 Recurrent Neural Networks

Recurrent Neural Networks (RNN) are neural networks in which the outputs of the neural elements of subsequent layers have synaptic connections with the neurons of previous layers. This leads to the possibility of taking into account the results of information transformation by the neural network in the previous stage for processing the input vector in the next stage of the network. Recurrent networks can be used to solve prediction and control problems. The three most famous architectures are the standard recurrent network, the recurrent unit with gates GRU, and the recurrent network with memory cells LSTM. LSTM is used in this chapter, and for that reason, more attention is paid to it.

3.3 LSTM

The LSTM network (Fig. 4) was introduced in 1997 by Hochreiter and Schmidhuber as a way to solve the problem of vanishing and exploding gradients. The model is similar to a regular recurrent neural network, but instead of a perceptron, it has a so-called LSTM cell in the recurrent neural network layer. An LSTM cell can be thought of as a repeating "sub-network". Each such subnet has its own system of control circuits, consisting of an input gate, a learning circuit, and an output circuit. Together, these circuits regulate the flow of information to and from the so-called internal state of the cell.

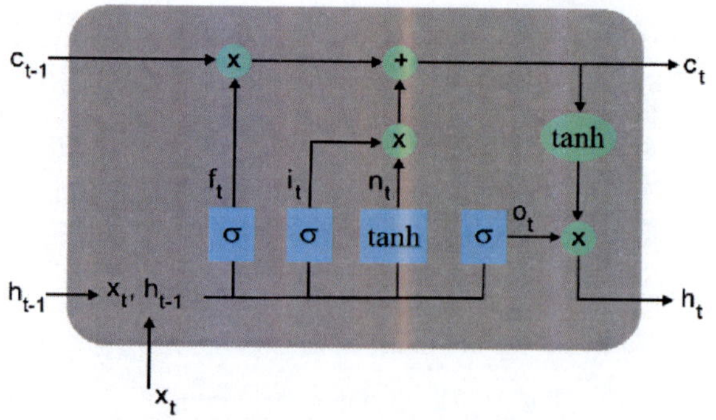

Fig. 4. Basic LSTM architecture.

In the process of calculating the next element, they pass two values instead of one, namely c_t and h_t. LSTM has three gates, namely *it* which scales the influence of the newly calculated cell $\tilde{C}_t, f_t$ which determines whether the previously calculated h_{t-1} will forget part of the information, and gate o_t, which determines which information will be passed to the next layer as variable h_t.

$$\begin{aligned}
f_t &= \sigma\left(W_f\left[h_{t-1}; x_t\right] + b_f\right) \\
i_t &= \sigma\left(W_i\left[h_{t-1}; x_t\right] + b_f\right) \\
o_t &= \sigma\left(W_o\left[h_{t-1}; x_t\right] + b_o\right) \\
\tilde{C}_t &= tanh\left(W_c\left[h_{t-1}; x_t\right] + b_c\right) \\
C_t &= i_t \cdot \tilde{C}_t + f_t \cdot C_{t-1} \\
h_t &= o_t \cdot tanh(C_t)
\end{aligned} \tag{3}$$

LSTM was developed much earlier than GRU, but both architectures are still commonly used today. GRU is one and therefore faster, but LSTM has more power for more complex problems.

4 Methods

The impact of vibrations on humans is being investigated both in operational and laboratory conditions. Greater attention is paid to the tests in laboratory conditions because, in this way, the stability of the micro-environment (noise, thermal loads) and the reproducibility of the results are ensured.

A hydrodynamic pulsator, HP-2007, and a 01dB-Metravib PRO-132 acquisition system were used for excitation. The hydrodynamic pulsator, HP-2007, is designed to provide excitation in the frequency range of 0.1–31 Hz and amplitude of 0–50 mm. The maximum possible load of the pulsator is 200 kg. With its help, it is possible to set one-component or two-component excitation, that is, types of excitation such as harmonic (sine, triangular or rectangular) excitation and stochastic excitation. Twenty male subjects participated in the experiment (Table 1).

Subjects were exposed to single-component random broadband excitation in the vertical (z) direction while sitting in an upright position with hands on thighs. They were exposed to whole-body vibrations for an excitation value of 0.45 m/s^2 r.m.s., in the frequency range of 0.1–20 Hz. Also, at the same time, the angle of inclination of the seat back was varied, with the values being 90° and 100° in relation to the seat part. Figure 5 describes the setup scheme of the measuring installation.

The duration of each experiment was 60 s, and the number of repetitions was 2. The parameters of the experiment were: sampling rate of 51,200 Hz; the duration of the block was 80 ms; the number of repetitions was 2; the number of shifts = 09 samples; 1 order of samples was 096, 1 period. Signal overlap is 75%. The frequency step $\Delta f =$ 0.390625 Hz, and the bandwidth is B = 39 Hz.

Table 1. Anthropometric characteristics of male subjects

Number of subjects	Gender	Years	Body height (cm)	Body height in sitting position (cm)	Weight (kg)	BMI
1	M	26	182	85	76	22.9
2	M	27	183	87	80	23.9
3	M	36	184	96	115	34
4	M	39	185	94	109	31.8
5	M	26	184	89	87	25.7
6	M	25	181	86	83	25.3
7	M	27	190	87	82	22.7
8	M	29	189	85	86	24.1
9	M	38	188	92	96	27.2
10	M	40	186	91	95	27.5
11	M	32	182	91	93	28.1
12	M	31	184	92	96	28.4
13	M	28	186	93	100	28.9
14	M	29	185	90	98	28.6
15	M	18	173	78	68	22.7
16	M	21	175	80	74	24.2
17	M	38	188	92	96	27.2
18	M	37	183	89	91	27.2
19	M	33	178	81	80	25.2
20	M	36	179	82	83	25.9
Mean		***30.8***	***183.25***	***88***	***89.4***	***26.57***
St dev.		***6.22***	***4.43***	***4.97***	***11.71***	***2.96***

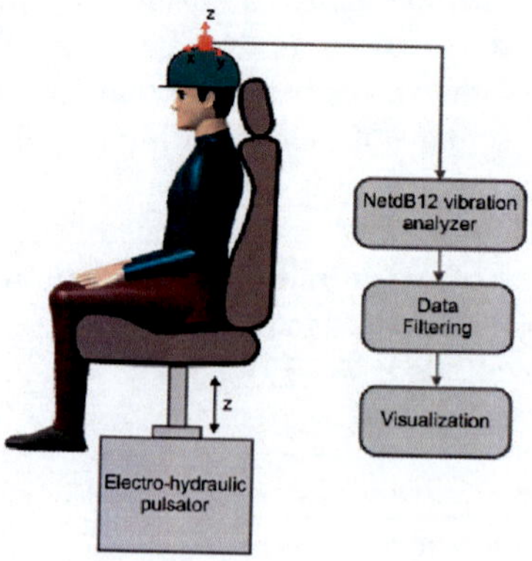

Fig. 5. Schematic view of the measuring installation.

5 Results

5.1 Experimental Measurements

The STHT function is observed in two directions-horizontal and vertical, which are significant in this experiment because they represent head movements in those directions. Figures 6 and 7 show the results of the STHT functions of all subjects in the direction of the horizontal and vertical axis for the case of an excitation of 0.45 m/s^2 and a sitting angle of 90°.

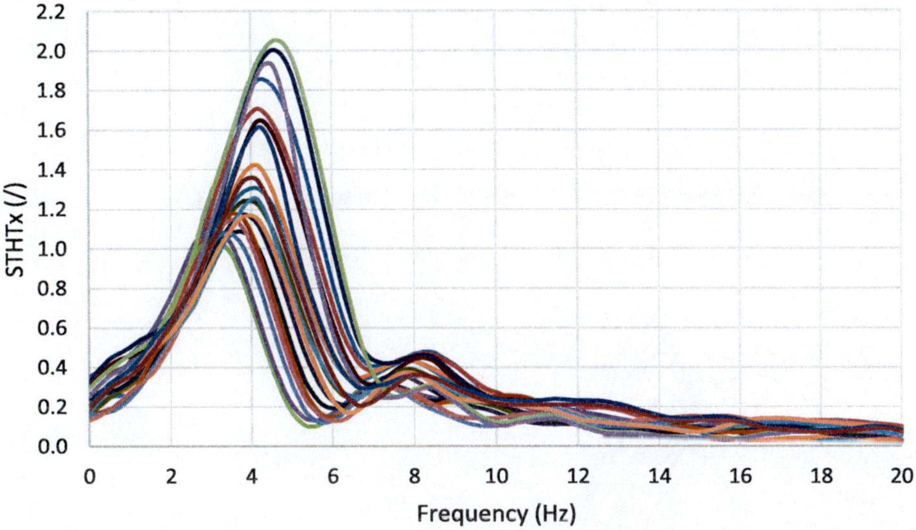

Fig. 6. STHT function for twenty male subjects in the horizontal direction (x) with a 90° sitting angle.

The results of the figure show that the resonant frequencies range from 3.71–4.81 Hz, with the lowest value of the STHT function being recorded at 1.07, and the highest at 2.04. A secondary slight peak can be observed in the range of 7.3–8.5 Hz.

The results of the figure in the case when the STHT function was observed in the vertical direction show that the resonant frequencies were in the range of 3.56–5.66 Hz, where the lowest value of the transfer function was 1.22, and the highest was 1.75. A secondary slight peak can be observed in the range of 7.3–8.5 Hz. In contrast to the previous figure, here it is characteristic that in addition to the first peak, there is also the appearance of a second significant peak that occurs between 9.61 Hz and 10.77 Hz. Then, we observed what happens when the seat angle is increased to a value of 100°.

The results of Fig. 8 show that now the resonant frequencies are in the range of 2.98–4.33 Hz, where the lowest value of the STHT function was recorded as 1.01, and the highest was 1.97, so it can be concluded that there is a decrease in the resonant values with the increase in the sitting angle when observing STHTx. It was also observed that for a seatback tilt angle of 100°, there was a slight increase in frequency response functions

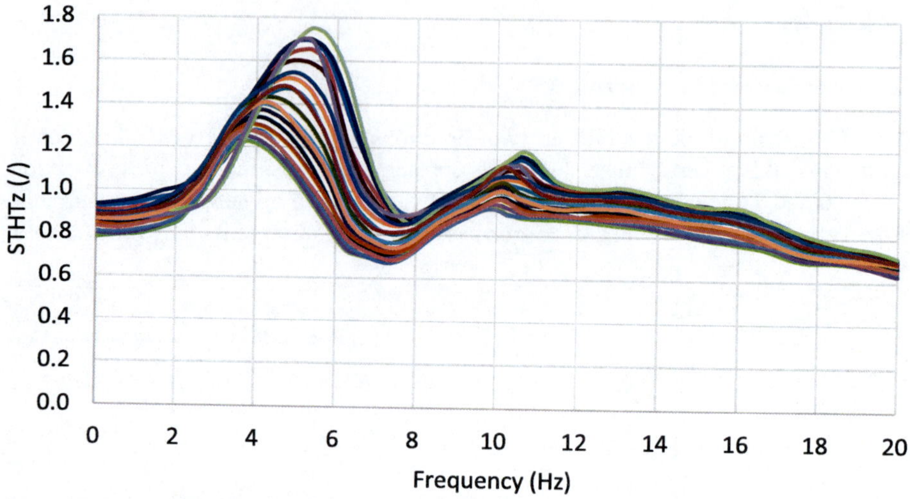

Fig. 7. STHT function for twenty male subjects in the vertical direction (z) with a 90° sitting angle.

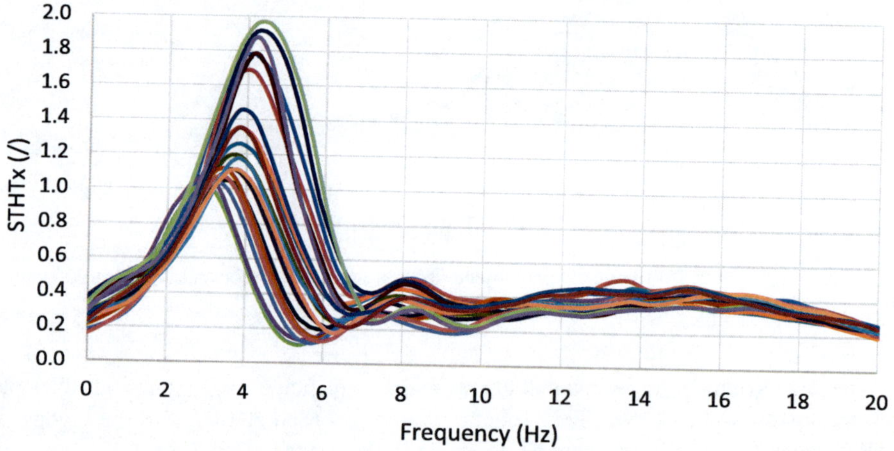

Fig. 8. STHT function for twenty male subjects in the horizontal direction (x) with a 100° sitting angle.

in the frequency range of 11–16 Hz, which was not observed in the mean STHT response values in the horizontal direction for a seatback tilt angle of 90°.

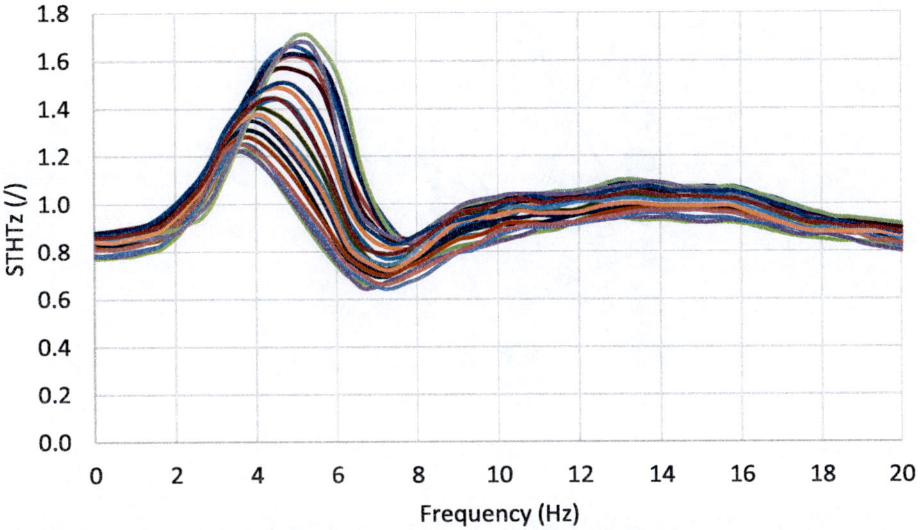

Fig. 9. STHT function for twenty male subjects in the vertical direction (z) with a 100° sitting angle.

Figure 9 shows that the resonant frequencies range from 3.43–5.27 Hz, with the lowest value of the STHT function of 1.19 and the highest of 1.71. Also, in the case of STHTz, an increase in the value of STHT functions was observed in the frequency range from 12.2 to 17.75 Hz, which was not the case with STHT functions when STHTz was observed for a sitting angle of 90°.

In general, it can be concluded that an increase in the angle of inclination of the seat back when the human body is exposed to vertical vibrations leads to lower resonance values and, therefore, to the value of the amplitude of the STHT frequency response function. These results were the input data of the ANN model, which had the task of predicting the response of the human body after the training phase.

5.2 Development of ANN Prediction Models

This part of the chapter deals with the application of three methods in order to determine which of them gives the best results in predicting the frequency response function of STHT based on measured experimental results. The time series forecasting method, which represents an important area of machine learning, was used in the framework of the ARIMA and Facebook Prophet models. The third model is based on recurrent neural networks using an LSTM cell. For the purposes of building the model and training the entire data set, the high-level programming language Python was used.

The data for training the neural network was obtained from experimental measurements of 20 subjects exposed to vertical excitation. Also, the parameter of the angle of inclination of the seat back is varied. Each subject was characterized by BMI, height, weight, sitting height, gender, and age. Figure 10 shows the scheme of the neural network used.

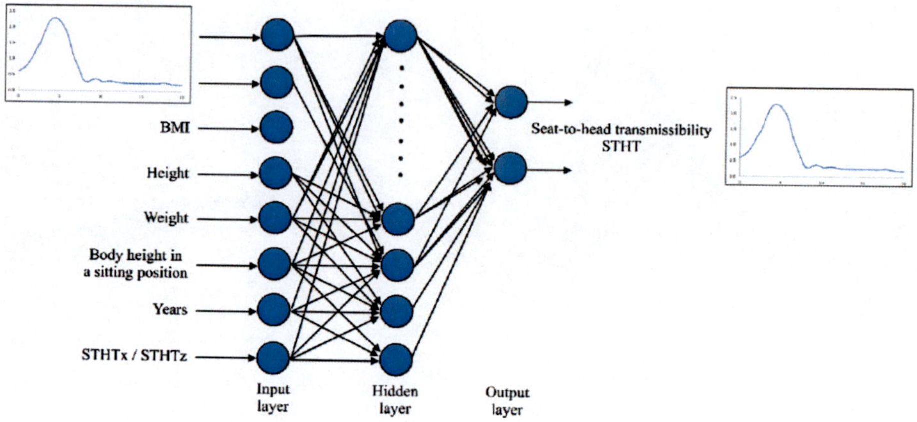

Fig. 10. Schematic of a recurrent neural network.

To test the robustness of the model, the dataset is divided into training, validation, and test sets, where eighteen randomly selected subjects are used for training, one for validation, and one for the testing phase.

The supervised learning problem is framed by predicting the STHT function at time (t) of a given STHT measurement and its data at the previous time step. After this transformation step, the eight input variables (input string) and one output variable (STHT value at the current moment), which was in the 8th position, are represented as:

$$var1(t-1), var1(t-1), \ldots, var8(t) \tag{4}$$

where *var1* is height (cm), *var2* is weight (kg), *var3* is BMI (kg/m^2), *var4* is sitting body height (cm), *var5* is age, *var6* is 90° inclination, *var7* is 100° inclination, and *var8* is output.

First, it started with the traditional approach - ARIMA [62]. The general model can be written as:

$$y'_t = c + \phi_1 y'_{t-1} + \ldots + \phi_p y'_{t-p} + \theta_1 \varepsilon_{t-1} + \ldots + \theta_q \varepsilon_{t-q} + \varepsilon_t \tag{5}$$

where y'_t is differentiated sequence of size y_t predicted at time t, φ is AR lag coefficient, θ is MA lag coefficient, *c* is mean value of changes between consecutive points. If *c* is positive, then the mean value of the changes leads to an increase in the value of y_t. Thus, y_t will tend to move upward. However, if *c* is a negative number, y_t will tend to move downward. The "predictors" on the right-hand side of the previous equation include both the lagged values of yt and the lagged errors. This is called the ARIMA (p, d, q) model, where the parameter *p* is the order of the autoregressive part, *d* is the degree of involvement of the first difference, and *q* is the order of the moving average part. The same stationarity and invertibility conditions used for the autoregressive and moving average models also apply to the ARIMA model. In our case, this model resulted in insufficiently good prediction results. The correlation factor R of the training phase had a value of 0.39, while the values of this factor for validation and test were 0.43 and 0.61, respectively. After that, two more advanced approaches were tested, Facebook Prophet [63] and recurrent neural networks (LSTM) [64]. Facebook Prophet usually produces very good predictions. It is highly customizable and accessible to data analysts without prior expertise in time series data. Essentially, Prophet is an additive regression model. This means that the model is a simple sum of several (optional) components. The Prophet forecasting model can be broken down into three main components: trend, seasonality and holidays. They are combined in the following equation:

$$y(t) = g(t) + s(t) + h(t) + \epsilon_t \tag{6}$$

where: $g(t)$ is a piecewise-linear or logistic growth curve for modeling non-periodic changes in time series, *s(t)* is periodic change, *h(t)* is effects of irregularly scheduled holidays and ε_t is an error that takes into account all unusual changes that the model does not accept. Using time as a regressor, Prophet attempts to fit several linear and nonlinear functions of time as components. Modeling seasonality as an additive component used the same approach applied by exponential smoothing in the Holt-Winters technique. In fact, the forecasting problem boils down to a curve-fitting exercise rather than explicitly looking at the time dependence of each observation. A trend is modeled by fitting a piecewise linear curve over the trend or non-periodic part of the time series. In our case, compared to the ARIMA model, Facebook Prophet gave slightly better results. The correlation factor R of the training phase had a value of 0.57, while the values of this factor for validation and test were 0.6 and 0.7, respectively. Finally, the LSTM model was used. An LSTM is defined with 30 neurons in the first hidden layer and one neuron in the output layer to predict the STHT for the selected subject. The model was trained with 50 training epochs while the data series size was 20. The coefficient in training, validation, and testing was over 98% (Table 2). This shows that the accuracy of the model was within an acceptable range.

Table 2. Correlation coefficient R in the phases of training, validation and testing

R correlation coefficient	Train	Validation	Test
ARIMA	0.39	0.43	0.61
Facebook Prophet	0.57	0.6	0.7
LSTM	0.98	0.99	0.98

Forecasts of time series on the test set of male subjects for the case of vertical excitation are shown in Figs. 11 and 12.

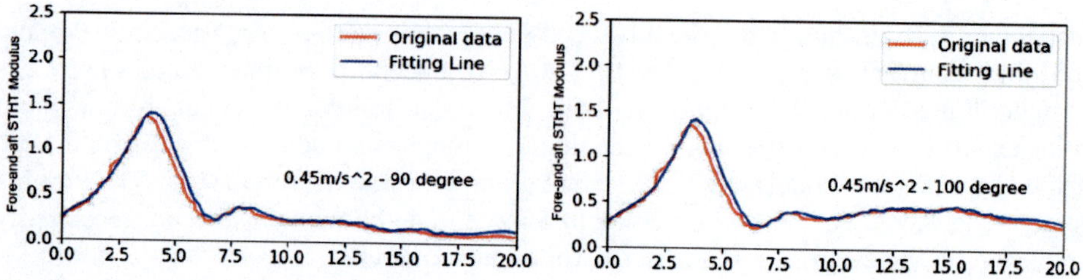

Fig. 11. Subject's original and predicted values - STHT function in the horizontal direction (frequency in Hz is shown on the x-axis).

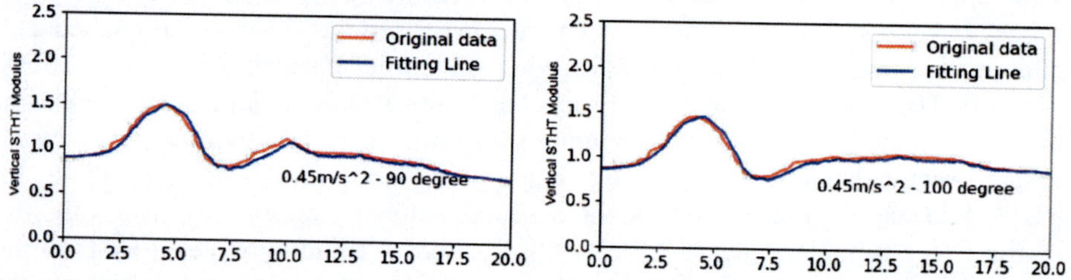

Fig. 12. Subject's original and predicted values - STHT function in the vertical direction (frequency in Hz is shown on the x-axis).

Loss curves during the training and validation phases are shown in Fig. 13. The root mean square error of the predicted STHT functions was 0.046.

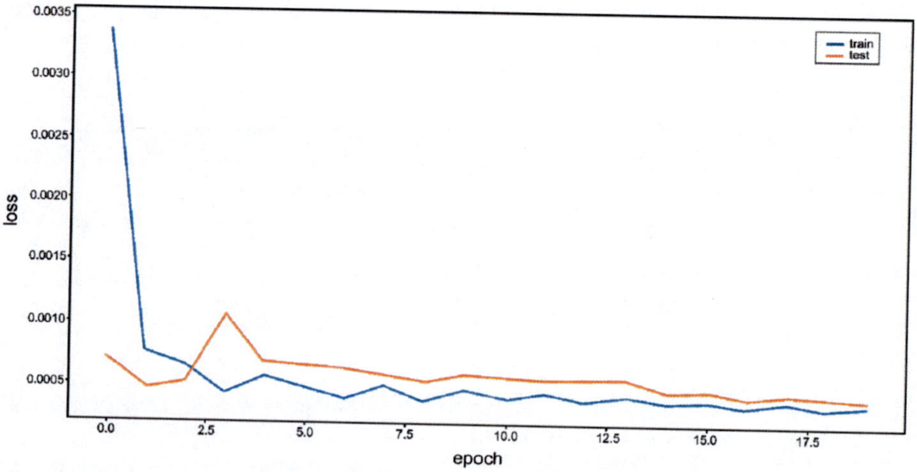

Fig. 13. Loss curves during training and validation phases for vertical excitation.

The developed model of an artificial neural network shows that it has adequate precision for biodynamic modeling. The main feature of the ANN model is the consideration of human height, weight, sitting height, BMI, and age during whole-body vertical vibration exposure. Despite the complexity of the achieved model, its predictive property makes it suitable for modeling and predicting the STHT function in the frequency domain. The ANN model shows an accuracy of 98%. Based on the presented results, it can be seen that machine learning applied to data that has the form of a time series, in this case recurrent neural networks, is an efficient and effective way to analyze and predict data.

6 Conclusions

Oscillatory comfort of a vehicle is a complex problem influenced by several factors: road characteristics, vehicle mechanical characteristics, and vehicle speed. The examination of the impact of vibrations on the human body is equally investigated both in operational and laboratory conditions. However, the laboratory conditions are of greater importance because they can provide conditions in which there is stability in the microenvironment and reproducibility of results.

The resonance frequencies of vertical vibrations were in the range of 3.71–4.81 Hz for the STHT function in the horizontal direction for a sitting angle of 90°, while for a sitting angle of 100° the range was 2.98–4.33 Hz. When observing the STHT function in the vertical direction, for a sitting angle of 90°, the resonant frequencies were in the range of 2.56–5.66 Hz, and for an angle of 100°, in the range of 3.43–5.27 Hz. The last part of the chapter shows the application of artificial intelligence methods and machine learning methods in order to predict the frequency response function based on measured experimental results. Three prediction methods were used. The first two methods were based on time series forecasting. ARIMA and Facebook Prophet are the first two examples. The third model was based on recurrent neural networks using an LSTM cell. Data for neural network training was obtained based on experimental data conducted on 20 subjects, for the case of one excitation direction, one vibration amplitude value, and two seatback tilt angles. Also, each subject was characterized by BMI, height, weight, body height in a sitting position, and age. A recurrent neural network was used after the insufficiently good accuracy of the first two approaches (ARIMA and Facebook Prophet). Its results showed great accuracy, which was reflected in high values of the R correlation factor: 0.98 (training phase), 0.99 (validation phase), and 0.98 (testing phase). The obtained results are shown for each combination of amplitude and angle of inclination of the seat back. The root mean square error of the predicted STHT responses of the LSTM model was 0.046.

The developed model of artificial neural network shows that it has adequate precision of biodynamic modeling. The main feature of the ANN model is to consider height, weight, sitting height, BMI, age and gender during the whole-body vibration exposure. Despite the complexity of the achieved model, the predictive property makes it suitable for modeling and predicting the STHT response in the frequency domain.

Acknowledgments. This research was supported by the Ministry of Education, Science and Technological Development of the Republic of Serbia through Grant TR35041.

References

1. ISO-2631-1. Evaluation of Human Exposure to Whole-Body Vibration. Part 1: General Requirements. International Organization for Standardization, Geneva (1997)
2. Chaffin, D.B., Andersson, G.: Chapter 9: guidelines for seated work. In: Occupational Biomechanics, pp. 289–323. John Wiley and Sons, New York (1984)
3. Griffin, M.J.: Vertical vibration of seated subjects: effects of posture, vibration level, and frequency. Aviat. Space Environ. Med. **46**(3), 269–276 (1975)
4. Fairley, T.E., Griffin, M.J.: The apparent mass of the seated human body: vertical vibration. J. Biomech. **22**(2), 81–94 (1989)
5. Kjellberg, B.O.: Wikström, Subjective reactions to whole-body vibration of short duration. J. Sound Vib. **99**, 415–424 (1985)
6. Gregory, P., Slota, M.S.: Changes in the Natural Frequency of the Trunk During Seated Whole-Body Vibration. Virginia Polytechnic Institute & State University, School of Biomedical Engineering & Sciences, The Kevin P. Granata Musculoskeletal Biomechanics Laboratory (2008)
7. .Brammer, J.: Human response to shock and vibration. In: Piersol, A.G., Harris, G.M. (eds.) Shock and Vibration Handbook, vol. 41, pp. 41–48. McGraw-Hill Professional, Cornwall (2010)
8. Mandapuram, S.C.: Biodynamic responses of the seated occupants to multi-axis whole-body vibration. Ph.D. thesis, The Department of Mechanical and Industrial Engineering, Montreal (2012)
9. Mansfield, N.J., Maeda, S.: The apparent mass of the seated human exposed to single axis and multi-axis whole-body vibration. J. Biomech. **40**, 2543–2551 (2007)
10. Demic, M., Lukic, J.: Investigation of the transmission of fore and aft vibration through the human body. Appl. Ergon. **40**, 622–629 (2009)
11. DeShaw, J., Rahmatalla, S.: Effect of lumbar support on human-head movement and discomfort in whole-body vibration. Occup. Ergon. **13**(1), 3–14 (2016)
12. Azizan, M., Fard, M., Azari, F., Jazar, R.: Effects of vibration on occupant driving performance under simulated driving conditions. Appl. Ergon. **60**, 348–355 (2017)
13. Popovic, V., Vasic, B., Petrovic, M., Mitic, S.: System approach to vehicle suspension system control in CAE environment. Stroj. Vestnik/J. Mech. Eng. **57**(2), 100–109 (2011)
14. Mačužić, S., Lukić, J., Ružić, D.: Three-dimensional simulation of the McPherson suspension system. Tehnički vjesnik **25**(5), 1286–1290 (2018)
15. Abdulrazzaq, A., Abdullah, A.A., Al-Rajihy, A.: Vibration control of automobile suspension system using smart damper, Int. J. Eng. Technol. **19**(1), 1–14 (2019)
16. Nawayseh, N., Griffin, M.J.: Tri-axial forces at the seat and backrest during whole-body vertical vibration. J. Sound Vib. **277**, 309–326 (2004)
17. Rakheja, S., Stiharu, I., Boileau, P.E.: Seated occupant apparent mass characteristics under automotive postures and vertical vibration. J. Sound Vib. **253**, 57–75 (2002)
18. Mandapuram, S., Rakheja, S., Ma, S., Demont, R.: Influence of back support conditions on the apparent mass of seated occupants under horizontal vibration. Ind. Health **43**, 421–435 (2005)
19. Mandapuram, S., Rakheja, S., Boileau, P.E., Maeda, S., Shibata, N.: Apparent mass and seat-to-head transmissibility responses of seated occupants under single and dual axis horizontal vibration. Ind. Health **48**, 698–714 (2010)
20. Bovenzi, M., Hulshof, C.: An updated review of epidemiologic studies on the relationship between exposure to whole-body vibration and low back pain. J. Sound Vib. **215**, 595–613 (1998)
21. Griffin, M.J.: Handbook of Human Vibration. Academic Press, London (1990)

22. Hulshof, C., Veldhuijzen van Zanten, B.: Whole-body vibration and low back pain- A review of epidemiologic studies. Int. Arch. Occup. Environ. Health **59**, 205–220 (1987)
23. Qiu, Y., Griffin, M.J.: Biodynamic responses of the seated human body to single-axis and dual-axis vibration. Ind. Health **48**, 615–627 (2010)
24. Rakheja, S., Dewangan, K.N., Dong, R.G., Pranesh, A.: Whole-body vibration biodynamics - a critical review: II. Biodynamic modelling. Int. J. Veh. Perform. **6**(1), 52–84 (2020)
25. Wang, W., Rakheja, S., Boileau, P.E.: The role of seat geometry and posture on the mechanical energy absorption characteristics of seated occupants under vertical vibration. Int. J. Ind. Ergon. **36**, 171–184 (2006)
26. Mandapuram, S., Rakheja, S., Marcotte, P., Boileau, P.E.: Analyses of biodynamic responses of seated occupants to uncorrelated fore-aft and vertical whole-body vibration. J. Sound Vib. **330**(16), 4064–4079 (2011)
27. Tsukahara, Y., Iwamoto, J., Iwashita, K., Shinjo, T., Azuma, K., Matsumoto, H.: What is the most effective posture to conduct vibration from the lower to the upper extremities during whole-body vibration exercise? Open Access J. Sports Med. **7**, 5–10 (2016)
28. Paddan, G.S., Griffin, M.J.: Transmission of yaw seat vibration to the head. J. Sound Vib. **229**(5), 1077–1095 (2000)
29. Rakheja, S., Dewangan, K.N., Dong, R.G., Marcotte, P.: Whole-body vibration biodynamics - a critical review: I. Experimental biodynamics. Int. J. Veh. Perf. **6**(1) (2020)
30. Kumbhar, P.B.: Simulation-based virtual driver fatigue prediction and determination of optimal vehicle seat dynamic parameters, Ph.D. thesis, Texas Tech University (2013)
31. Griffin, M.J., Lewis, C.H., Parsons, K.C., Whitham, E.M.: The biodynamic response of the human body and its application to standards. In: AGARD Conference Proceedings, vol. 253, Paris, France (1979)
32. Hinz, B., Seidel, H.: The nonlinearity of human body's dynamic response during sinusoidal whole-body vibration. Ind. Health **25**, 169–181 (1989)
33. Paddan, G.S., Griffin, M.J.: The transmission of translational seat vibration to the head-I. Vertical seat vibration. J. Biomech. **21**(3), 191–197 (1988)
34. Wang, W., Rakheja, S., Boileau, P.E.: Effect of back support condition on seat to head transmissibilities of seated occupants under vertical vibration. J. Low Freq. Noise Vib. Act. Control **4**, 239–259 (2006)
35. Hinz, B., Menzel, G., Blüthner, R., Seidel, H.: Seat-to-head transfer function of seated men - determination with single and three axis excitations at different magnitudes. Ind. Health **48**, 565–583 (2010)
36. Nawayseh, N., Griffin, M.: Tri-axial forces at the seat and backrest during whole-body fore-and-aft vibration. J. Sound Vib. **281**(3), 921–942 (2005)
37. Mertens, H.: Nonlinear behavior of sitting humans under increasing gravity. Aviat. Space Environ. Med. **49**(2), 287–298 (1978)
38. Griffin, M.J.: Human response to vibration. J. Sound Vib. **84**(4), 615–617 (1982)
39. Parsons, K.C., Griffin, M.J., Whitham, E.M.: Vibration and comfort III. Translational vibration of the feet and back. Ergonomics **25**(8), 705–719 (1982)
40. Griffin, M.J., Whitham, E.M.: Individual variability and its effect on subjective and biodynamic response to whole-body vibration. J. Sound Vib. **58**(2), 239–250 (1978)
41. Dewangan, K.N., Rakheja, S., Marcotte, P., Shahmir, A.: Comparisons of apparent mass responses of human subjects seated on rigid and elastic seats under vertical vibration. Ergonomics **56**(12), 1806–1822 (2013)
42. Hinz, B., Seidel, H., Menzel, G., Blüthner, R.: Effects related to random whole-body vibration and posture on a suspended seat with and without backrest. J. Sound Vib. **253**, 265–282 (2002)
43. Stein, G.J., Múčka, P., Chmúrny, R.: Preliminary results on an x-direction apparent mass of human body sitting in a cushioned suspended seat. J. Sound Vib. **298**, 688–703 (2006)

44. Stein, G.J., Múčka, P., Hinz, B., Blüthner, R.: Measurement and modelling of the y-direction apparent mass of sitting human body–cushioned seat system. J. Sound Vib. **322**(1–2), 454–474 (2009)
45. Wang, W., Rakheja, S., Boileau, P.E.: Effects of sitting postures on biodynamic response of seated occupants under vertical vibration. Int. J. Ind. Ergon. **34**(4), 289–306 (2004)
46. Patra, S.K., Rakheja, S., Nelisse, H., Boileau, P.E., Boutin, J.: Determination of reference values of apparent mass responses of seated occupants of different body masses under vertical vibration with and without a back support. Int. J. Ind. Ergon. **38**, 483–498 (2008)
47. Pranesh, A.M., Rakheja, S., Demont, R.: Influence of support conditions on vertical whole-body vibration of the seated human body. Ind. Health **48**, 682–697 (2010)
48. Toward, M.G.R., Griffin, M.J.: Apparent mass of the human body in the verticaldirection: effect of a footrest and a steering wheel. J. Sound Vib. **329**, 1586–1596 (2010)
49. Zimmermann, C.L., Cook, T.M.: Effects of vibration frequency and postural changes on human responses to seated whole-body vibration. Int. Arch. Occup. Environ. Health **69**, 165–179 (1997)
50. Porter, J.M., Gyi, D.E., Tait, H.A.: Interface pressure data and the prediction of driver discomfort in road trials. Appl. Ergon. **34**(3), 207–214 (2003)
51. Eletter, S.F., Yaseen, S.G., Elrefae, G.A.: Neuro-based artificial intelligence model for loan decisions. Am. J. Econ. Bus. Adm. **2**(1), 27–34 (2010)
52. Won, S.H., Song, L., Lee, S.Y., Park, C.H.: Identification of finite state automata with a class of recurrent neural networks. Neural Net. IEEE Trans **21**, 1408–1421 (2010)
53. Lukić, J., Mačužić Saveljić, S.: Ann driver model based on seat to head transmissibility. In: 10th International Automotive Technologies Congress, OTEKON 2020, Bursa, Türkiye, 6–7 September 2021
54. Macuzic Saveljic, S., Arsic, B., Saveljic, I., Lukic, J., Filipovic, N.: Artificial neural network for prediction of seat-to-head frequency response function during whole body vibrations in the fore-and-aft direction. Technical Gazette **29**(6) (2022)
55. Widrow, B., Greenblatt, A., Kim, Y., Park, D.: The No-Prop algorithm: a new learning algorithm for multilayer neural networks. Neural Netw. **37**, 182–188 (2013)
56. . Gohari, M., Rahman, A.R., Raja, P., Tahmasebi, M.: New biodynamical model of human body responses to vibration based on artificial neural network. In: 14th Asia Pacific Vibration Conference, Dynamics for Sustainable Engineering, Hong Kong Polytechnic University, Hong Kong SAR, China (2011)
57. Wan, Y., Schimmels, J.M.: A simple model that captures the essential dynamics of a seated human exposed to whole body vibration. Adv. Bioeng. **31**, 333–334 (1995)
58. Muksian, R., Nash, C.D.: A model for the response of seated humans to sinusoidal displacements of the seat. J. Biomech. **7**, 209–215 (1974)
59. Allen, G.: A critical look at biomechanical modeling in relation to specifications for human tolerance of vibration and shock. In: AGARD Conference Proceedings, vol. 253, pp. 6–10 (1978)
60. Gohari, M., Rahman, R.A., Raja, R.I., Tahmasebi, M.: A novel artificial neural network biodynamic model for prediction seated human body head acceleration in vertical direction. J. Low Freq. Noise Vib. Active Contr. **31**(3), 205–216 (2012)
61. Gohari, M., Rahman, R.A., Tahmasebi, M., Nejat, P.: Off-road vehicle seat suspension optimisation, Part I: derivation of an artificial neural network model to predict seated human spine acceleration in vertical vibration. J. Low Freq. Noise Vib. Act. Control **33**(4), 429–441 (2014)
62. Box, G.E., Jenkins, G.M., Reinsel, C., Ljung, G.M.: Time Series Analysis: Forecasting and Control. John Wiley & Sons (2015)
63. Taylor, S.J., Letham, B.: Forecasting at scale. Am. Stat. **72**(1), 37–45 (2018)
64. Mandic, D.P., Chambers, J.A.: Recurrent Neural Networks for Prediction: Learning Algorithms, Architectures and Stability, Wiley (2001)

A Review of the Application of Artificial Intelligence in Medicine: From Data to Personalised Models

Anđela Blagojević[1,2](✉) and Tijana Geroski[1,2]

[1] Faculty of Engineering, University of Kragujevac, Sestre Janjić 6, 34000 Kragujevac, Serbia
{andjela.blagojevic,tijanas}@kg.ac.rs

[2] Bioengineering Research and Development Center (BioIRC), Prvoslava Stojanovića 6, 34000 Kragujevac, Serbia

Abstract. Artificial intelligence leverages sophisticated computation and inference to generate insights, enables the system to reason and learn, and empowers decision making of clinicians. Starting from data (medical images, biomarkers, patients' data) and using powerful tools such as convolutional neural networks, classification, and regression models etc., it aims at creating personalized models, adapted to each patient, which can be applied in real clinical practice as a decision support system to doctors. This chapter discusses the use of AI in medicine, with an emphasis on the classification of patients with carotid artery disease, evaluation of patient conditions with familiar cardiomyopathy, and COVID-19 models (personalized and epidemiological). The chapter also discusses model integration into a cloud-based platform to deal with model testing without any special software needs. Although AI has great potential in the medical field, the sociological and ethical complexity of these applications necessitates additional analysis, evidence of their medical efficacy, economic worth, and the creation of multidisciplinary methods for their wider deployment in clinical practice.

Keywords: Image processing · Deep learning · Data mining · Medical expert systems

1 Introduction

Advances in the computational power paired with massive amounts of data generated in healthcare systems make many clinical problems suitable for artificial intelligence (AI) applications. Artificial intelligence has been successfully applied in the automation of the process of analysis of medical data, shortening the time for diagnosis, as well as ensuring high accuracy and repeatability of results. Algorithms can be applied to automatically diagnose diseases based on MRI/CT/X-ray images, predict patient survival rates more accurately, estimate treatment effects on patients using data from randomized trials and automate the task of labeling medical datasets using natural language processing. Algorithms in medicine have so far demonstrated several potential benefits to both physicians and patients.

N. Filipovic (Ed.): AAI 2022, LNNS 659, pp. 271–305, 2023.
https://doi.org/10.1007/978-3-031-29717-5_17

2 Application of Artificial Intelligence in Medicine

AI has found application in several fields of medicine with the aim to improve the processes of stratification and diagnosing of patients, discovery and development of drugs, as well as communication between doctors and patients, transcription of medical documents, such as prescriptions and remote treatment of patients. Therefore, this chapter deals with wide ranges of AI applications in different branches of medicine. Different aspects of AI tasks will be covered, such as segmentation, classification, feature extraction, disease progression as well as model integration on the cloud platform. The main focus are different branches of medicine and application of AI in cardiovascular field, as well as the response to modern problems such as COVID-19. These field have been particularly examined due to the importance of the investigated problems, as they pose an urging solution.

2.1 Stratification of Patients with Carotid Artery Disease

Arterial stenosis is one of the most common cardiovascular diseases and it occurs because of the plaque deposition within the coronary vessel. Therefore, if this disease is not discovered on time and adequately treated, it may have critical consequences, such as a stroke and even death. In order to avoid these scenarios, it is obvious that early detection is the most important task. Since the manual annotation of the atherosclerotic plaque is a time-consuming process, the automatization of segmentation process is necessary. Paper presented by Dašić et al. [1] represents a model that identifies and segments plaque components such as fibrous and calcified tissue and lipid core, by using Convolutional Neural Network (CNN).

2.1.1 Dataset

The dataset used in this research [1] was collected during the project “A multidisciplinary approach for the stratification of patients with carotid artery disease – TAXINOMISIS” [2]. The acquisition of a dataset that contains original and annotated US images was the first step in the development process of plaque segmentation method. The original dataset includes captured common carotid artery, carotid bifurcation and branches in transversal and longitudinal projections of 108 patients. Ultrasound examination was done in both B mode and Color doppler mode, so the dataset consists of both types of images. From this dataset, only ultrasound common carotid artery (CCA) images in transversal projections were used, because they gave the clearest view of the atherosclerotic plaque. Finally, this resulted in 67 images [1].

2.1.2 Methodology

Atherosclerotic plaque components segmentation is defined as multiclass segmentation problem (semantic segmentation), where U-net model is used to detect the following four classes: background (part of the image outside the carotid artery media), calcified, lipid and fibrous plaque components. U-net architecture is a commonly used method in the literature for the purposes of segmentation and it has been showed that U-net achieves great results in solving problems of segmentation on biomedical images [3]. Proposed U-net architecture is simpler than the original U-net model. Encoder path consists of five blocks, where each block contains two convolutional layers with kernel of size 3 × 3 pixels followed by 2 × 2 max pooling layer [1]. Encoder blocks use convolutional layers with 16, 32, 64, 128 and 256 filters respectively. Decoder path is a structure symmetric to the encoder path. In each decoder block, 2 × 2 up convolution and skip connection are followed by two more convolutional layers with 3 × 3 filters, and the last decoder block produces the segmentation mask with 1 × 1 convolution and sigmoid activation function. For each convolutional layer, activation function is rectified linear unit (ReLU). Padding was used for all convolutional layers so that the resulting segmentation map preserves the same height and width. Because of that, the resulting segmentation map has the same resolution as the input image. In this chapter, the particular importance is given to the prevention of overfitting. Datasets in medicine are very often small due to the high costs of screening, a lack of time of healthcare workers and in general, due to the overload of the whole healthcare system. For this reason, overfitting is a problem that often occurs during the development and testing of the ML models. To prevent this, the loss function has to be monitored, also, other methods for overfitting prevention such as dropout layers, data augmentation, regularization, early stopping and many other are often introduced. Also, every other convolutional layer was followed by a dropout layer as a way of overfitting prevention [1]. For the training phase, 100 epochs were used, with batch size 16. Different optimizers were tested, but the results were similar, so Adam was used in the final model. As a loss function, custom weighted loss function that combines categorical focal and dice losses was used. Weights were estimated according to the number of pixels for each class, resulting in classes lipid plaque and calcified plaque having larger weights values due to lower number of pixels belonging to these two classes.

2.1.3 Results

For the evaluation of the segmentation results, Jaccard similarity coefficient (JSC) is often used. JSC is computed as Intersection over Union between the annotated image (ground truth) and the segmentation mask predicted by U-net model. Mean JSC value, as well as JSC values for each class, are shown in Table 1.

Table 1. Class-wise and mean JSC scores.

Classes	JSC score values [%]
Background class	95.94
Fibrous plaque class	67.34
Lipid plaque class	25.17
Calcified plaque class	26.54
Mean	53.75

Looking at Table 1, it can be concluded that the proposed model segmented well the fibrous component. On the other hand, the lipid and calcified plaque components were segmented less accurately. This is related to the imbalanced classes. Additionally, looking at the Fig. 1, another problem can be spotted. Instead of a couple of large segments of different plaque components (as seen on annotated image on Fig. 1b), predicted segmented mask is filled with a lot of smaller segments (Fig. 1c). This is probably due to the small size of the patches. The occurrence of patch/block artifacts in the output image is a disadvantage of processing each patch independently. There are methods for decreasing the appearance of these artifacts, one of which is to utilize overlapping patches rather than non-overlapping patches during the image processing step [4]. Larger patches of size 32x32 pixels were tested as well, but the results did not improve.

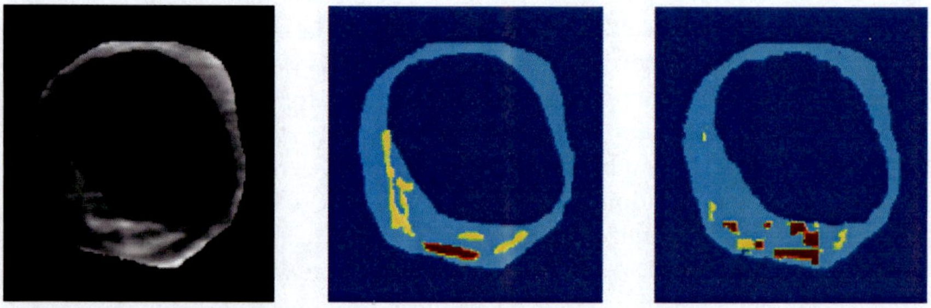

Fig. 1. Original ultrasound Media (a), annotated plaque components (b), U-net segmentation mask (c) [1].

Since there are no other papers that deal with the topic of semantic segmentation of plaque components in ultrasound imaging data, it is hard to compare the results. Also, it should be noted that there is a lot of previously discussed works that achieved better results but on different imaging data (e.g., MRI) with larger sets of images.

2.1.4 Discussion

There are several papers on the issue of plaque segmentation that address the problem in various ways using diverse imaging data. Clarke et al. [5] used the minimal distance classification approach to produce decent results in MRI images. Hofman et al. [6] examined different supervised learning algorithms on MRI images, however all the models had low accuracy for calcification components. Unfortunately, there hasn't been much study done on plaque segmentation on carotid artery ultrasound images due to their low picture quality with substantial noise, artifacts, and so on [7]. Therefore, its importance in assessing susceptible plaques is often underestimated, since most evaluations focus on the limits of intima-media thickness (IMT) [8–10]. Some study publications were effective in localizing plaque segments but were unable to characterize plaque composition [11, 12]. Lekadir et al. [13] introduced a convolutional neural network (CNN) model for automated plaque composition categorization that demonstrated high accuracy. The problem with their technique is that the CNN model is unable to function with a US picture of the entire carotid wall, but only with a patch image of each plaque section individually. This is cumbersome since it necessitates a significant amount of hand plaque segment extraction. There is also a lack of clear visual depiction of carotid wall plaque makeup because their model only does the classification and not the segmentation. Nonetheless, this publication and a slew of other studies have shown that convolutional neural networks are cutting-edge in image segmentation.

2.2 Assessment of Patient Condition with Familiar Cardiomyopathy

Cardiomyopathies are structural and functional abnormalities of the ventricular myocardium that are unexplained by flow limiting coronary artery disease or abnormal loading conditions [14]. Familial cardiomyopathies (FCM) are most diagnosed through in vivo imaging, with either echocardiography or, increasingly, cardiac magnetic resonance imaging (MRI). The treatment of symptoms of FCM by established therapies could only in part improve the outcome, but novel therapies need to be developed to affect the disease process and the time course more fundamentally. This is interesting area for AI application in terms of analysis of patient-specific data and development of patient-specific models for monitoring and assessing patient condition with familiar cardiomyopathy. During the project "In Silico trials for drug tracing the effects of sarcomeric protein mutations leading to familial cardiomyopathy – SILICOFCM" [15], several models were developed for monitoring and assessment of patient condition from the current through the progression of the disease [16, 17]. The first step in controlling and treating cardiovascular diseases is the accurate diagnosis using different diagnostic imaging systems (i.e., magnetic resonance imaging, echocardiography, angiography). Echocardiography has become one of the preferred medical imaging modalities that are used to depict and visualize heart left ventricle (LV). This is primarily due to the portability and low cost of ultrasound imaging devices [18]. Segmentation of the heart left ventricle (LV) is a very important step when setting up an adequate diagnostic using quantitative measurements such as end diastolic and end systolic volumes, left ventricular mass, etc. [19]. Therefore, SILICOFCM integrates the model for left ventricle segmentation in ultrasound images using the dataset from real patients with cardiomyopathy and an automatic extraction of clinically relevant parameters necessary for estimating the condition of the patient. Such methods could be helpful in the development of computer-based diagnostic tools that clinicians can use for decision making process and for setting up an accurate diagnosis [15].

2.2.1 Dataset

The dataset was collected during the project SILICOFCM [15] by the Institute of Cardiovascular Diseases, Vojvodina – Sremska Kamenica (12 patients) and Newcastle University and Newcastle upon Tyne Hospitals NHS Foundation Trust (6 patients). These recordings consisted of 153 2D image ultrasound apical view.

In order to increase the instances of the DICOM dataset, mirroring effect was applied in order to increase the number of training images. Other standard augmentation techniques were not investigated, as augmented images in such way would not have a physical meaning (upside down LV, zooming in would result in enlarged heart, which is already enlarged due to cardiomyopathy disease etc.). It should be emphasized that brightness and contrast had to be improved as part of the preprocessing to low signal-to-noise ratio in ultrasound images. Although novel data augmentation techniques were not investigated as part of this research, we aim to improve this step in the future by using i.e., Generative Adversarial Networks (GANs) for data augmentation. The total dataset was randomly split into training and validation set of 142 images and the remaining testing set of 11 images (without data augmentation). For both databases, the resolution was 1016 x 708 pixels. All the images were in DICOM format. The results of the automatic segmentation process were compared with the mean of the output segmented by the radiologists. Since the inter-observer variability has been statistically significant in manual segmentation, automatic segmentation emerges as the solution to reduce this variability.

As already mentioned, clinicians always inspect more than one view in diagnosing cardiomyopathy. Therefore, in automatic diagnosis of the patient status with cardiomyopathy, more than one view should also be analyzed. On one hand, apical view in order to estimate Left Ventricular Length, Diastolic, 2D - LVLd [cm] and Left Ventricular Length, Systolic, 2D – LVLs [cm] (in abbreviations lower letter *d* and *s* correspond to diastole and systole phases) [20]. On the other hand, M-mode were analyzed because it is crucial view in estimating Interventricular Septum Thickness, Diastolic, M-mode - **IVSd [cm]**, LV Internal Dimension, Diastolic, M-mode - **LVIDd [cm]**, Left Ventricular Posterior Wall Thickness, Diastolic, M-mode - **LVPWd [cm]**, Interventricular Septum Thickness, Systolic, M-mode - **IVSs [cm]**, LV Internal Dimension, Systolic, M-mode - **LVIDs [cm]** and Left Ventricular Posterior Wall Thickness, Systolic, M-mode - **LVPWs [cm]** [20]. All these parameters, once determined, are necessary to calculate further parameters [20]:

- Left Ventricle Volume, Diastolic, M-mode, Teicholz - **EDV** [ml]
- Left Ventricle Volume, Systolic, M-mode, Teicholz - **ESV** [ml]
- Ejection Fraction, M-mode, Teicholtz - **EF** [%]
- LV Fractional Shortening, M-mode - **FS** [%]
- LV Stroke Volume, M-mode, Teicholtz - **SV** [ml]
- LV Mass, Diastolic, M-mode - **LVd mass** [g]
- LV Mass, Systolic, M-mode - **LVs mass** [g]
- LV Mass Index, Diastolic, M-mode - **LVd Mass Index** [g/m^2]
- LV Mass Index, Systolic, M-mode - **LVs mass Index** [g/m^2]

Additionally, 53 patients were collected from the Clinical Centre of Kragujevac, Serbia, which were manually annotated by an expert thus the discussed parameters of interest were extracted for the purposes of setting the ground truth for comparison with the proposed developed workflow. As a result, 66 images with the 4-chamber view, 32 images with the 2-chamber view and 53 images with the M-mode view were available as training dataset for the automatic extraction of parameters. In addition, 14 patients (14 images) from ICVDV and 6 patients (6 images) from UNEW in M mode were collected for testing purposes (Table 2).

Table 2. Description of the utilized datasets.

Name of the submodule	Name of the partner	Number of patients	Number of images
Apical view	ICVDV	12	120
	UNEW	6	33
	CCKG	53	98
M-mode view	ICVDV	14	14
	UNEW	6	6
	CCKG	53	53

2.2.2 Methodology

The proposed methodology description is divided in two sections – section A: on the methods used to analyze apical view and section B: section on the methods used to analyze M-mode view. Full workflow is shown in Fig. 2. DICOM image format is used as the input to the system. The end user (expert) selects which view is best represented by the image and feed it to the algorithm.

Three alternatives are provided by the SILICOFCM tool: 4-chamber, 2-chamber, or M-mode [15]. The SILICOFCM Tool will further analyze the images depending on the view mode:

1. Apical 4-chamber view analysis includes segmentation of the LV using U-net previously trained and calculating the bordering rectangle as shown in Fig. 2 (left side), based on which parameters LVLd [cm] and LVLs [cm] A4C will be calculated. The user should define if the view represents the systolic or diastolic phase.
2. Apical 2-chamber view analysis includes segmentation of the LV using U-net previously trained, and calculating the bordering rectangle as shown in Fig. 2 (right side), based on which parameters LVLd [cm] and LVLs [cm] A2C will be calculated. The user should define if the view represents the systolic or diastolic phase.
3. M-mode view analysis includes bordering of the characteristic areas of LV – septum in diastolic phase, diameter in diastole, LV wall in diastole, septum in systole, diameter in systole and LV wall in systole (Fig. 2- middle). Based on these areas, parameters IVSd [cm], IVSs [cm], LVIDd [cm], LVIDs [cm], LVPWd [cm], LVPWs [cm] will be calculated. The user should define that the view is M-mode.

If the user has all three views in systolic and diastolic phase available (which should be the case when imaging the patient), then all relevant parameters are calculated from these three views and the automatic calculation of relevant cardiomyopathy diagnostic parameters can be further performed (i.e. – EF [%], ES [%], SV [ml], LVd mass [g], LVs mass [g], etc.).

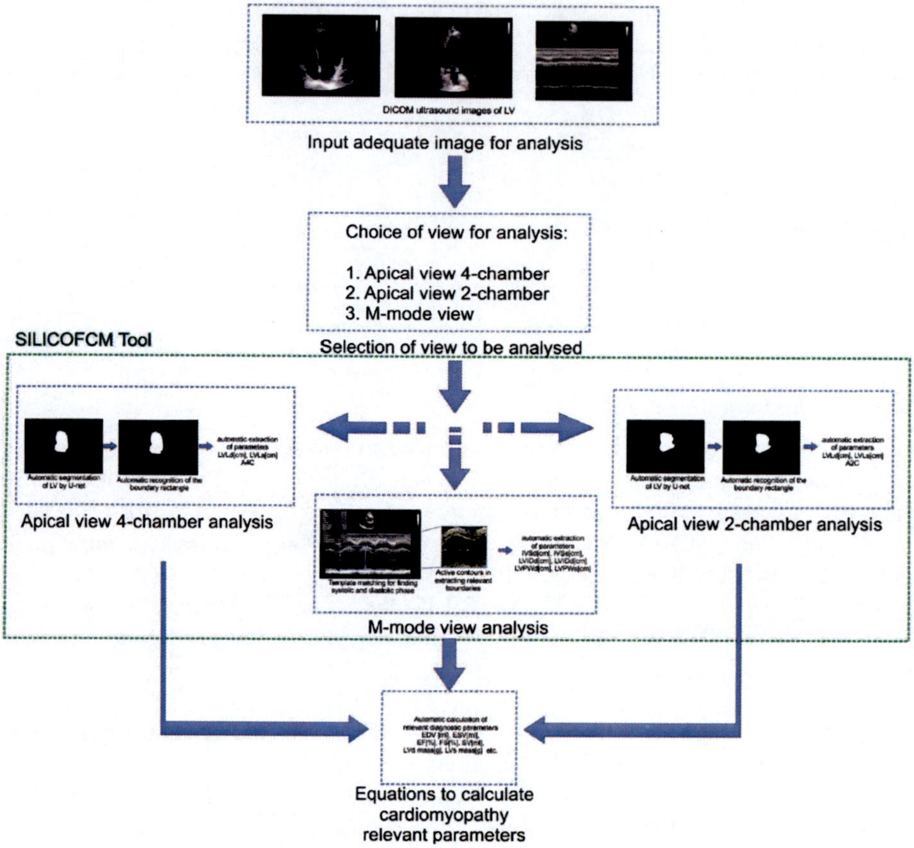

Fig. 2. Workflow of the proposed methodology for the automatic heart ultrasound segmentation and geometric parameter extraction.

2.2.3 Apical View

As mentioned, for the segmentation of the LV in the ultrasound images U-net neural network was used [17, 21]. The proposed architecture consists of two 3 × 3 convolutional layers and 2 × 2 max pooling in the contraction path. The expansion path consists of consecutive 2 × 2 up-conv and two 3 × 3 convolutional layers. After each up-conv, we have concatenation of feature maps. In the proposed network, all convolutions had the filter size of 3 × 3. The network requires a fixed size input of 128 × 128 pixels, therefore all input ultrasound images of the heart were resized to the required size. This is due to the fact that the search space had to be reduced, images were recalled from the size 708 × 708 pixels to size 128 × 128 pixels. Also, higher resolution (256 × 256) analysis has been applied, but the results did not show improvement with the downside being the higher processing demand.

Additionally, as the level of details in images is not important (only one region is the region of interest), reduction in image resolution is justified. Pixel intensities of the ultrasound images used as masks were rescaled to a range [0, 1]. The training process lasted for 10 epochs, stochastic gradient descent was used with the learning rate of 0.002 and regularization factor of 0.005. ReLU activation function was used. The data is fed to the network, which then propagates along the described paths (contraction, expansion and concatenation). The final result is a binary segmented image. In order to be able to generalize and evaluate the performance of the trained model, the dataset was divided into the training, validation and testing dataset. The training dataset included 120 images, the validation dataset included 22 images and testing was performed on unknown/blind 11 images. Augmented images of the original images were added in the training and validation phases. Subsequently, 10 epochs of training and validation are performed. Validation data was used to provide an unbiased evaluation of a model learned from the training dataset and for fine-tuning of the hyperparameters. It should be emphasized that validation and testing datasets are different, as testing dataset did not include any artificially augmented images, but only real unseen images.

Segmentation accuracy of the proposed automatic method was compared to the manual segmentation. In the validation phase, 22 of the 153 images were randomly selected, excluded from training and used for validation. In the testing phase, 11 images from the dataset were selected to calculate the evaluation metrics.

Dice similarity coefficient was utilized in order to calculate the overlapping regions between the automatic segmentation and the ground truth. Other evaluation metrics can be used, i.e., Hausdorff distance H, which is calculated in millimeters. Also, Jaccard coefficient (JC) is calculated similarly to Dice coefficient and is generally used to compare the similarity and diversity of two segmented areas. It is defined as the number of pixels of the intersected area, divided by the number of pixels that represent the union area.

After segmentation, the output image is forwarded to the system for drawing the bordering rectangle (Fig. 3). The output result corresponds to the longer side length, which has a meaning of LVLd [cm] and LVLs [cm] A4C. The same methodology is applied for the way 2-chamber view, except the final outputs are LVLd [cm] and LVLs [cm] A2C.

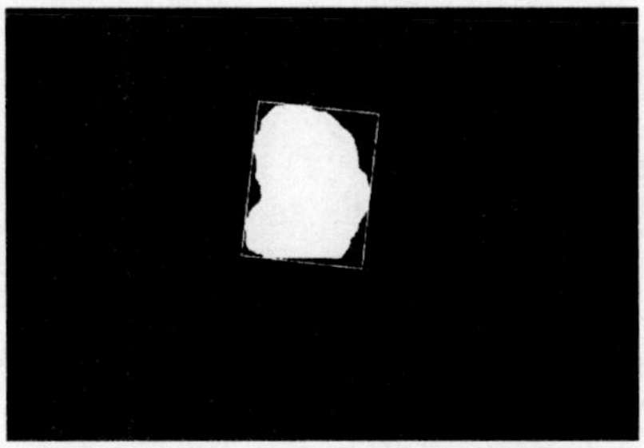

Fig. 3. LV segmentation and visualization of the LV height calculation – LVLd [cm] and LVLs [cm] A4C.

2.2.4 M-Mode

As far as the methodology for M-mode analysis is concerned, the first step is the template matching. Templates to be matched include all the relevant described borders (wall thickness, septum thickness and heart interior). One example of such template is given in Fig. 4. It can be seen that ultrasound image quality is very low (even from a small patch), as noise is very large, which shows the challenges of image processing and feature detection.

Fig. 4. Template for the dataset used.

It should be emphasized that for every new dataset, the template should be extracted manually just once for each dataset – that specific ultrasound machine. After this, no further manual action is required. Felzenszwalb and Huttenlocher's graph-based image segmentation algorithm is a tool widely used in computer vision, both because of the simple algorithm and the easy-to-use and well-programmed implementation provided by Felzenszwalb [22]. Recently, the algorithm has frequently been used as pre-processing tool to generate over-segmentations or so-called superpixels - groups of pixels perceptually belonging together. Therefore, we propose that the found area matching the template will be extracted on the analyzed image, after which Felsenszwalb's efficient graph-based image segmentation algorithm [22] will overtake to perform segmentation in order to distinguish between the heart wall, septum and heart interior. In this line of work, the algorithm is frequently used as baseline for state-of-the-art superpixel algorithms. Best parameters of this algorithm were found to be:

- scale = 250,
- sigma = 0.7,
- min_size = 2000.

Some additional erosion in 4 iterations and dilatation in 12 iterations are added, after which the binarization threshold is performed. Output values after binarization are IVSd [cm], IVSs [cm], LVIDd [cm], LVIDs [cm], LVPWd [cm], LVPWs [cm]. Due to great deal of noise present in the ultrasound images, the same described procedure is repeated for the upper half of the image and the lower half of image. This was found to reduce mean square error between the manually extracted values and those automatically determined by the algorithm.

In addition, in order to convert the extracted parameters from unit *pixels* to the unit *cm*, it was required to extract from DICOM info data the information about the conversion scale. In the available images, DICOM tags (0018,602C) Physical Delta X and (0018,602E) Physical Delta Y contained the adequate values. Other DICOM tags such as patient's height, weight etc. can be extracted as well to suit different purposes.

2.2.5 Results

The proposed methodology results have shown that U-net neural network can learn to recognize the heart's left ventricle from ultrasound images. In the presented scenario in Fig. 5, the network has performed very well, with some external additional areas, that were removed in a fine-tuning stage.

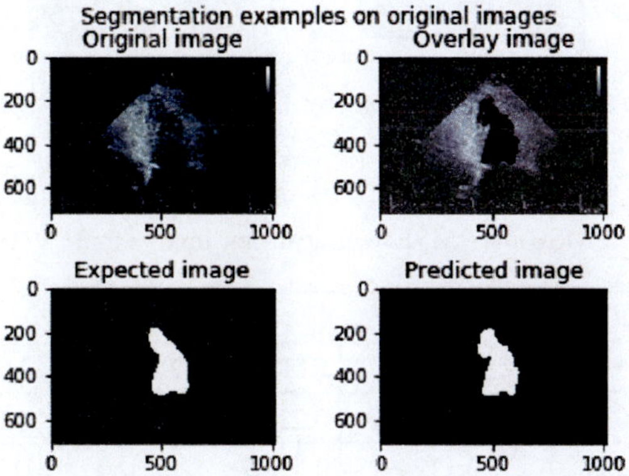

Fig. 5. Comparison of the LV segmentation by U-net and manual segmentation (a case scenario).

A kernel of 10x10 size was used for dilatation in the images, as the trend was that some external additional small surfaces were detected that were not so much connected to the main surface. This modification of the output image improved the accuracy by 1%.

In the worst-case scenarios, the marked area had smaller surface than the expected, leading to the area underestimation (Fig. 6).

The loss function of the U-Net had a falling trend as expected both in training and validation during the 5 epochs and the accuracy of the training and validation data was increasing up to 88.79% and 80.35% respectively.

The dice similarity coefficient was calculated to compare manual segmentation with the automatic segmentation performed by U-net. Test accuracy on 128 × 128 images was 83.49%, test accuracy on 1016 × 708 images without kernel was 82.39% and test accuracy on 1016 × 708 images with kernel was 83.40%. The time for preprocessing of results was 3.75 s per epoch for the training data to be finished, validation data took 0.43 s and test data took 0.58 s to be finished. The inference runtime of the network was 94.4 ± 4.2 s per epoch, but this time it may be reduced by optimizing the network architecture and computation graph, as well as using better computer or GPU configuration.

The results for automatic extraction of parameters on Apical view images in the form of mean absolute error (MAE), mean square error (MSE) and root mean square error (RMSE) are presented in Table 3.

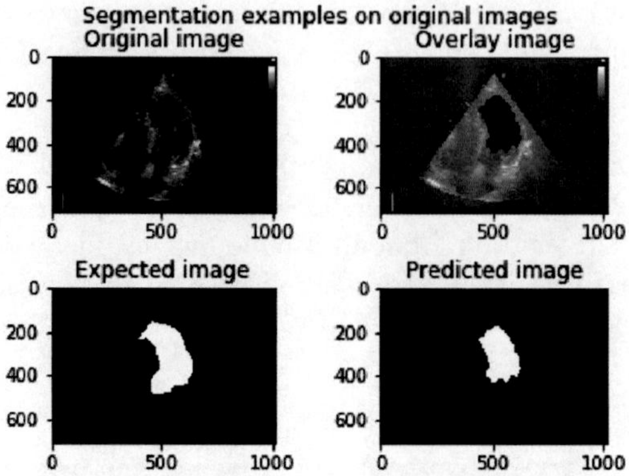

Fig. 6. Comparison of the LV segmentation by U-net and manual segmentation (worst case scenario).

Table 3. Values of MAE, MSE and RMSE for apical view images for ICVDV and UNEW datasets.

Parameter name	ICVDV			UNEW		
	MAE	MSE	RMSE	MAE	MSE	RMSE
LVLd[cm] A4C	0.1897	0.0459	0.2143	0.2442	0.0815	0.2855
LVLs[cm] A4C	0.2820	0.1124	0.3352	0.5180	0.2683	0.5180
LVLd[cm] A2C	0.0627	0.0088	0.0939	0.5314	0.7532	0.8679
LVLs[cm] A2C	0.2443	0.0823	0.2869	0.0340	0.0012	0.0340

The proposed method with U-net has been shown to segment LV successfully in a fully automatic manner and with robustness when it comes to different imaging conditions (imaging conditions were different in two hospitals from our dataset – contrast difference, signal to noise ratio, overlay with blood flow etc.). The method was also shown to be robust even if a small dataset is used for training (even a training set of 30 images it has been shown to produce competitive results). In addition to that, automatic extraction of LVLd[cm] and LVLs[cm] A4C has shown to perform well, with root mean square error of 0.3052 cm for all parameters, combined datasets. Although there could be some improvements, it can be concluded that the results are promising and can further be tested on other available datasets.

On the other hand, further automatic extraction of the parameters IVSd[cm], IVSs[cm], LVIDd[cm], LVIDs[cm], LVPWd[cm], LVPWs[cm] are reported in Table 4 in the form of mean absolute error (MAE), mean square error (MSE) and root mean square error (RMSE) for available dataset.

Although there could be some differences in the manual and automatic segmentation that could be classified as large, it should be emphasized that a small difference in the number of pixels would mean greater difference in the centimeters. It should be noted that not every image had the same conversion pixel-centimeters, as it depends on

Table 4. Values of MAE, MSE and RMSE for CCKG dataset.

Parameter name	CCKG		
	MAE	MSE	RMSE
IVSd[cm]	0.8921	1.3754	1.1728
IVSs[cm]	2.1119	5.6667	2.3805
LVIDd[cm]	1.0543	2.2327	1.4942
LVIDs[cm]	1.4824	3.1217	1.7668
LVPWd[cm]	0.4679	0.3660	0.6050
LVPWs[cm]	0.6062	0.5030	0.7093

the machine used, but also on the calibration of the same machine on different days. Therefore, extracting DICOM tag with this information for each image was necessary. Some of the conversions were for apical view 1 pixel = 0.02895143 cm, 1 pixel = 0.034221 cm, 1 pixel = 0.0362321 cm etc. and for the M mode view examples are 1 pixel = 0.08874659999999 cm, 1 pixel = 0.0593407 cm, 1 pixel = 0.03579973 cm. This means that even a 5 difference in pixels (which is expected due to unclear boundaries, fast and approximate measuring by the expert) can mean up to 0.5 cm difference, which for the wall and septum thickness is a lot, taking into consideration that thickness is usually around 2 cm. This part of the system will be further investigated to improve the results. Additionally, the models can be integrated as part of the platform workflow starting from images to further automatically generate a parametric model of the left ventricle from patient-specific ultrasound images [23].

2.2.6 Discussion

Some of the main problems in developing an algorithm for automatic LV segmentation are specific characteristics of ultrasound images such as low signal-to-noise ratio, weak echoes, more than one anatomical structure in the image, etc. [24]. As a result, many authors have attempted to solve the segmentation problem using a variety of approaches, including active shape, active contours, layout methods, and machine learning methods [25–28]. The literature shows that these approaches are not so sensitive to the initial conditions, and their main limitations are the image conditions. In contrast, deformable templates are robust to image conditions, however, they are very sensitive to initialization conditions [29]. In comparison to other methods applied for automatic extraction of LV, it can be seen that low-level methods have the assumption that myocardium is displayed brighter in the images, therefore the LV blood pool is represented with darker structures in images [30]. However, when this assumption is violated, LV will be incorrectly segmented. Unsupervised learning models such as deformable templates have also been introduced to solve some issues that are present when using active contours. However, deformable templates have the main limitation that they need to initialize the optimization function, meaning that the mostly manual step has to be introduced [31]. Since the limitations of these methods were noted and in order to create a method that will deal with different imaging conditions, as well as work on small datasets, we have moved away from traditional methods and turned to convolutional deep neural networks such as U-net. Because of the specific LV shape and size, which is characteristic for the cardiomyopathy, no papers in the literature dealing with implementation of deep neural networks could be compared with the previously presented results.

Deep neural networks are introduced to address some of the drawbacks of the previously listed standard approaches. Oktay et al. [32] discussed the use of neural networks in left ventricular segmentation in terms of 3D image segmentation. Oktay et al. overcame the issue of limited training data by regularizing training with an anatomical 3D model of the heart based on a large database of manually annotated cardiac magnetic resonance images. Carneiro et al. [31] successfully performed LV segmentation in echocardiographic images utilizing a 4-chamber view using a deep learning strategy that decouples rigid and nonrigid classifiers. Zyuzin and colleagues [33] used 2D 4-chamber view echocardiography images, and implemented U-net to segment the LV. Some writers, such as Smistad et al. [28], have also suggested using a pre-trained U-net. They propose using a pre-trained U-net and the Kalman filter, then comparing the results. The findings indicated that the Dice coefficients of the Kalman filter and the suggested U-net approach were similar, while the Hausdorff distance of the proposed U-net method was significantly superior. However, no application of U-net in LV segmentation on images of patients with cardiomyopathy has been found in the literature, so it cannot be stated that previously proposed methods in the literature will successfully segment LV in images, owing to the asymmetrical pattern of LV hypertrophy present in patients with cardiomyopathy.

With the exception of two studies, no study has dealt with the automated identification of LV boundaries in M-mode images in terms of automatic segmentation and extraction of important information. Unser et al. propose a method for automatic extraction of myocardial borders in M-mode echocardiograms that involves multistep processing algorithms, including noise reduction, border enhancement using appropriate filters, and final border extraction by searching for optimal paths along the time axis [34]. They do, however, evaluate typical M-mode echocardiograms from healthy people and do not publish the number of patients or statistical parameters such as accuracy, false positive rate, true negative rate, etc. They do, however, evaluate typical M-mode echocardiograms from healthy people and do not publish the number of patients or statistical parameters such as accuracy, false positive rate, true negative rate, and so on. The second study, published more recently, promises automated contour identification in M-mode pictures [35]. Their model begins with a manual candidate contour and then moves each candidate contour towards the required borders, functioning as active contours. The active contours method is known to be less successful in the face of increasing levels of noise, and it is unknown how this methodology would behave if the images were tested on patients with higher levels of noise.

With all of this said, it is evident that a fully automatic LV segmentation and automatic extraction of the characteristics of interest is required. This is especially the case because there is no research that deals with both Apical and M-mode view in medical image processing. A fully automatic LV segmentation system based on two views has the potential to streamline the clinical work-flow and reduce the inter-user variability.

2.3 Personalized COVID-19 Model

Artificial intelligence algorithms have proven to be an adequate method for many purposes, including the analysis of COVID-19 disease caused by SARS-Cov-2 virus [36, 37]. This disease was investigated during the project "Use of Regressive Artificial Intelligence (AI) and Machine Learning (ML) Methods in Modelling of COVID-19 Spread – COVIDAI" [38] and some new knowledge about this disease is extracted. Nevertheless, researchers all around world are still making the effort to find the adequate solution that will eradicate the virus [39]. Knowledge extracted from personalized AI model is published in the work of Blagojević et al. [40] and could help doctors to determine whether the patients will develop a critical condition, or they can be treated at home with a mild condition. This type of research could also be helpful to hospital managers in their decision-making process in order to avoid inappropriate allocation of intensive care beds.

2.3.1 Dataset

The dataset is obtained during the project "Use of Regressive Artificial Intelligence (AI) and Machine Learning (ML) Methods in Modelling of COVID-19 Spread" - COVIDAI [38]. During this project, data were collected in two hospitals – the Clinical Center of Kragujevac, Serbia and the Clinical Center of Rijeka, Croatia. Finally, the dataset consisted of blood test analyses from 105 patients (44 female and 61 male patients), with age distribution given in the form mean ± standard deviation – 52.77 ± 16.63. Also, the dataset was divided in three groups: demographic data (gender and age), symptoms (fever, cough, fatigue, chest pain, muscle pain, headache, dyspnea, loss of taste or smell) and blood analysis (complete blood count, coagulation, kidney function, hepatic function, enzymes, electrolytes, oxygenation and acid-base balance, inflammation indices, carbohydrate metabolism). Finally, based on values of blood biomarkers, patients are divided in four classes: mild, moderate, severe and critical clinical condition. The distribution is given in the following: 31.80% of total patients belongs to class of mild clinical condition, 50.90% to moderate, 13.88% to severe and 3.42% belongs to class of critical clinical condition.

2.3.2 Methodology

The idea of the personalized model is to predict the blood analysis in advance and based on that prediction, determine the severity of the clinical condition. For the prediction of each blood biomarker, several machine learning methods (KNN, SVM, Decision tree, Extra tree and Gradient Boost regressor) were used and their performances have been compared in order to select the final regression model. After evaluation of the proposed methods performances, as a final decision for the prediction of blood test analysis, Gradient boosting regressor (GBR) was chosen.

As previously mentioned, the main task of this model is the assessment of the clinical condition of patients with the aim to assess how the COVID-19 develops over time. This was achieved by firstly assessing the values of biomarkers using the methods explained in the previous section, after which the patients were classified into one of 4 classes (mild, moderate, severe and critical). For the described classification task, rule-based decision model of extreme gradient boosting (XGBoost) was constructed. XGBoost as well as GBR, was trained with optimal hyperparameters obtained by the grid search method.

2.3.3 Results

Ten blood biomarkers (LDH, urea, creatinine, CRP, WBC, albumins, percentage of lymphocytes, HGB, RDW, MCHC) were selected as the best features that had the greatest contribution to the classification task and most reliably described the development of COVID-19 disease in patients. Selection of the best features was implemented by method based on the K highest scores.

Due to the small number of patients' data available in time, 34 patients were selected as a test set. After prediction of the blood analyses values, it is possible to predict the patients' clinical condition in advance by the XGBoost classification algorithm. The classification model was tested on 34 patients and achieved an accuracy of 94% in predicting the patients' condition on 14th day. In addition to accuracy, precision, specificity, sensitivity, F1-score, area under curve (AUC) and precision recall (PR) values were computed. The average value of precision is 0.95, the average value of the specificity is 0.98, the average F1-score is 0.96, the value of AUC is 0.99 and the average PR score is 0.98. Added value of this study is also connected to the interpretability, as the best results in this study are achieved by XGBoost, which is an algorithm based on decision trees. An example of one tree which is a part of XGBoost model is presented in Fig. 7. This is especially important for data sets that are unbalanced or biased [41]. Other prediction models such as ANNs do not provide clinically useful interpretable rules that could explain the reasoning process behind their predictions. They just produce the different scores (i.e., accuracy, precision, recall, etc.) which represent the probability that a patient would get infected by the SARS-COv-2 virus.

2.3.4 Discussion

Although there are many research papers related to AI-based prediction of COVID-19 disease, most of these works perform binary classification in terms of severity of disease or positive-negative samples [42–44]. The main advantage of the proposed methodology is the usage of multiclass classification, as well as disease tracking in terms of worsening or bettering of clinical condition. Added value of this study is also connected to the model interpretability, since the best results are achieved by algorithm based on decision trees. Prediction models such as neural networks do not provide interpretable rules that could explain the reasoning process behind their predictions. It is only possible to track the models' performances and produce the final score, which represent the possibility that a patient would get infected. In addition to achieving the highest accuracy of the model, a good prediction model should clearly show the decision processes to clinicians.

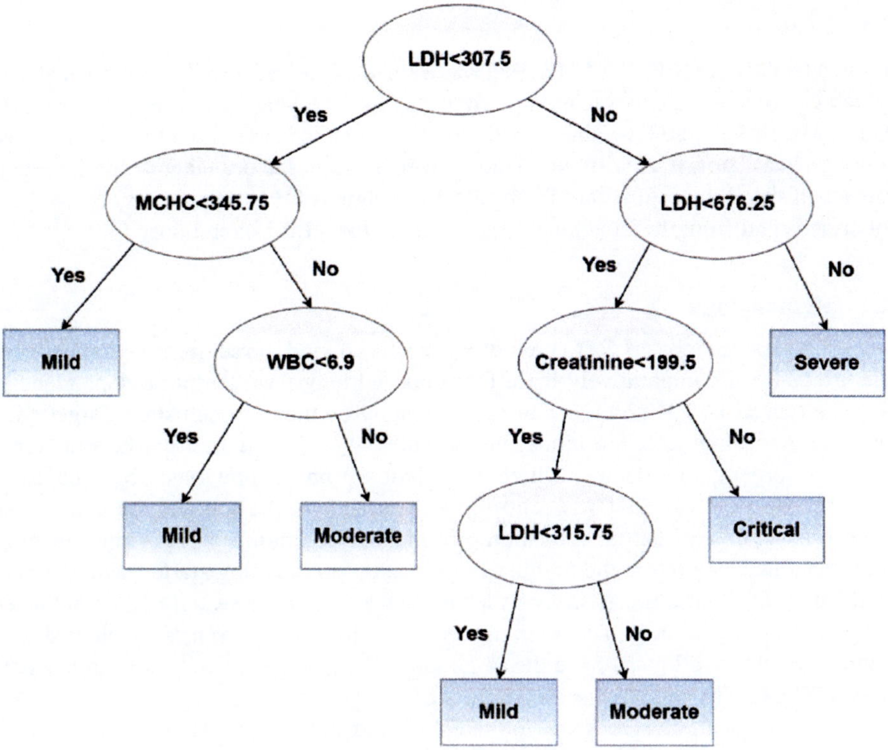

Fig. 7. An example of tree which is a part of XGBoost classification model [40].

2.4 Epidemiological COVID-19 Model

Artificial intelligence also plays a significant role in monitoring several infectious diseases, in the meantime, it has been proven to be adequate methodology for the development of an epidemiological model in order to track the number of people infected with COVID-19. Deep learning methods such as recurrent neural networks (RNNs) and long short-term memory networks (LSTMs) are well known as a method suitable for modeling temporal sequences [45]. Considering this fact, LSTM networks were the preferable method for analyzing epidemiological situation and monitoring COVID-19 during the project COVIDAI [38]. Prognostic epidemiological model developed during this project is published in the work of Šušteršič et al. [21] and it can be helpful in terms of predicting epidemic peaks.

2.4.1 Dataset

The encoder decoder (ED)-LSTM model was trained and tested on official statistical data of Benelux countries. Data was monitored during the one-year period from March 15th, 2020 to March 15th, 2021. Values for the real numbers of cases infected, hospitalized but not on ventilator, ICU with ventilator as well as deceased are taken from the report statistics of the Belgian institute for health for Belgium [46], statistic reports [47] for Netherlands and from the Luxembourgish data platform for Luxembourg [48].

2.4.2 Methodology

Long short-term memory (LSTM) represents a special kind of recurrent neural network structure that can comparatively learn the proposed long-term temporal dependencies [49]. The typical LSTM block is configured mainly by memory cell state, forget gate, input gate, and output gate. The input gate I_t decides which information can be transferred to the cell, forget gate f_t decides which information from the previously cell should be neglected. The control gate $\overline{C}_t$ is controlling the update of the cell and the output gate O_t controls the flow of output activation information. The number of the input features is presented as x_t and H_t is the number of hidden units. Learning started with the zero initial values of C_0 and H_0. Also, parameters such as bias b and weight W are adjusted during the learning process. The internal memory of the unit is given as C_t and it should be emphasized that all gates have the same dimension as the size of your hidden state [21, 50]. The LSTM cell is presented in Fig. 8.

As the purpose of this work was to predict second and third peaks based on the position of previously peaks and monitor the course of pandemic, encoder-decoder LSTM (ED-LSTM) was taken into consideration in order to choose an adequate model for this task [21]. The encoder-decoder LSTM network was developed as a sequence-to-sequence neural network to effectively map a fixed-length input to a fixed-length output. The advantage of these neural networks is that the mentioned two lengths of inputs and outputs do not have to be the same. The ED-LSTM has two implementation pathways: the first pathway – encoding and the second pathway – decoding. The purpose of the encoding is to encode an input sequence into a fixed-length vector representation and prepare the initial states for the decoder path (defined as encoder state in Fig. 8). On the other hand, the purpose of the decoding phase is to decode the vector representation and define a distribution of the output sequence. The architecture of the defined neural network is given in Fig. 8.

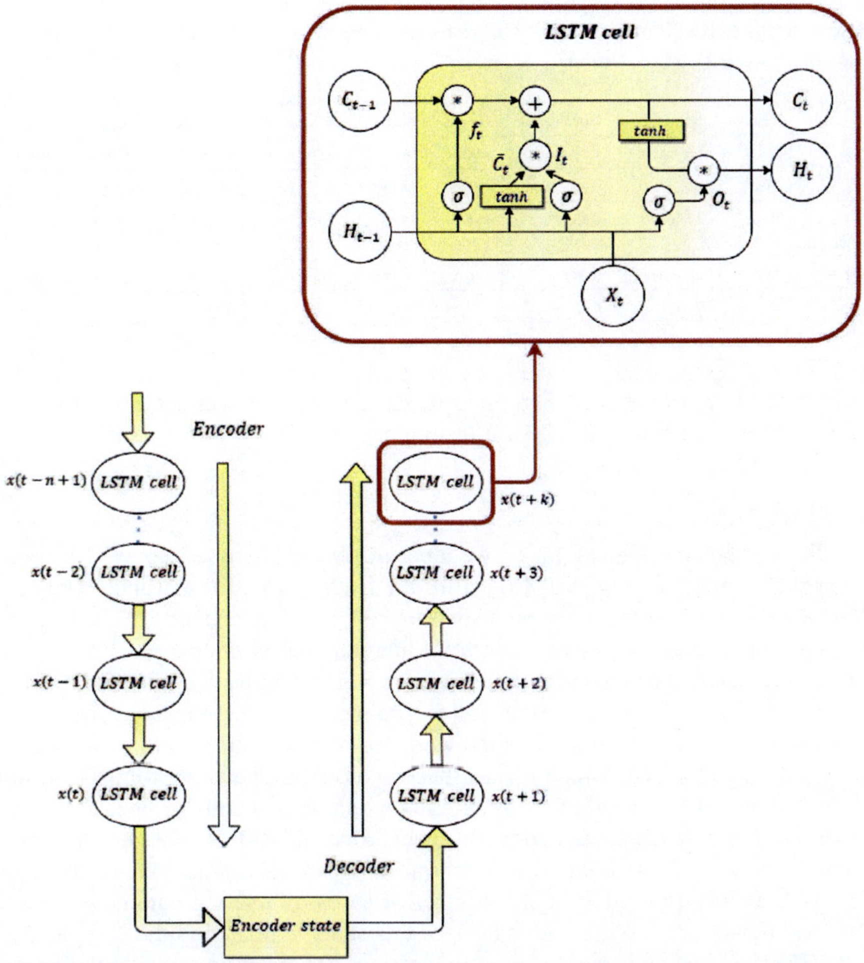

Fig. 8. Architecture of ED-LSTM, including the graphical representation of LSTM cell.

2.4.3 Results

In this section, the results of the applied LSTM-ED model were presented [21]. As evaluation metrics, RMSE, MAE and R^2 score were used. Average values of the mentioned metrics for all five iterations are given in Table 5.

The moving average smoothing technique was applied in order to remove the variation between time steps in real values of the test dataset. Actually, a new series, where the values were comprised of the average of five days of observations from the real data, were created.

During the validation process, it is concluded that LSTM-ED is capable of predicting the number of deceased cases in a period of 100 days [21]. In this one-year period, two peaks have appeared, so data from October 20th, 2020, up to March 15th, 2021

Table 5. Values of RMSE, MAE and R^2 score for the infected, severe, critical and deceased cases for the countries of Benelux Union.

Country	Belgium	Netherlands	Luxembourg	Country	Belgium	Netherlands	Luxembourg	Country	Belgium
Metrics	RMSE	MAE	R^2 score	RMSE	MAE	R^2 score	RMSE	MAE	R^2 score
Infected	535.93	440.47	0.76	434.28	362.96	0.82	25	20.59	0.76
Severe (hospitalized)	20.42	17.16	0.83	94.6	79.43	0.78	12.24	10.45	0.83
Critical (ICU)	38.97	28.15	0.6	37.61	30.06	0.65	3.17	2.63	0.66
Deceased	8.72	7.54	0.73	5.23	4.18	0.66	0.38	0.31	0.77

were included as a testing set. Therefore, the dataset division was set in the following manner – 58% for training and 42% for testing process.

2.4.4 Discussion

By predicting epidemiological peaks, we tend to achieve "flattened curve" of the disease spread in order to prevent it [51]. But due to the lack of the official data about COVID-19 epidemic, some early published models had the tendency to overfit, or in some cases, parameters were overtaken from literature based on less precise evidence. During this research, data related to the epidemics were carefully compiled from reliable sources: state, regional and local level and incorporate them into the developed model. In general, for the three countries which were considered, official and simulated values showed a good match, which means that the model is showing promising results. Also, LSTM network showed that it is able to predict second and third peaks based on the position of previously peaks with the lower values of RMSE. The match between simulated and real values, as well as higher values of RMSE can be affected by several things, such as underreporting of the number of cases, estimating parameters, setting initial conditions, etc. The main drawback of this study is related to the complexity of the COVID-19 epidemic spread, current model does not consider behavioral responses to the epidemic, re-infection – no immunization, variants of the virus etc. Therefore, future research will include different complex phenomena, especially medical intervention and asymptomatic infection, in order to better describe the COVID-19 spread and development.

2.5 Integration of Different Models into Multiscale Platform

An integration platform is a unified collection of integration software (middleware) components that allow users to achieve the following:

- Create, protect, and manage integration flows that connect various applications, systems, services, and data repositories.
- Allow for quick API generation and lifecycle management to fulfill a variety of hybrid integration requirements.

To put it another way, an integration platform gives enterprises the integration capabilities they need to integrate their systems, applications, and data throughout their environment. Therefore, we provide an example of such integrated platform with multiscale models to investigate cancer, cardiovascular, bone disorders and tissue engineering [52]. Such platform enables the users to access the models without the access to certain computer specifications or operating systems requirements.

2.5.1 Architecture

The platform is developed under the project "Increasing scientific, technological and innovation capacity of Serbia as a Widening country in the domain of multiscale modelling and medical informatics in biomedical engineering – SGABU" [52]. The architecture is presented in Fig. 9 with the modules and their corresponding engines and tools. The SGABU framework can be defined as a hierarchical multilayer schema [53]. The framework is comprised of five layers.

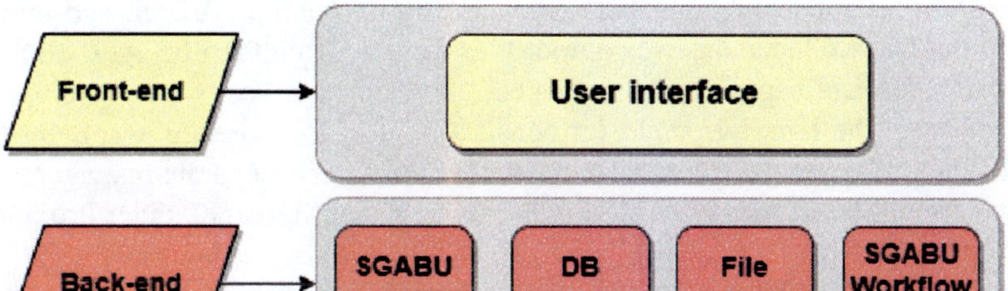

Fig. 9. Architecture of SGABU platform.

At the very bottom is the hardware layer, which houses CPUs, RAM, and VMs. Above the hardware layer is the security layer, which includes additional methods for platform user access management, encrypted communication, and user authentication. The next layer relates to the workflow layer, which serves as the platform's main service. The engines of the workflow layer are as follows: i) the workflow engine, (ii) the Docker container engine, (iii) the data quality control engine, and (iv) the 3D visualizer.

Each of these engines communicates directly with the SGABU tools and modules at the back-end layer. The last layer is the front-end layer, which is responsible for the user interface.

DBServer includes a relational MySQL (MariaDB) database. Functional Engine Server (FES), which oversees executing CWL (Common Workflow Language) [54] compliant scientific workflows, is the most resource-intensive backend component. When the reverse proxy receives a request from the frontend, it forwards it to the FES, which returns a file (or many files) based on the kind of request. All external HTTP traffic is SSL encrypted and routed through a central NGINX reverse proxy.

The 3D visualizer is represented by a modified Kitware Paraview picture. It is inextricably linked with FES. Its goal is to conduct 2D/3D simulation data preparation, post-processing, and transformation activities. Because of this close connection, the user is relieved of the task of dealing with massive workflow output files. A front-end-integrated modified Paraview Glance receives output files directly from the FES API, eliminating massive transfers until specifically requested.

Hardware: The hypervisor cluster consists of 96 CPU cores, 376 GB RAM and 1.89 TB SSD, interconnected with 1 Gbps internal network, separate management network and the uplink. It runs KVM (Kernel-based Virtual Machine) under Proxmox VE 7 environment.

Backend: Laravel [55] is a free and open-source PHP framework that provides a set of tools and resources to build modern PHP applications. It is used as a backend framework for SGABU platform. It follows a model-view-controller design pattern, which generally makes it a lot easier to start creating and after that, maintaining the functionality of the platform. On the SGABU platform, it also provides important built-in features such as authentication, sessions, routing, migration system etc.

FES API: The FES (*Functional Engine Server*) API is a REST API developed in Python whose main purpose is to run and manipulate various workflows [56]. FES-API implements the required executor for running workflows. The FES in the background employs Common Workflow Language (CWL) which provides portability of the workflows. It is possible to manage the entire lifecycle of a single workflow using the functional server interface, including creation, handling inputs and outputs. FES is in charge of providing input files to the workflow as well as storage space where the workflow will store its outputs. The significant advantage of this system is an asynchronous execution of workflows as background processes, which allows multiple workflows to run simultaneously. The workflow is created in the initial step of the workflow cycle. The service must be given with a "familiar" CWL template and input parameters when constructing a process. It is supplied the space where the input data will be saved, as well as the place where the workflow will keep the outcomes. Another useful feature is the ability to download the whole workflow results. To remove a workflow, it should be deleted from the system, excluding the workflow template.

Frontend: As one of the most popular software development tools available, Angular is used for building single-page client applications using HTML and TypeScript [57]. This framework implements core and optional functionality as a set of TypeScript libraries that are imported into applications.

The key value proposition of Angular is the ability to create apps that can run on practically any device, whether mobile, web, or desktop. SGABU's frontend is designed to be responsive. It indicates that diverse devices (desktop computers, laptops, tablets, and smartphones) can render the given web pages appropriately. As a result, responsive web platforms are automatically adapted to different web browsers and screen sizes.

Some models or datasets in the SGABU platform will produce outputs intended for visualization with a goal of better understanding. Plotly.js [58] is implemented for interactive data visualization and it supports various graphs like line charts, bar charts, scatter plots, area plots, histogram, etc. Plotly employs JavaScript to create interactive graphs that allow users to zoom in on the graph or add extra information such as data on hover. It enables versatile graph customization, making charts more relevant and intelligible for consumers.

2.5.2 Use Case - Model

The majority of the SGABU platform's simulation modules are built as CWL processes. This technique is an obvious choice since it employs Docker containerization and a standardized means of encoding inputs, outputs, and intermediate outcomes, resulting in intrinsic findability, accessibility, interoperability, and reusability (FAIR principles). The effort involved in offering CWL-type processes is divided into two main actions:

1. developing CWL implementation on FES (*Functional Engine Service*) backend and
2. developing an appropriate UI.

The first step is to interview the module supplier and discuss workflow inputs, outputs, ranges and constraints, technologies, how the results should be presented to the user, and so on. The majority of the integrated processes are written in Matlab, C++, and Fortran, with some help from other libraries. It was discovered that they could all be converted into portable CWL workflows that ran within Docker containers. Due to their vast size and licensing concerns, only Matlab-based modules are not containerized and use the Matlab executable directly. The procedures given are reusable by any other party due to a standardized and completely FAIR methodology. The second activity was to design the UI, which resulted in proper workflow input forms with stringent verification of numeric values, file kinds, and so on, as well as output visualization tabs with tabular views, interactive diagrams, 3D views, and animations. The primary dashboard on the SGABU platform provides access to each module, as seen in the picture below (Fig. 10).

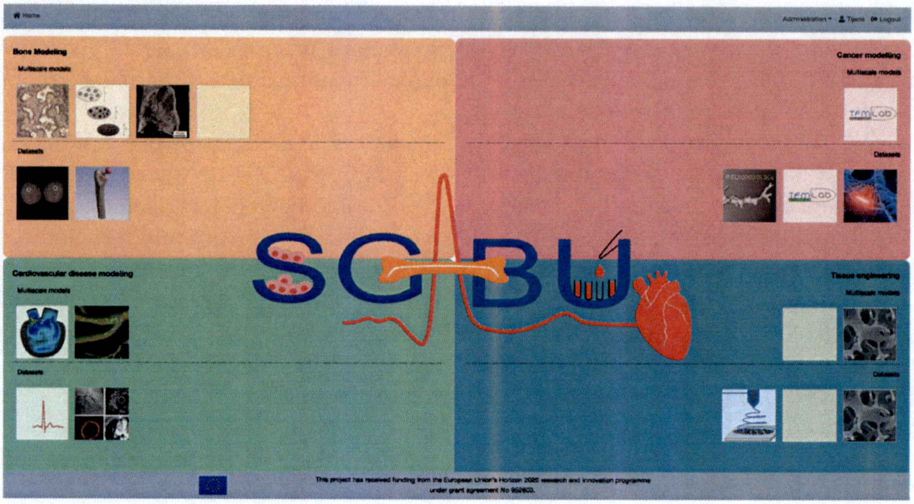

Fig. 10. Main platform dashboard

After selecting one of the modules, the help menu appears, directing the user to the instructions for using a certain model/dataset, as well as theoretical background and references for additional reading. The combined module interface will be demonstrated using the **ParametricHeart** workflow (Fig. 11). In the upper section of the window, users can see given titles and statuses of the previously ignited workflows. Possible statuses of any workflow are:

- Not yet executed
- Terminated
- Running
- Finished OK
- Finished Error

The user is free to start the workflow any time by filling required fields in *Add new workflow* form and clicking *Run*. Thanks to asynchronous workflow execution implementation, multiple workflows can be executed simultaneously. In the specific case of ParametricHeart workflow, a user is expected to fill out the following fields:

- Workflow name
- Left section:
  - o Base division
  - o Connection division
  - o Aortic division
  - o Wall division
  - o Valves division
  - o Mitral division

- Right section:

  - o IVS-diastolic [cm]
  - o LVID-diastolic [cm]
  - o LVPW-diastolic [cm]
  - o IVS-systolic [cm]
  - o LVID-systolic [cm]

- Inlet Velocity time function
- Outlet velocity time function

The division fields control the number of finite elements in each component of the model. **Base division** controls the number of finite elements along the height of the left ventricle base. **Connection division** controls the number of finite elements along the height between the base and valves. **Valves division** controls the number of finite elements along the width of valves. **Aortic/mitral division** controls the number of finite elements along the height of the aortic/mitral valve. **Wall division** controls the number of finite elements along the width of the heart wall. Regarding the dimensions that need to be filled in, **IVS-diastolic/systolic [cm]** represents the interventricular septum (IVS) in diastole/systole, **LVID-diastolic/systolic [cm]** represents left ventricular internal diameter (LVID) in diastole/systole and LVPW-diastolic/systolic [cm] represents left ventricular posterior wall (LVPW) in diastole/systole.

All fields except for **Workflow name** are numerical and the allowed value ranges are provided and verified on execution.

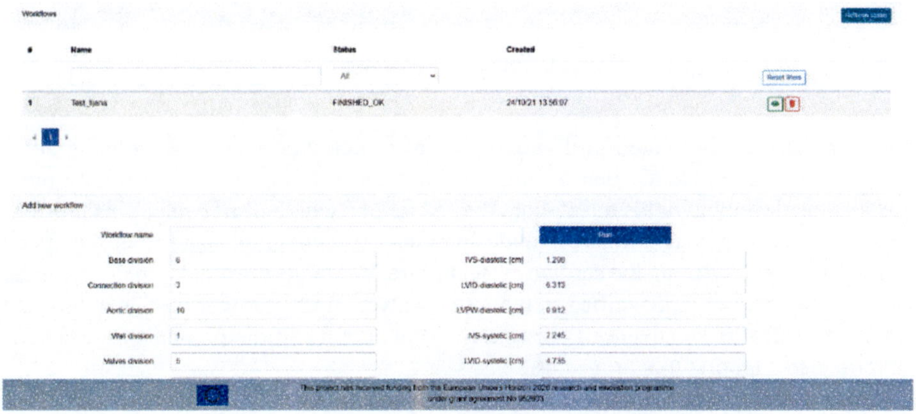

Fig. 11. User Interface for ParametricHeart module.

Figure 12 depicts the specified velocity functions over the course of a cardiac cycle. The mitral valve of the left ventricle is given an inlet velocity function, whereas the aortic valve gets an outlet velocity function. Figure 12 also includes interactive graphs with inlet and exit velocity time curves.

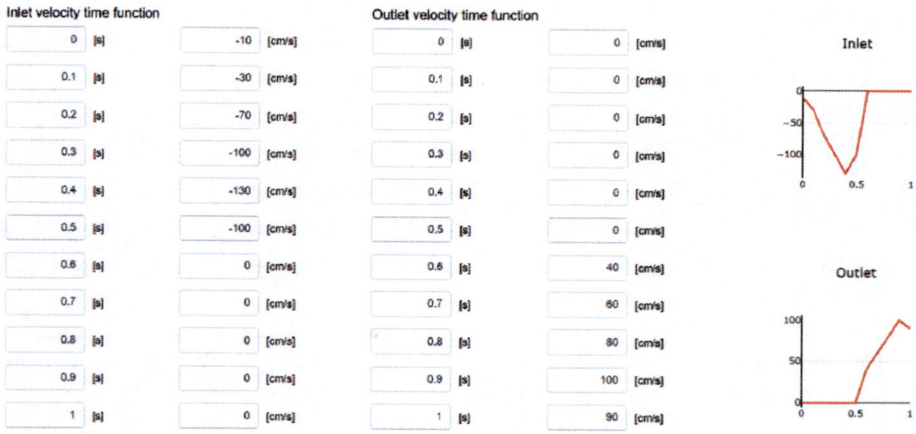

Fig. 12. Prescribing inlet and outlet time function for ParametricHeart module.

The SGABU platform's user interface incorporates exception management (Empty forms, non-numerical forms, out of range values). Once everything is filled out correctly, the workflow may begin. In the left panel, the user may view the current status of the workflow (Fig. 13).

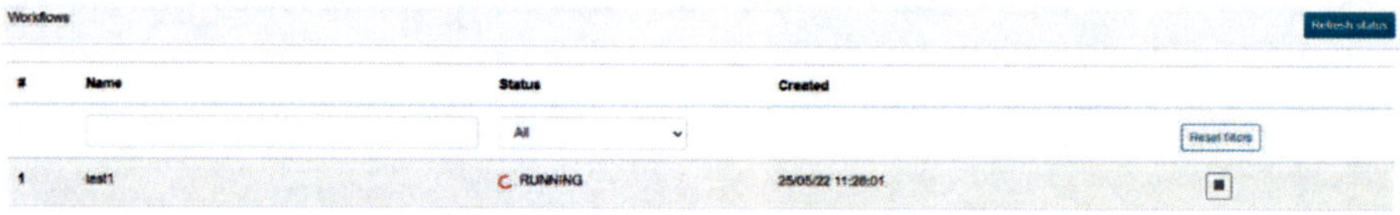

Fig. 13. Running the workflow.

The results are presented in a variety of ways, including tables, statistics, graphs, video, and 3D views. The simulation output in ParametricHeart consists of velocity, pressure, and displacement fields throughout the course of a whole heart cycle. This process, in addition to raw physical fields, includes pressure-volume graphs and representations of myocardial work during the cardiac cycle. Figure 14 depicts the input divisions, model size, and inlet velocity prescribed to the mitral valve. Binding a workflow's inputs and outputs is critical if an advanced user plans to conduct a parametric analysis. SGABU platform also supports this usage case.

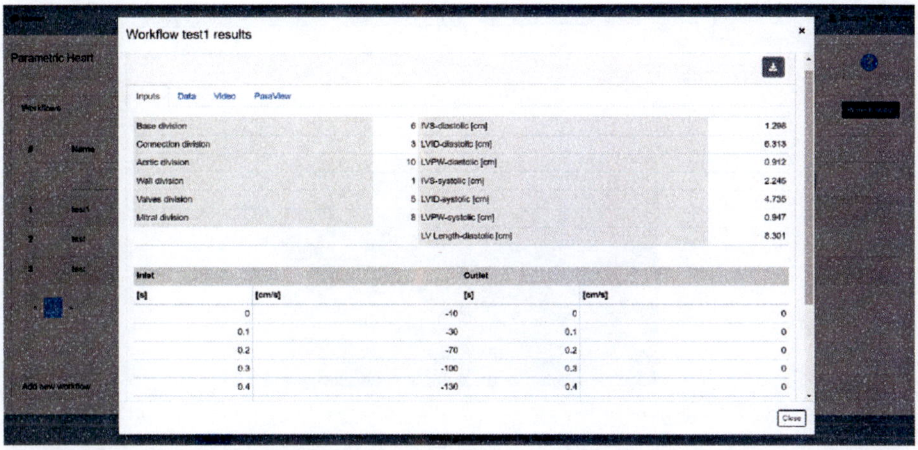

Fig. 14. Inputs tab of the results section.

The results can be downloaded in the form of csv files for any further offline investigation. In Fig. 15, we show a PV diagram, ejection fraction and global work efficiency obtained after executing this finite element simulation.

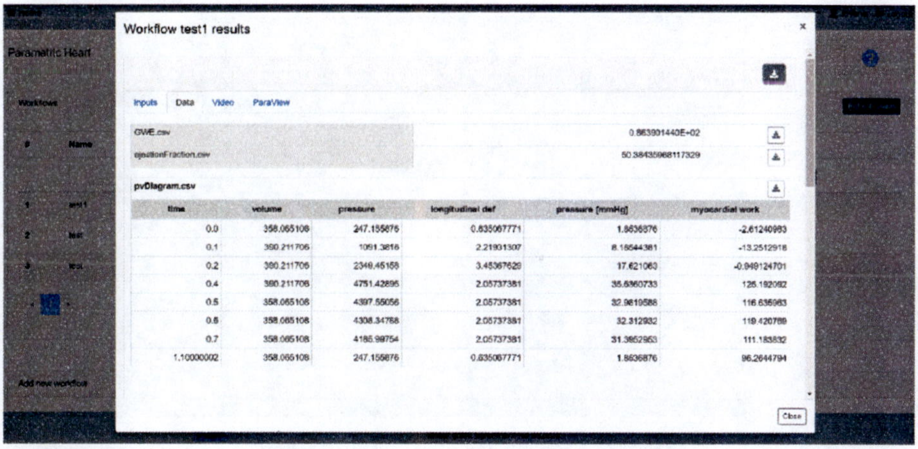

Fig. 15. Data tab of the results section.

Figure 16 presents a displacement field within the left ventricle model in the form of a video file. The displacements are the highest during the systolic phase of the cycle.

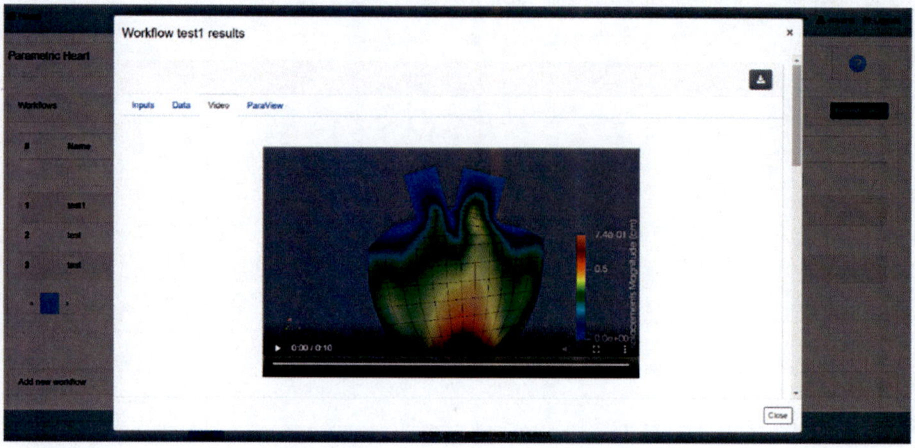

Fig. 16. Video tab of the results section.

Clicking the Paraview tab will open a new browser tab with full 3D representation of the workflow outputs. In Fig. 17 we show a geometry of the left ventricle.

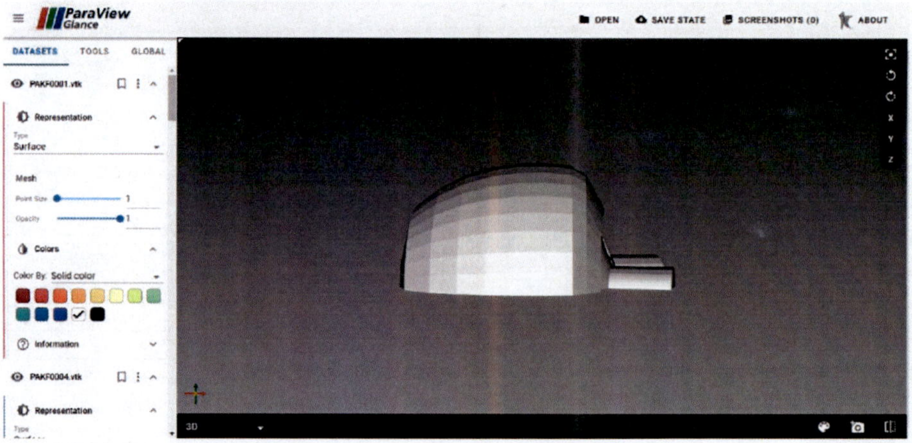

Fig. 17. Paraveiw tab of the results section.

The velocity field is seen in Fig. 18. It can be seen that inlet velocities increase during diastole and are equal to zero during systole. Furthermore, during systole, muscles contract and force blood out of the left ventricle. The velocities are greatest near the bottom of the heart at the start of systole. The aortic valve has the maximum velocities in the middle of systole.

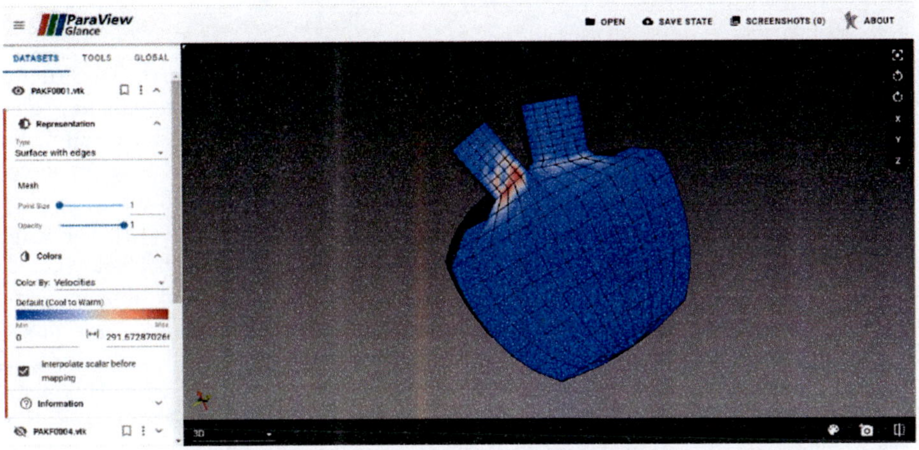

Fig. 18. Paraview velocities show of PAKF0001.vtk.

2.5.3 Use Case – Dataset

A single dataset integration work often begins with an interview with the dataset supplier on how he or she wants the individual dataset to be displayed in the platform. The next job is to put the agreed-upon UI into action. The team then goes back to the data supplier with the request for testing and iterative UI adjustment until the provider's criteria are met. Due to the tabular nature of certain datasets, the activity is straightforward, however the majority of datasets required further customization by front-end developers using technologies such as Angular, Plotly.js, Paraview Glance, and so on. Hip joint (UKG) - This procedure is included in the bone modeling field (Fig. 19). This file includes 301 DICOM pictures from a CT scanner.

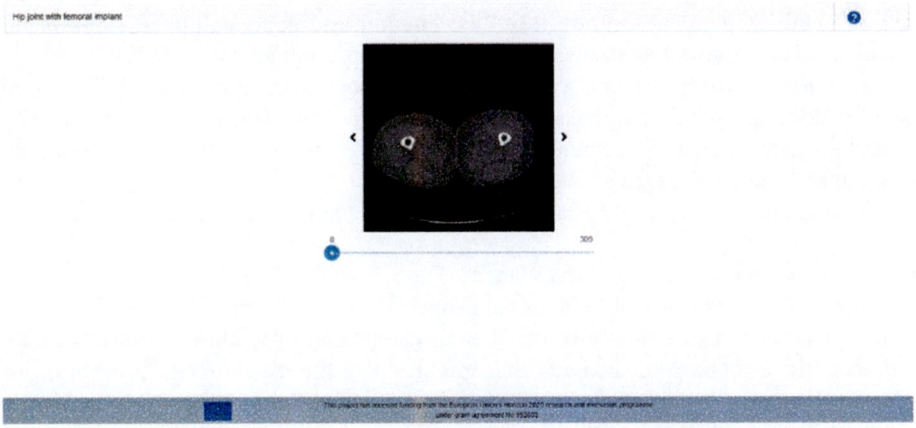

Fig. 19. User Interface for Hip joint with femoral Implant dataset.

2.5.4 Discussion

There are several instances of a successful biomedical research platforms. PANBioRA, for example, is a modular platform that standardizes biomaterial assessment and opens the door to pre-implantation, individualized diagnostics for biomaterial-based applications [59]. The SILICOFCM platform is cloud-based and represents a novel in silico clinical trials solution for the design and functional optimization of whole cardiac performance as well as monitoring the success of pharmacological therapy [15]. The Bioengineering and Technology (BET) platform fosters innovation, supports research, and federates the larger multidisciplinary community involved in translational bioengineering [60]. However, no platform integrates models from several medical fields and shows the most advanced solutions in terms of open data and open models, where users can change different parameters and discuss the differences in obtained results. This platform is developed not only for the researching purposes, but with the goal of integrating teaching materials for various fields. SGABU is a web-based platform that is available for users to access services and tools for biomedical research.

3 AI in Medicine – Current Limitations and Future Trends

AI's influence on healthcare is quite promising, and it has the potential to fundamentally change the healthcare professions in the very near future. AI may be used in a variety of health-related fields, ranging from hospital care and clinical research to medication development and diagnostic prediction. The rapidly expanding availability and cheap cost of strong computer resources is propelling healthcare into the digital age. The incorporation of contemporary technology into a physician's everyday practice offers safe real-time data access and big data analytics. This enhances overall treatment quality by increasing collaboration among specialists [61].

Total public and private sector investment in healthcare AI is expected to increase in the next years, reaching $6.6 billion by 2021. By 2026, the US healthcare business may save $150 billion per year as a result of this. The improvements will have a significant influence on automated operations, precision surgery, preventative medical intervention, and diagnostics in the healthcare setting. Experts expect that the impact will be larger on the operational and administrative aspects of healthcare than on the therapeutic side [62]. Artificial intelligence advancements are expected to provide clients with high-quality personalized and data-driven services.

Although the potential of AI in medicine is huge, there are certain limitations:

- At this stage, it needs human surveillance – Though artificial intelligence has advanced significantly in the medical field, human supervision is still required. Surgery robots, for example, work rationally rather than empathetically. Health professionals may detect critical behavioral insights that might aid in the diagnosis or prevention of medical issues. As this field develops, there will be greater contact between healthcare practitioners and technology specialists. To be effective, AI requires human input and assessment.
- AI may overlook social variables - Patients' demands frequently extend beyond their immediate physical circumstances. Social, economic, and historical considerations

can all play a role in making suitable suggestions for certain individuals. For example, an AI system may be able to assign a patient to a certain care facility based on a diagnosis. This approach, however, may not take into consideration the patient's financial constraints or other customized preferences. When an AI system is used, privacy becomes a concern. When it comes to gathering and utilizing data, companies like Amazon have complete control. Hospitals, on the other hand, may encounter certain difficulties when attempting to transmit data from Apple mobile devices, for example. These legislative and social constraints may limit AI's ability to aid in medical procedures.

- AI may lead to unemployment - While AI may help lower expenses and relieve clinical strain, it may also make certain positions obsolete. This variable may result in the displacement of experts who have committed time and money in healthcare education, posing equality issues. According to World Economic Forum research from 2018, AI will produce 58 million employments by 2022. However, according to the same estimate, AI would displace or eliminate 75 million jobs by the same year. The main reason for the loss of job possibilities is that as AI becomes increasingly integrated across industries, occupations requiring repetitive activities will become obsolete.
- Inaccuracies are still possible - Medical AI is primarily reliant on diagnostic data from millions of cataloged instances. Misdiagnosis is very feasible when there is limited data on specific illnesses, demographics, or environmental variables. This becomes especially crucial when giving specific medications. There is also a problem of missing data. In the case of prescriptions, some information on specific populations and treatment reactions may be missing. This incidence may cause difficulties in diagnosing and treating individuals from specific populations. To accommodate for data gaps, AI is always developing and improving. It is vital to highlight, however, that certain groups may still be excluded from current domain knowledge.
- Susceptible to security risks - AI systems are vulnerable to security issues since they rely on data networks. With the advent of Offensive AI, enhanced cyber security will be essential to maintain the technology's long-term viability. According to Forrester Consulting [63], 88% of security decision-makers believe aggressive AI is a growing concern. As artificial intelligence (AI) leverages data to make systems smarter and more accurate, cyberattacks will include AI to get wiser with each success and failure, making them more difficult to forecast and avoid. When malicious threats outmaneuver security systems, the attacks become far more difficult to counter.

AI unquestionably has the ability to enhance healthcare systems. The automation of time-consuming procedures can free up clinician schedules to allow for greater patient interaction. Improving data accessibility aids healthcare providers in taking the necessary precautions to avoid sickness. Real-time data can help diagnosis to be made more accurately and quickly. Artificial intelligence is being used to decrease administrative mistakes and save valuable resources. As SMEs get more involved in AI development, the technology becomes more relevant and well-informed. AI is rapidly being used in healthcare, and limitations and obstacles are being tackled and solved.

The total public and private sector investment in healthcare AI is expected to grow in the coming years, reaching a $6.6 billion investment by 2021. This may result in annual savings of $150 billion for the US healthcare economy by 2026. Changes will fundamentally reshape the healthcare landscape and affect automated operations, precision surgery, preventive medical intervention, and diagnostics. Experts forecast a more remarkable impact on the operational and administrative sectors of healthcare rather than the clinical part of it. The development of AI is expected to provide customers with high quality personalized and data-driven services [61].

4 Conclusions

This chapter deals with application of AI in medicine, with the focus of the work in the area of stratification of patients with carotid artery disease, assessment of patient condition with familiar cardiomyopathy and COVID-19 models (personalized and epidemiological). The chapter also covers the topic of integration of the models into the cloud-based platform, to deal with the access of models without any specific software requirements. In the end, current limitations and future trends are discussed. The incredible ability of artificial intelligence to analyze vast amounts of data, make sense of the visuals, and identify patterns that even the most skilled human eye misses, has spurred hope that the technology may enhance medicine. Finally, AI bears the potential of "humanizing health care" by bringing the physician closer to the patient through the creation of tailored models.

Acknowledgement. The research was funded by Serbian Ministry of Education, Science, and Technological Development, grant [451-03-68/2022-14/200107 (Faculty of Engineering, University of Kragujevac)]. This research is also supported by the project that has received funding from the European Union's Horizon 2020 research and innovation programmes under grant agreement No 952603 (SGABU project). This article reflects only the author's view. The Commission is not responsible for any use that may be made of the information it contains. T. Geroski (maiden name Sustersic) also acknowledges the support from L'OREAL-UNESCO awards "For Women in Science" in Serbia.

References

1. Dašić, L., Radovanović, N., Šušteršič, T., Blagojević, A., Benolić, L., Filipović, N.: Patch-based convolutional neural network for atherosclerotic carotid plaque semantic segmentation. IPSI Trans. Internet Res. **19**(1), 57–62 (2022)
2. TAXINOMISIS project: A multidisciplinary approach for the stratification of patients with carotid artery disease. [Online]. https://taxinomisis-project.eu/
3. Ronneberger, O., Fischer, P., Brox, T.: U-net: convolutional networks for biomedical image segmentation. In: International Conference on Medical Image Computing And Computer-Assisted Intervention (2015)
4. Zhang, D., Wang, Z.: Image information restoration based on long-range correlation. IEEE Trans. Circuits Syst. Video Technol. **12**(5), 331–341 (2002)
5. Clarke, S.E., Hammond, R.R., Mitchell, J.R., Rutt, B.K.: Quantitative assessment of carotid plaque composition using multicontrast MRI and registered histology. Magn. Reson. Med. **50**(6), 1199–1208 (2003)

6. Hashimoto, B.E.: Pitfalls in carotid ultrasound diagnosis. Ultrasound Clin. **6**(4), 462–476 (2011)
7. Loizou, C.P., Pattichis, C.S., Pantziaris, M., Tyllis, T., Nicolaides, A.: Snakes based segmentation of the common carotid artery intima media. Med. Biol. Eng. Compu. **45**(1), 35–49 (2007)
8. Golemati, S., Stoitsis, J., Sifakis, E.G., Balkizas, T., Nikita, K.S.: Using the Hough transform to segment ultrasound images of longitudinal and transverse sections of the carotid artery. Ultrasound Med. Biol. **33**(12), 1918–1932 (2007)
9. Cheng, J., et al.: Fully automatic plaque segmentation in 3-D carotid ultrasound images. Ultrasound Med. Biol. **39**(12), 2431–2446 (2013)
10. Zhou, R., et al.: Deep learning-based carotid plaque segmentation from B-mode ultrasound images. Ultrasound Med. Biol. **47**(9), 2723–2733 (2021)
11. Jain, P.K., Sharma, N., Giannopoulos, A.A., Saba, L., Nicolaides, A., Suri, J.S.: Hybrid deep learning segmentation models for atherosclerotic plaque in internal carotid artery B-mode ultrasound. Comput. Biol. Med. **136**, 104721 (2021)
12. Lekadir, K., et al.: A convolutional neural network for automatic characterization of plaque composition in carotid ultrasound. IEEE J. Biomed. Health Inform. **21**(1), 48–55 (2016)
13. Mechrez, R., Goldberger, J., Greenspan, H.: Patch-based segmentation with spatial consistency: application to MS lesions in brain MRI, Int. J. Biomed. Imaging **2016**, 1–13 (2016). ID 7952541
14. Elliott, P., et al.: ESC guidelines on diagnosis and management of hypertrophic cardiomyopathy the task force for the diagnosis and management of hypertrophic cardiomyopathy of the European society of cardiology (ESC). Eur. Heart J. **35**, 2733–2779 (2014)
15. SILICOFCM project: In Silico trials for drug tracing the effects of sarcomeric protein mutations leading to familial cardiomyopathy. [Online]. https://silicofcm.eu/
16. Šušteršič, T., Blagojević, A., Simović, S., Velicki, I., Filpović, N.: Development of machine learning tool for segmentation and parameter extraction in cardiac left ventricle ultrasound images of patients with cardiomyopathy. In: 17th International Symposium on Computer Methods in Biomechanics and Biomedical Engineering and 5th Conference on Imaging and Visualization (CMBBE2021), Bonn, Germany (2021)
17. Šušteršič, T., Blagojević, A., Simović, S., Velicki, L., Filpović, N.: Automatic detection of cardiomyopathy in cardiac left ventricle ultrasound images. In: 11th International Conference on Information Society and Techology (ICIST), Kopaonik, Serbia (2021)
18. Lang, R., et al.: Recommendations for chamber quantification: a report from the American Society of Echocardiography's Guidelines and Standards Committee and the Chamber Quantification Writing Group, developed in conjunction with the European Association of Echocardiograph. J. Am. Soc. Echocardiogr. **18**(12), 1440–1463 (2005)
19. Noble, A., Boukerroui, D.: Ultrasound image segmentation: a survey. IEEE Trans. Med. Imaging **25**(8), 987–1010 (2006)
20. Systems, G.M.: Technical Publications, Vivid I, Reference Manual. General Electric Co (2005)
21. Šušteršič, T., et al.: Epidemiological predictive modeling of COVID-19 infection: development, testing, and implementation on the population of the Benelux union. Front. Public Health **9**, 1567 (2021)
22. Felzenszwalb, P., Huttenlocher, D.: Efficient graph-based image segmentation. Int. J. Comput. Vision **59**(2), 167–181 (2004)
23. Filipovic, N., et al.: In silico clinical trials for cardiovascular disease. J. Visual. Exp. Jove **183**, e63573 (2022)
24. Moradi, S., et al.: MFP-UNet: a novel deep learning based approach for left ventricle segmentation in echocardiography. Physica Medica, **67**, 58–69 (2019)

25. Noble, J., Boukerroui, D.: Ultrasound image segmentation: a survey. IEEE Trans. Med. Imaging **25**, 987–1010 (2006)
26. Ghelich Oghli, M., Mohammadzadeh, A., Kafieh, R., Kermani, S.: A hybrid graph-based approach for right ventricle segmentation in cardiac MRI by long axis information transition. Phys Medica **54,** 103–116 (2018)
27. Ghelich Oghli, M., Mohammadzadeh, M., Mohammadzadeh, V., Kadivar, S., Zadeh, A.: Left ventricle segmentation using a combination of region growing and graph based method. Iran J. Radiol. **14**(2), e42272 (2017)
28. Smistad, E., Ostvik, A., Haugen, B., Lovstakken, L.: 2D left ventricle segmentation using deep learning. IEEE International Ultrasonics Symposium, pp. 1–4 (2017)
29. Carneiro, G., Nascimento, J.: Combining multiple dynamic models and deep learning architectures for tracking the left ventricle endocardium in ultrasound data. IEEE Trans. Pattern Anal. Mach. Intell. **35**(11), 2592–2607 (2013)
30. Paragios, N.: A level set approach for shape-driven segmentation and tracking of the left ventricle. IEEE Trans. Med. Imaging **22**(6), 773–776 (2003)
31. Carneiro, G., Nascimento, J., Freitas, A.: The segmentation of the left ventricle of the heart from ultrasound data using deep learning architectures and derivative-based search methods. IEEE Trans. Image Process. **21**(3), 968–982 (2011)
32. Oktay, O., et al.: Anatomically constrained neural networks (ACNNs): application to cardiac image enhancement and segmentation. IEEE Trans. Med. Imaging **37**(2), 384–395 (2017)
33. Zyuzin, V., et al.: Identification of the left ventricle endocardial border on two-dimensional ultrasound images using the convolutional neural network UNet. In: IEEE Ural-Siberian Conference on Biomedical Engineering, Radioelectronics and Information Technology, pp. 76–78 (2018)
34. Unser, M., Pelle, G., Brun, P., Eden, M.: Automated extraction of serial myocardial borders from M-mode echocardiograms. IEEE Trans. Med. Imaging **8**(1), 96–103 (1989)
35. Rabben, S.I., et al.: Semiautomatic contour detection in ultrasound M-mode images. Ultrasound Med. Biol. **26**(2), 287–296 (2000)
36. Ikemura, K., et al.: Using automated machine learning to predict the mortality of patients with COVID-19: prediction model development study, J. Med. Internet Res. **23**(2), e23458 (2021)
37. Sen, S., Saha, S., Chatterjee, S., Mirjalili, S., Sarkar, R.: A bi-stage feature selection approach for COVID-19 prediction using chest CT images. Appl. Intell. **51**(12), 8985–9000 (2021). https://doi.org/10.1007/s10489-021-02292-8
38. COVIDAI project: Use of Regressive Artificial Intelligence (AI) and Machine Learning (ML) Methods in Modelling of COVID-19 Spread. [Online]. http://www.covidai.kg.ac.rs/
39. Cabaro, S., et al.: Cytokine signature and COVID-19 prediction models in the two waves of pandemics. Sci. Rep. **11**(1), 1–11 (2021)
40. Blagojević, A., et al.: Artificial intelligence approach towards assessment of condition of COVID-19 patients – Identification of predictive biomarkers associated with severity of clinical condition and disease progression. Comput. Biol. Med. **138**, 104869 (2021)
41. Liu, L., et al.: An interpretable boosting model to predict side effects of analgesics for osteoarthritis. BMC Syst. Biol. **12**(6), 29–38 (2018)
42. Rahman, T., et al.: Mortality prediction utilizing blood biomarkers to predict the severity of COVID-19 using machine learning technique. Diagnostics **11**(9), 1582 (2021)
43. Yao, H., et al.: Severity detection for the coronavirus disease 2019 (COVID-19) patients using a machine learning model based on the blood and urine tests. Front. Cell Dev. Biol. **8**, 683 (2020)
44. Rahman, T., et al.: QCovSML: A reliable COVID-19 detection system using CBC biomarkers by a stacking machine learning model. Comput. Biol. Med. **143**, 105284 (2022)

45. Schmidhuber, J.: Deep learning in neural networks: An overview. Neural Netw. **61**, 85–117 (2015)
46. Infectious Diseases Data Explorations & Visualizations. [Online]. https://epistat.wiv-isp.be/covid/
47. Our World in Data. [Online]. https://ourworldindata.org/coronavirus/country/netherlands
48. The luxembourgish data platform. [Online]. https://data.public.lu/fr/
49. Hochreiter, S., Schmidhuber, J.: Long short-term memory. Neural Comput. **9**(8), 1735–1780 (1997)
50. Chandra, R., Jain, A., Chauhan, D.S.: Deep learning via LSTM models for COVID-19 infection forecasting in India. *arXiv preprint* arXiv:2101.11881 (2021)
51. Jin, Y., et al.: Virology, epidemiology, pathogenesis, and control of COVID-19. Viruses **12**(4), 372 (2020)
52. SGABU project: Increasing scientific, technological and innovation capacity of Serbia as a Widening country in the domain of multiscale modelling and medical informatics in biomedical engineering. [Online]. http://sgabu.eu/
53. Nikolić, J., Atanasijević, A., Živić, A., Šušteršič, T., Ivanović, M., Filipović, N.: Development of SGABU platform for multiscale modeling. IPSI Trans. Internet Res. **19**(1), 50–55 57–62
54. Common Workflow Languarge. [Online]. https://www.commonwl.org/. Accessed 6 Oct 2022
55. "Laravel," [Online]. https://laravel.com/. Accessed 6 Oct 2022
56. Ivanovic, M., Zivic, A., Tachos, N., Gois, G., Filipovic, N., Fotiadis, D.: In-silico research platform in the cloud-performance and scalability analysis. In: 2021 IEEE 21st International Conference on Bioinformatics and Bioengineering (BIBE) (2021)
57. Angular [Online]. https://angular.io/. Accessed 5 Oct 2022
58. Plotly [Online]. https://plotly.com/. Accessed 5 Oct 2022
59. "PANBioRa [Online]. https://www.panbiora.eu/. Accessed 4 Nov 2022
60. Bioengineering and Technology platform – BET [Online]. https://www.epfl.ch/research/facilities/ptbet/. Accessed 4 Nov 2022
61. Habuza, T., et al.: AI applications in robotics, diagnostic image analysis and precision medicine: current limitations, future trends, guidelines on CAD systems for medicine. Inform. Med. Unlock. **24**, 100596 (2021)
62. AI and Healthcare: A Giant Opportunity. Forbes [Online]. https://www.forbes.com/sites/insights-intelai/2019/02/11/ai-and-healthcare-a-giant-opportunity/?sh=721ccb44c682. Accessed 6 Oct 2022
63. Forrester at a Glance. [Online]. https://www.forrester.com/about-us/fact-sheet/. Accessed 6 Oct 2022
64. Carneiro, G., Nascimento, J., Freitas, A.: The segmentation of the left ventricle of the heart from ultrasound data using deep learning architectures and derivative-based search methods. IEEE Trans. Image Process. **21**(3), 968–982 (2011)

Digital Platform as the Communication Channel for Challenges in Artificial Intelligence

Jelena Živković(✉) and Đorđe Ilić

Faculty of Engineering, University of Kragujevac, 6 Sestre Janjić Street, 34000 Kragujevac, Serbia
jelenamarkovic0909@gmail.com

Abstract. The paper presents the implementation of a digital platform that could be used as a communication channel in Artificial Intelligence. Digital Communication uses artificial intelligence that can quickly find solutions to certain problems. When a person discovers a potential problem, artificial intelligence can access the platform, so they can send their solutions from their computer in any corner of the world. Problems and issues in the field of artificial intelligence can be solved through the described platform. Setting project tasks in the mentioned area include a detailed description of the problem, the deadline by which a possible solution is to be sent and an award for the best solution. The platform can be used both in industry and the education system. The industry can find new experts to cooperate with and the education system, that has been striving for digitalization in recent years, can enable a larger number of students to test themselves in the field of Artificial Intelligence in a simpler way, solving its tasks. Prizes can be set as points in the case of the education system or money in the case of industry.

Keywords: Digital Platform · Artificial Intelligence · Challenge

1 Introduction

As a consequence of the COVID-19 virus pandemic, the world's population has opted to use online platforms for learning and making new business connections [1]. Artificial intelligence as an area of wide application is among the most attractive areas in the world [2]. Artificial intelligence can produce many questions and challenges that need to be answered [3]. By using the SMART2M platform individual users can submit their ideas to the problem, which can be a potential solution. In this way, scientists around the world can share their knowledge gained through many years of experience. All registered users can respond to the challenges and thus join in solving the given problem. By not revealing the identity of the company, the platform preserves its privacy, and the potential solution to the problem is visible only to the company and its makers.

This way of communication prevents the use of work of other people, and an additional advantage of this way of communication is the large number of different solutions to one problem. The popularity of this field is growing each year, so the number of

N. Filipovic (Ed.): AAI 2022, LNNS 659, pp. 306–319, 2023.
https://doi.org/10.1007/978-3-031-29717-5_18

educational institutions that are involved in this field is increasing [4]. Researching Artificial intelligence in educational institutions includes constant interactions with students [5]. Professors can set tasks via the platform, while students as the registered users of the platform can send their examples of solutions and receive points. For a better result in education, our platform can enable cooperation between educational institutions, for example, competitions in this area or exchange experiences. By following the world's needs, the platform will be susceptible to changes in order to contribute to the growth and to open a new door of communication in this area in the future.

2 Platform Concept

Section Platform concept represents the way the platform was created. Primarily, the platform was created for educational purposes, but now it can also be used by companies. Adequate users are identified, as well as the functions of each type of user. The design of the platform is adaptable to all devices, and the way it is applied depends on the type of user. Using Angular with the programming languages HTML, CSS and TypeScript made it possible to create a simple user interface that represents the frontend of the platform [6] [7]. Font effects of text files, colors, the appearance of elements within the page, as well as "hyperlinks" it may contain, are achieved using the HTML programming language [6, 7]. The more readable stylization of the elements, as well as the adaptability of the user interface on all devices were made in the programming language CSS. Each information entry is accompanied by reactive forms that allow checking the correctness of the entered data. Paths and field positions within the reactive path are defined using HTML [6, 7], and the styling itself, which includes defining colors and sizes, is a combination of CSS and Bootstrap. Access to certain pages and data is limited depending on the type of user, so for example companies cannot access information and functions that are only available to platform administrators or challenges from other companies. The listed constraints represent authentication and authorization, which are defined using the TypeScript language. The operations that form the basis of using the platform are adding, viewing, modifying, and deleting data. These operations are enabled through Laravel [8], which is a PHP framework and data from frontend comes via defined API routes that support methods post, put, get and delete [9]. The methods that control data manipulation are written in the so-called controllers of the Laravel development environment [8]. Filtering and sorting methods are also defined in the controllers, which are executed by a combination of the "Eloquent" model and SQL queries [10], based on the received data from the frontend, the backend sends the returned information. The mentioned framework defines the backend platform. In addition to the mentioned validation in data entry fields (reactive forms), the backend provides the platform with additional validation using the "validate" function, which is part of the "Request" library. Validation is performed based on the set parameters that check the correctness of the received data. In addition, users are provided with tabular access to data that can be sorted and filtered. For communication between backend and frontend, API is used (Fig. 1).

The goal was to create a web platform where scientists, professors, students and other people exchange experiences about problems in artificial intelligence.

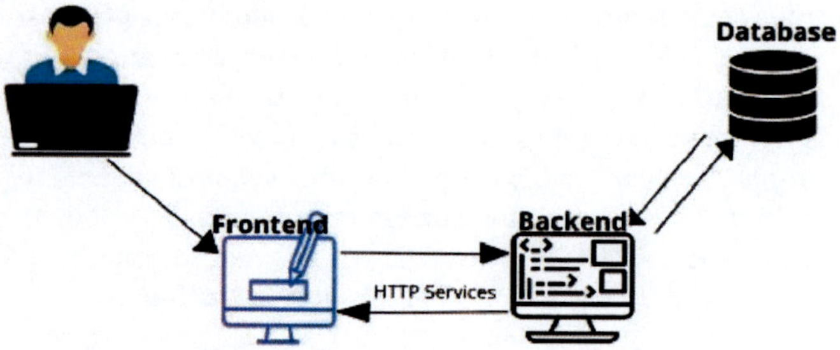

Fig. 1. Platform architecture scheme

3 Backend, Frontend and Discussion

In this section, we will describe a difference between frontend and backend. Elements which were used for creating the platform SMART2M are: Angular and Laravel (Fig. 2).

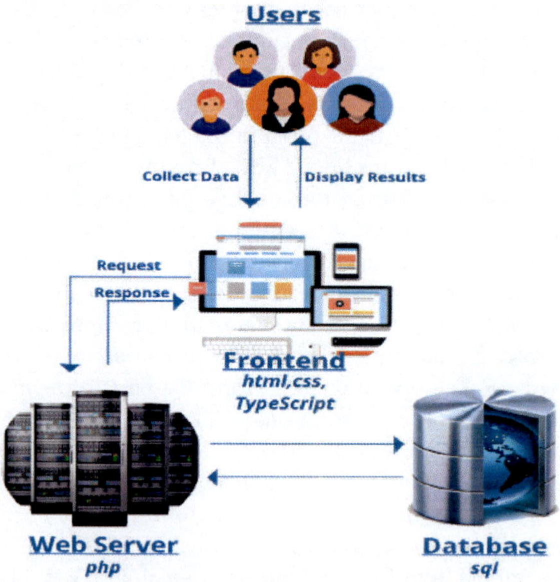

Fig. 2. Web Application Architecture Diagram

3.1 Backend

The business layer and data layer of the application were created using Laravel [8]. Laravel is framework of PHP [9] that follows MVC architecture. Laravel is simple for using [8], has excellent documentation, which is very important for developers. The platform SMART2M has built-in features like routing, migration system, CRUD operations, filtering, and sorting.

3.2 Frontend

Frontend is created by using Angular [11] with three basic technologies: HTML, CSS, TypeScript. HTML is required for site content [6,7], CSS for control of display and TypeScript for interaction. Any forms on the platform are created as reactive forms, and many validators are available in TypeScript. The site is accessible to users via a browser, and it is responsive.

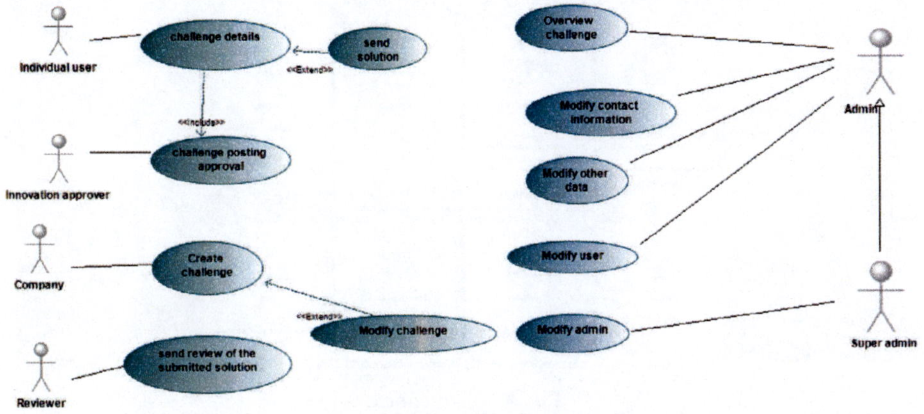

Fig. 3. Use case diagram

Use case diagram (Fig. 3) is the schematical representation of all possibilities of the actors. The platform SMART2M contains six types of actors (Individual user, Innovation approver, Company, Reviewer, Admin and Super admin). Prerequisites for any possibility are registration and login. A company creates a challenge or modifies an existing challenge, but the Innovation approver has the ability to approve the challenge. Individual user sends solutions, and Reviewer sends the review of the submitted solution. Admin and Super admin have many possibilities like overview challenge, modify contact information, other data or user, but Super admin has one feature more than the admin, and that is to modify the admin.

3.3 Public

The *Home page* is visible to everyone, and it shows four of the latest challenges and short instructions for using the platform. The short instructions are illustrated in cards (Fig. 4). Adding challenges or solutions is possible in four steps.

Setting up the challenge:

- First, the user needs to register as a company/investor or research group
- Then it is necessary to design and describe the challenge
- Review of the submitted solutions
- Selection of the most optimal

Sending a solution:

- First, the user needs to register as a solver
- Then it is necessary to read the challenge in detail
- Design and sending a solution

Fig. 4. Instructions

Possible questions can be eliminated by contacting our support team. In section Contact Us, people can fill in the fields and send their problem or suggestion for progress. The contact form has four fields (Full Name, Email, Phone Number, Message), where every field is mandarory. Messages must be concise and not confusing. The second way for contacting the support is sending an mail to one of the two emails, which are on the footer of site. The Contact Us form is shown in Fig. 5.

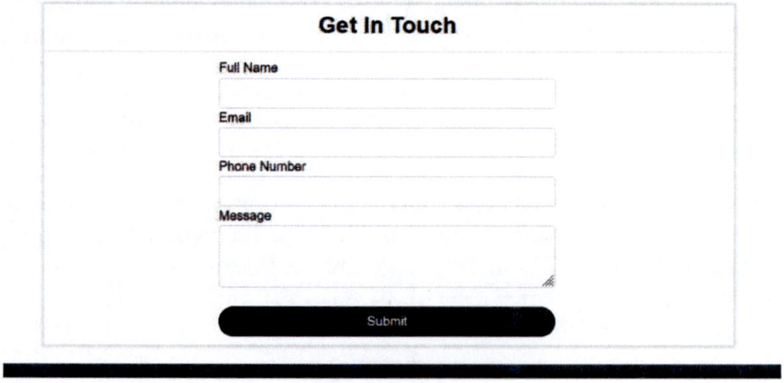

Fig. 5. Contact us form

3.4 Company, Investor or Research Group

After registration, when the user chooses Research group, Investor or Company, he/she can post their problem. The "problem" can be homework (for students), task (for competitions), or a real problem. It is necessary to fill out a form for challenges and select the type of award. Users must know that the abstract is public, and everybody can see them, but full details are only for logged-in users. It is possible to post images to describe the problem and to choose discipline. Figure 4 shows the form for adding a challenge, and the required fields are marked with an asterisk (*) (Research topic name, Award, Currency, Deadline, Abstract, Discipline, Type).

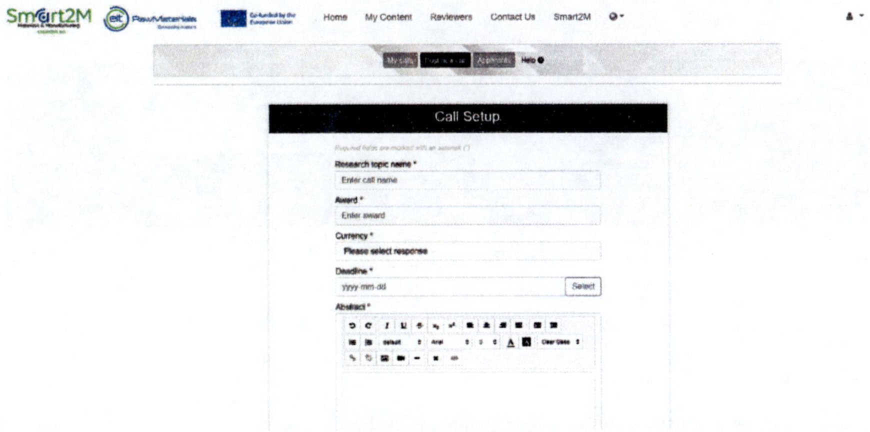

Fig. 6. Post new challenge

Awards can be in the value of money, points or corporation agreement. Companies can offer internship for the best solvers, and in that manner accomplish communication between industry and academia. This communication is very important, because for students this platform can open the door for an industry career, companies can get new people who are eager for new knowledge. Scientists from the entire world can reach their cognition by considering other people's opinions.

When the user posts a challenge, it will not be immediately visible, because it needs to be accepted by challenge's approver. When he accepts, the challenge will be visible. Challenge accepting will be described in some of the following chapters. The reason why this function is required, is to prevent hate speech or not respecting the rules of the platform. The platform will never discover the data of the challenge creator, because we don't want to endanger their position in relation to the competitors. The notification of the challenge status is sent via email to the user. The deadline may be up to several months, and it may change. If the deadline has passed before the invention solution, the user can update his challenge deadline extension. Later, the user can add reviewers on his challenge, they can be reviewers of the platform or anybody (by entering the email of the elected). Reviews help the challenges owner to make a decision about the best solution or more, and the access to applicants is shown in Fig. 6. The company can see the solutions, data of the solvers and reviews, later it can accept or reject the solution with feedback. In Fig. 5, the data is blurred for user privacy. Professors can see the time when students sent their solution, and thus determine their timeliness.

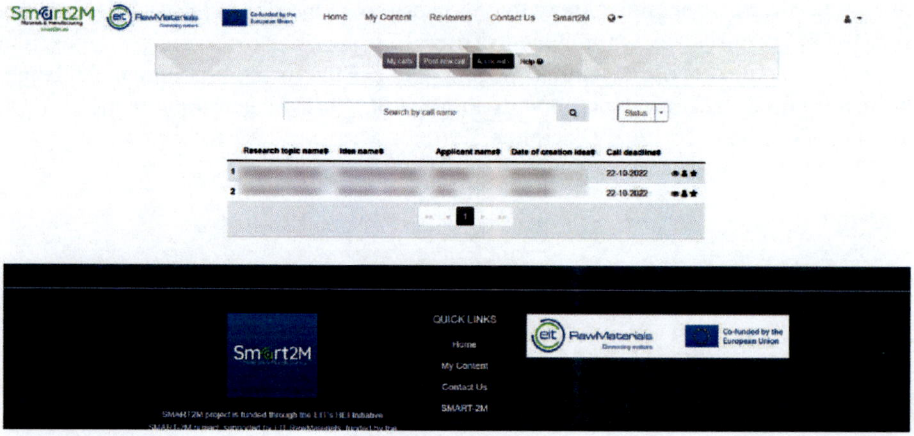

Fig. 7. Applicants

3.5 Solvers

The solvers are students, scientists or anybody who has knowledge about Artificial intelligence. In most cases challenges are very interesting and can force the solver to observe the task from another angle. By finding a challenge and coming up with potential solution, the solver develops his skills. The solver can send his solution by clicking a button at the bottom of the page with the details of the challenge. The solving implies filling out a form and sending an attachment. Users can send only one solution for one challenge, they cannot edit and delete it. If the solver is a student, he waits for feedback from the professor about his idea. On the platform SMART2M organizations can create competitions and in such a way, competitors wait for the winner announcement. The winner is chosen by reviewers and the task owner. Reviewers don't know anything about the solver's data, objectivity will be achieved in that competition. Some problems can form a new partnership between universities and companies. Students from different universities can exchange experiences in the Artificial intelligence area. For students, this is a chance to present their knowledge to people who will think about their solutions, although they don't have experience. The first step in finding a job is hard, but the platform can be that step and the solver can send solutions from his home because solvers can apply innovative solutions, from anywhere around the world. Sometimes students see different ways of solving the problem than the professionals, and maybe a better one. One solver can be the winner of multiple challenges, and one challenge can have multiple winners. When accepting the idea, an email will be sent to the solver about the status information. If the solution is rejected, the user receives the rejection reason in an email – feedback. While solvers writing their ideas do not know anything about the challenge owner, only when the winner is chosen, the company (or research group or investor) enters the process of identity disclosure. Companies hide their own identity so as not to reveal their vulnerabilities to competitors. Since the challenge owner can change the deadline and can decrease it, the solver must be fast in sending ideas.

Registration does not include the data of credit cards because the platform helps companies to find the solution for their shortcomings and solvers to show their knowledge but does not include transactions. By using the platform SMART2M, artificial intelligence gets new young scientists and a colorful range of thinking. Figure 7 shows the page for sending ideas (Fig. 8).

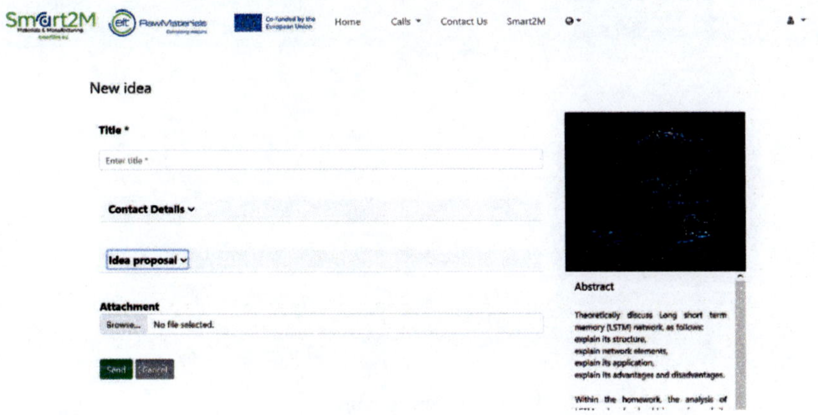

Fig. 8. New idea

Contact Details hides data like full name, email, phone number, Name of Affiliated Organization, Company or University, organization website. The data user fills the occasion registration but if he wants the data is subject to changes. Idea proposal includes Proposal brief, Capabilities, Description of idea and project plan, this information is required. A brief discussion of the solver's general ability to provide the resources listed above and information that might be useful to the challenge owner. Solvers are not expected to provide a complete solution to the problem, but, rather, to present an approach that could be pursued by the challenge owner. The Solver should explain what he/she can provide and what might be required of the Company and a brief overview of the proposed project (deliverables, timelines, milestones, and cost estimates). An attachment is not required, but it is desirable, formats for attachment are.docx or.pdf. After sending it, it is necessary to wait for the result and then the award process begins [12].

3.6 Reviewer

Before accepting the solution, there is a possibility of evaluating it and commenting on it. This possibility is desirable, but not necessary. In this way, reviewers help companies to accept the most optimal solution. A reviewer can be added by a platform administrator or by an organization. Chanceries of organizations choose reviewers and challenges whose solutions they will be in charge of. After a successful application, the page with the innovation on which it is selected is displayed. In the detailed overview of the innovation, there is a tabular representation of the solution, with a reminder of whether the solution has been evaluated (Fig. 9). In the model for displaying solutions, it is possible to enter comments and ratings, which are illustrated with stars (Fig. 10). It is important to indicate that the identity of the user who sent the solution was not revealed to the reviewer.

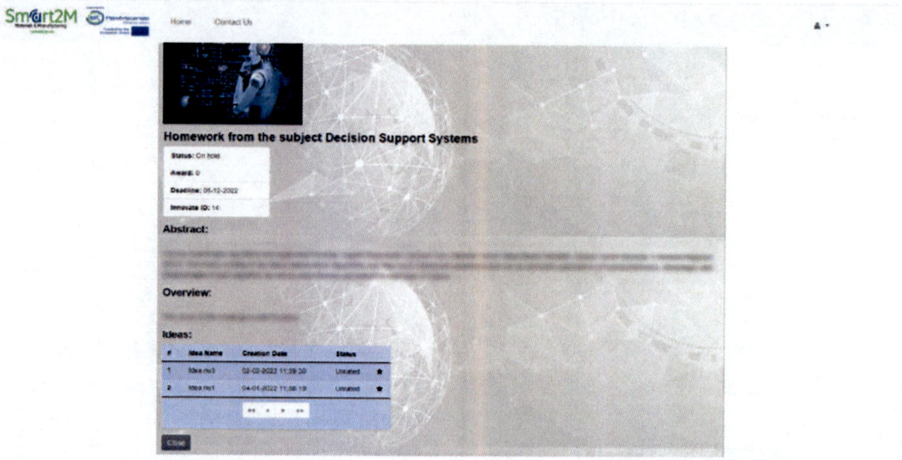

Fig. 9. Review solutions

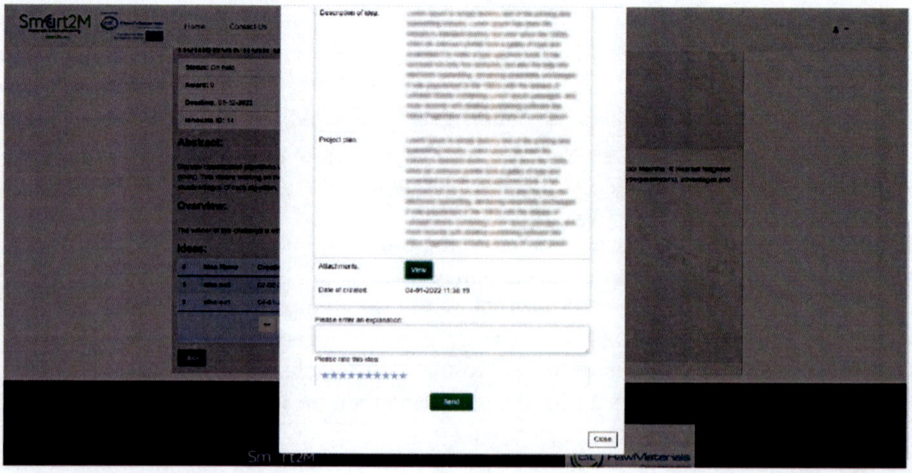

Fig. 10. Evaluation of solutions

3.7 Administrator

The administrator is the user with almost the highest privilege on the platform, and the user with the highest privilege is the super administrator who will be mentioned below. The administrator has a detailed insight into the data of all users, regardless of their type, with the exception of the super administrator. In addition, he has the ability to delete them, but not to modify their data. Users whose data the administrator has access to are: individual users who can submit solutions, companies, reviewers of solutions and users who approve challenges. The above information for the users is tabulated in the panels for administrators.

Primarily after a successful login, the administrator can see the statistics of registered users of the platform. More precisely, the administrator can see the number of challenges (regardless of their status), the total number of submitted solutions, the number of registered individual users and companies. Additionally, the best ones with the highest number of accepted solutions and the highest number of sent solutions are shown.

Depending on the selection of the item, the panels with accompanying information change. A table with their data is available on the panel of individual users. Image, username, first name, last name, email, and country are the data given in the table (Fig. 11). If necessary, the administrator can delete the user or just view his information.

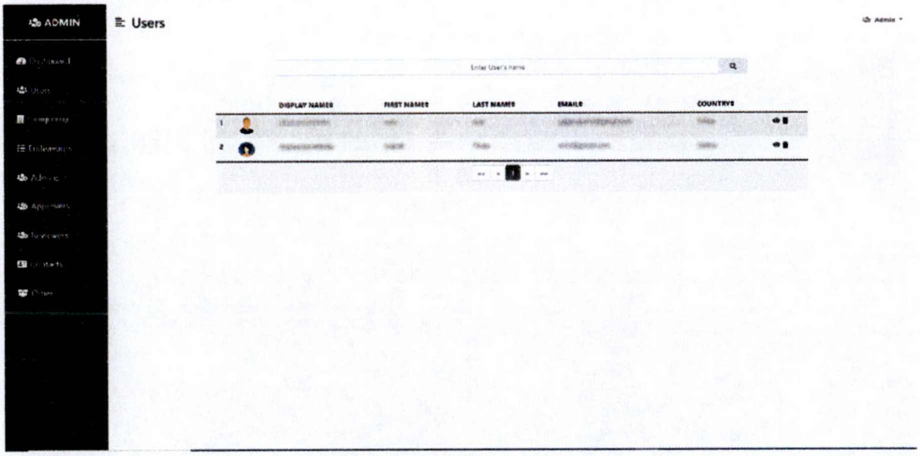

Fig. 11. Individual user view

Company data is displayed identically to that of an individual user, there is also the possibility of viewing all information, with the possibility of deleting, but not changing, the already existing data. In the table, the administrator can see the number of challenges per company. An overview of the detailed data in the module is shown in Fig. 12.

The administrator can see all challenges, those that are pending, active and those that have been returned for modification. The administrator does not have the ability to modify the challenge but he/she can delete it at any time. The table shows the challenges with the number of solutions submitted for that challenge.

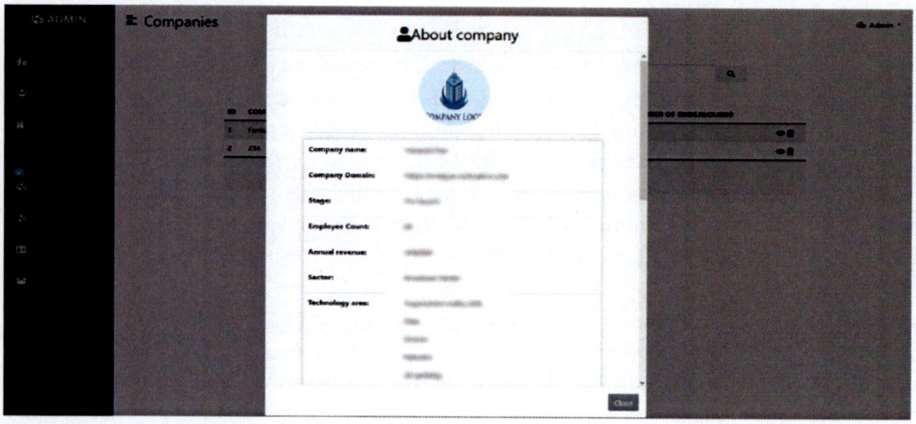

Fig. 12. Detailed view of company

If solution reviewers or challenge users need to be added, the administrator can do so by adding the user's email in the empty field (Fig. 13). It is necessary to enter a valid email with which the individual user is registered. The administrator, in addition to adding these two types of reviewers, can delete them and view their data [5].

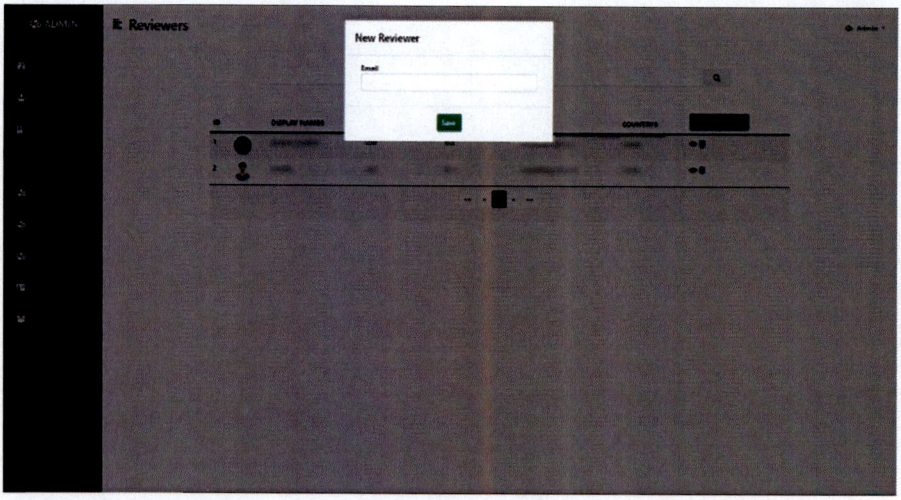

Fig. 13. Adding reviews

Contact information can be updated, deleted by the administrator, and new contact persons can also be added. When adding a new contact person, it is necessary to completely fill out the form with the required data. The data that needs to be entered are name, surname, email, and organization.

Fig. 14. Adding a contact of a person

The administrator can add, delete, and update items: sector, technology area, type and discipline (Fig. 14). The listed items are displayed in the form of a list, the sector and field of technology refer to companies and are entered by the user during registration, and the discipline refers to the challenges that are selected when adding a challenge. If the field for adding one of the listed items is validly filled, a warning message will be displayed, otherwise a success message will be displayed.

All data in the tables on the platform can be sorted in descending or ascending order. Filtering can also be performed by searching for certain parameters.

3.8 Super Administrator

Compared to the administrator, the super administrator has an additional option, which is a detailed insight into the administrator's data. In addition, he has the option of adding a new administrator by entering his email address (Fig. 15). Administrators can be deleted at any time.

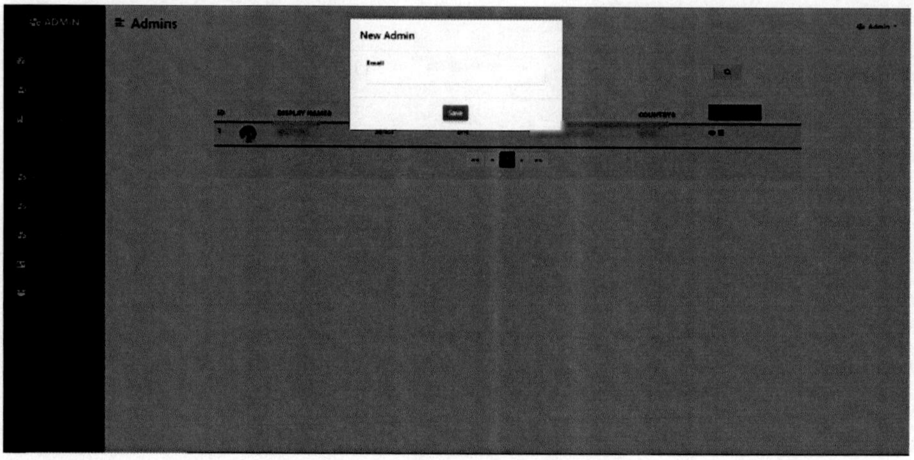

Fig. 15. Adding an administrator

4 Conclusion

The given paper represents the digitization of the methodology that is applied when solving problems within the organization. The problem-solving process itself will be accelerated because companies can set a date by which each idea should be sent. If it happens that the number of ideas is not enough or none of the ideas represents a solution to the problem, the duration of the challenge can be extended by the company. In this case, companies are not limited to one or more ideas, as if solving problems were approached by employees within the company. The platform can also be used by higher education institutions that will present their tasks/projects as challenges to students.

A simple display allows for easy use of the platform, and the storage of all information inside it prevents the loss or damage of essential information that the company needs. Additionally, all information can be accessed anytime from anywhere in the world. The data of the companies that should face the problem is hidden, so they can have the freedom to pose their problems without the fear of their competitors seeing their weaknesses. Making a decision about the most optimal solution is facilitated by reviews of users who are “in charge” of the company. Solution rating bias is prevented by hiding information about users who submitted solutions.

Users who will potentially submit solutions to challenges have an overview of all active challenges, they can access challenges daily to better pay attention to details, which can be crucial when creating solutions. In addition to the obligatory detailed description of the problem, users can immediately see the amount of the reward for the accepted solution. Access to information is enabled depending on the type of user and the type of information that needs to be accessed. Therefore, each user is protected from a hierarchically higher user. Administrators, as well as the super administrator, do not have the ability to change the user’s personal data, change the submitted solutions or change the set challenges.

The result of the described paper is successfully dealing with the problem that arose in the company. The advantages of this approach is the rapid discovery of innovative solutions that resulted from looking at the challenges from different angles, ideas are more numerous in most challenges, and the number of challenges that can be set by the company is unlimited.

Acknowledgement. This paper is funded through the EIT's HEI Initiative SMART-2M project, supported by EIT Raw Materials, funded by the European Union.

References

1. Huang, Y., Tu, Y., Wu, H., Wan, C., Yeh, C., Lu, L., Tsai, T.: Applying an innovative blended model to develop cross-domain ICT talent for university courses. In: Proceedings - Frontiers in Education Conference, FIE, Institute of Electrical and Electronics Engineers Inc. (2019)
2. Singhal, S., Ahuja, L., Monga, H.: State of the art of machine learning for product sustainability. In: Proceedings - IEEE 2020 2nd International Conference on Advances in Computing, Communication Control and Networking, ICACCCN 2020, pp. 197–202. Institute of Electrical and Electronics Engineers Inc. (2020)
3. Androutsopoulou, A., Karacapilidis, N., Loukis, E., Charalabidis, Y.: Transforming the communication between citizens and government through AI-guided chatbots. In: Government Information Quarterly, pp. 358–367. Elsevier Ltd. (2019)
4. Cheredniakova, A., Lobodenko, L., Lychagina, I.: A study of advertising content in digital communications: the experience of applying neuromarketing and traditional techniques. In: Proceedings of the 2021 Communication Strategies in Digital Society Seminar, ComSDS 2021, pp. 9–13. Institute of Electrical and Electronics Engineers Inc. (2021)
5. Kengam, J.: Artificial Intelligence in Education (2020). https://doi.org/10.13140/RG.2.2.16375.65445
6. J. Duckett, HTML and CSS: Design and Build Websites, John Wiley & Sons, 2011
7. Frain, J.: Responsive web design with HTML5 and CSS: develop future-proof responsive websites using the latest HTML5 and CSS techniques, Packt Publishing (2020)
8. Stauffer, M.: Laravel: Up & Running: A Framework for Building Modern PHP Apps, 2nd edn., O'Reilly Media, Sebastopol (2019)
9. Lerdorf, R., Tatroe, K.: Programming PHP, O'Reilly Media, Sebastopol (2002)
10. Viescas, J.: SQL Queries for Mere Mortals: A Hands-on Guide to Data Manipulation in SQL, 4th edn., Addison-Wesley Professional, New York (2018)
11. Freeman, A.: Pro Angular 9: Build Powerful and Dynamic Web Apps, 4th edn., Apress, New York (2020)
12. Mitrevski, N., Ilić, D., Marković, J., Šušteršič, T., Živić, F., Grujović, N., Filipović, N.: SMART2M digital platform as the communication channel for academia - industry collaboration. In: ICIST 2022 Proceedings, pp. 214–217 (2022)

Mathematical Modeling of COVID-19 Spread Using Genetic Programming Algorithm

Leo Benolić[1(✉)], Zlatan Car[2], and Nenad Filipović[1,3]

[1] Bioengineering Research and Development Centre (BioIRC), 6 Prvoslava Stojanovića Street, 3400 Kragujevac, Serbia
{leo.benolic,fica}@kg.ac.rs

[2] Faculty of Engineering, University of Rijeka, 58 Vukovarska Street, 51000 Rijeka, Croatia
car@riteh.hr

[3] Faculty of Engineering, University of Kragujevac, 6 Sestre Janjića Street, 34000 Kragujevac, Serbia

Abstract. This paper analyses the possibilities of using Machine learning to develop a forecasting model for COVID-19 with a publicly available dataset from the Johns Hopkins University COVID-19 Data Repository and with the addition of a percentage of each variant from the GISAID Variant database. Genetic programming (GP), a symbolic regressor algorithm, is used for the estimation of new confirmed infected cases, hospitalized cases, cases in intensive care units (ICUs), and deceased cases. This metaheuristics method algorithm was used on a dataset for Austria and neighboring countries Czechia, Hungary, Slovenia, and Slovakia. Machine learning was done to create individual models for each country. Variance-based sensitivity analysis was initiated using the obtained mathematical models. This analysis showed us which input variables the output of the obtained models is sensitive to, like in the case of how much each covid variant affects the spread of the virus or the number of deceased cases. Individual short-term models have achieved very high R2 scores, while long-term predictions have achieved lower R2 scores.

Keywords: Artificial intelligence · COVID-19 · genetic programming · mathematical prediction models · variants

1 Introduction

COVID-19 began to spread in 2019 in Wuhan, China, within three months, it started to spread in every province of mainland China and eventually to the other 27 states [1]. The virus comes from the coronavirus family of diseases from bats, it was renamed to COVID-19 to distinguish it from coronavirus Severe Acute Respiratory Syndrome (SARS) and Middle East Respiratory Syndrome (MERS), where SARS and MERS share 79% and 50% of the genome sequence respectively [2]. The characteristics of the virus have led to the fact that today, after more than two years, we have over 500 000 000 cumulative confirmed cases of COVID-19 infection [1]. The most recent variant

N. Filipovic (Ed.): AAI 2022, LNNS 659, pp. 320–331, 2023.
https://doi.org/10.1007/978-3-031-29717-5_19

is omicron, which has higher transmissibility but decreased level of symptom severity and mortality, indicated by the fact that in Europe, most countries have 20000–50000 cumulative confirmed cases per million, which means that almost half of the people were naturally exposed to the virus and vaccination of 75% [3]. After the end of the Omicron wave, many EU countries are returning to normal. Although the interest devoted to the coronavirus is currently declining, collected data regarding the virus should be used as much as possible so that we can better prepare for future similar threats by creating an *in silico* epidemiological model.

This paper investigates the possibility of using machine learning techniques, more precisely symbolic regression genetic programming (GP) algorithm, which is actively used in medical research. D'Angelo et al. [4] published a proposal for distinguishing between bacterial and viral meningitis using genetic programming and decision trees which revealed that the GP shows good results only with a few false positives on the trained dataset of blood and cerebrospinal parameters. Ain et al. [5] show the use of Genetic Programming for feature selection and construction for skin cancer image classification. Authors conclude that GP is providing a much better or comparable performance in most cases of classification. The same authors also have another publication where it implements the use of local and global image-extracted features to the same problem and they conclude that using GP provides better results[6]. Tan et al. [7] show a GP approach to oral cancer prognosis. With a data set of 31 cases, it is able to achieve average scores of 83.87% accuracy and an AUC score of 0.8341 for the classification task.

Similar attempts of COVID-19 estimation have been made in the literature. For instance, this algorithm was used for that with a short dataset of only the beginning of the pandemic [8]. This research is based on the COVID-19 dataset from Austria and its neighboring countries, Czech, Hungary, Slovakia, and Slovenia. Time-series variables were used as input data for obtaining a predictive model of the future state in the form of a mathematical equation of newly confirmed cases, hospitalized, cases in the intensive care unit, and deceased cases. The goal is to obtain satisfactory accuracy for a longer period of time that could be used to plan lockdowns and increase the capacity of COVID hospitals, along with analyzing the importance of input model variables such as the percentage of each covid variant.

2 Materials and Methods

In order to facilitate and accelerate machine learning, characteristics of the spread and behavior of viruses over time have to be determined. A paper by Wang, F. et al. (2020) has defined the timeline of COVID-19 cases in the first month of hospital admission, which is useful to adjust the input to facilitate and speed up machine learning [9]. For ML, the dataset will be used from COVID-19 Data Repository by the Center for Systems Science and Engineering (CSSE) at Johns Hopkins University due to its standardized form [3]. Also "GISAID EpiCoV" database will be used, where they tested the percentage of each variant [10].

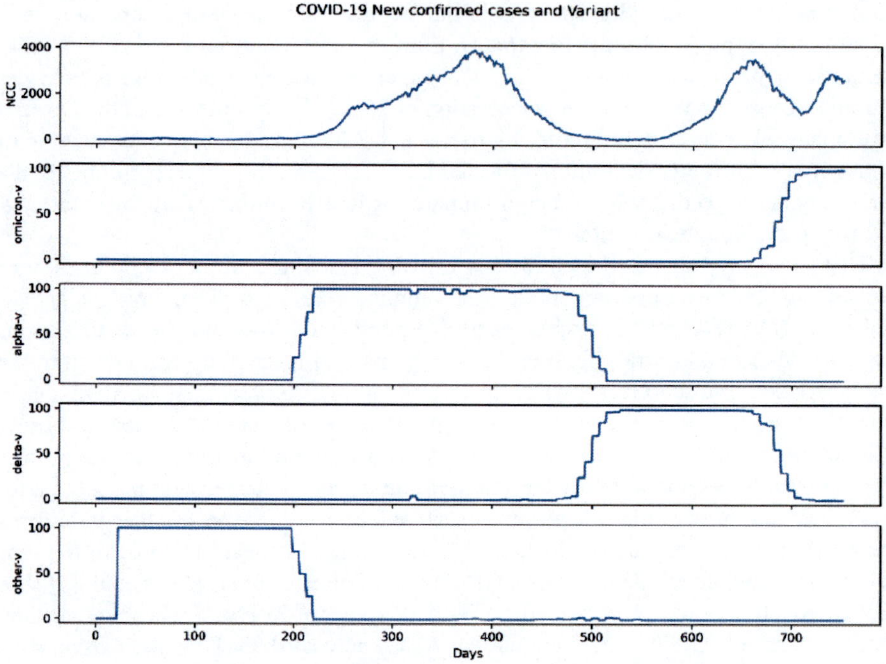

Fig. 1. Example of number of newly infected and percentage of each variant

Figure 1 shows that after increasing the percentage of new variants, the number of newly infected patients also increases. With that increase, an increase in new variants is also shown to be dominant. This is a result of more infections and the spread of the virus. It was necessary to insert the percentage of each variant as an input to increase the accuracy of the model because each variant has its own characteristics in terms of transmissibility, hospitalization, and mortality [10].

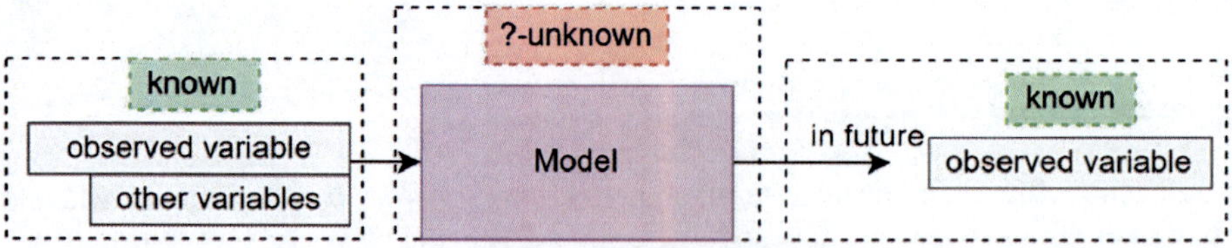

Fig. 2. Black-box model

Figure 2 displays the input and output data, such as positive rate, reproduction rate, population size, variant, newly confirmed cases, deceased cases, etc. In the time-dependent model, the future output value depends on the past value. To find the approximation of the model machine learning will be used, and for that, the dataset is divided into 70% for training and 30% for testing.

GP algorithm was chosen as the ML learning method as the GP algorithm is a metaheuristic method inspired by Charles Darwin's theory of natural evolution [11]. The GP symbolic regressor is able to create a symbolic mathematical function that best describes the given data. The particularly strong point of GP is that the result is universal, generally understandable and can be easily transferred to another program/environment. In the GP the mathematical function is written in the form of a tree, where the functions are nodes and the sheets are variables or constants. The next figure is an example of equation = (X0*3.4 + sin(X1) in the form of a tree (Fig. 3).

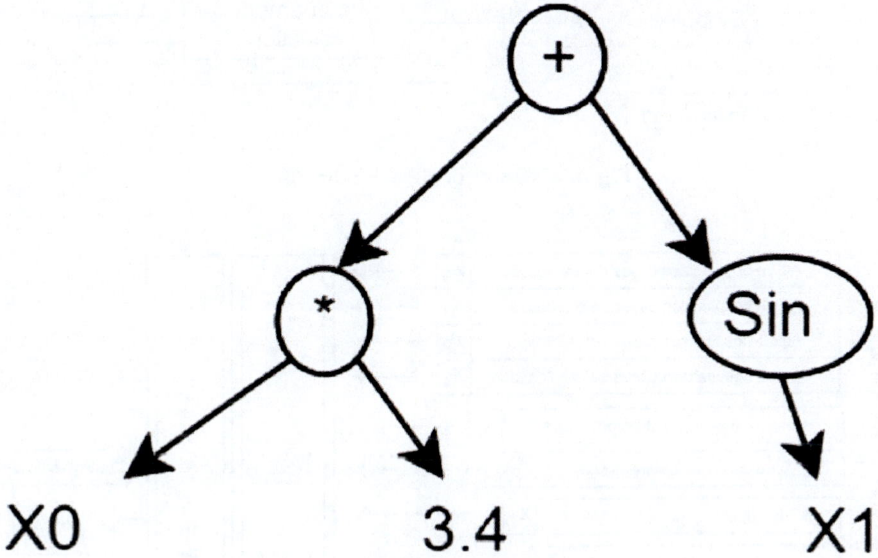

Fig. 3. Example of equation tree

Nodes have possibility to be different function from the list -function set [add, sub, mul, div, sqrt, log, abs, neg, inv, max, min, sin, cos, tan], leaves -determined in the terminal set for constant values of defined range or variables [model input X0 X1 X2 Xx3, etc.]. Nodes and leaves are primarily obtained randomly, they are altered by the process of mutation reproduction and crossover. After the implementation of the genetic operation, the offspring population is evaluated to assess the quality of the results and to select the best results with tournament selection that will participate in the next iteration of the genetic algorithm. The algorithm terminates this loop after reaching the stopping criteria or the maximum number of generations. The working principle is shown in Fig. 4 [12].

GP is not sequential nor time dependent, and doesn't have memory because it is an algorithm that sets the past input values of time series in multiple points and other variables for prediction of the future value. Figure 4 shows the configuration of inputs for the prediction of the death of people one week after hospitalization.

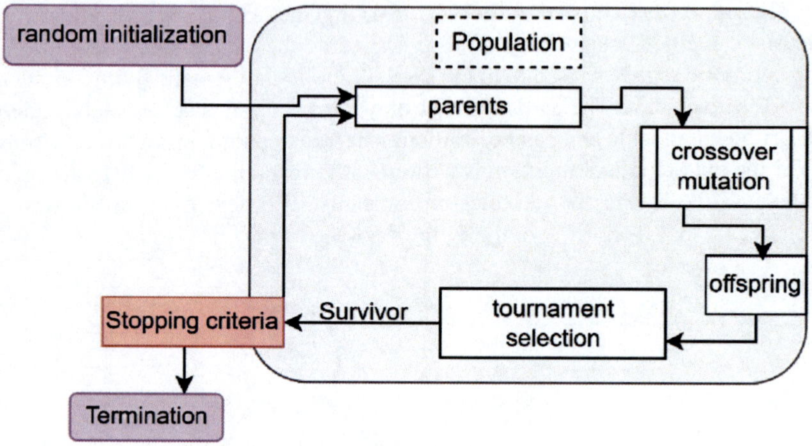

Fig. 4. Example of GP algorithm flow

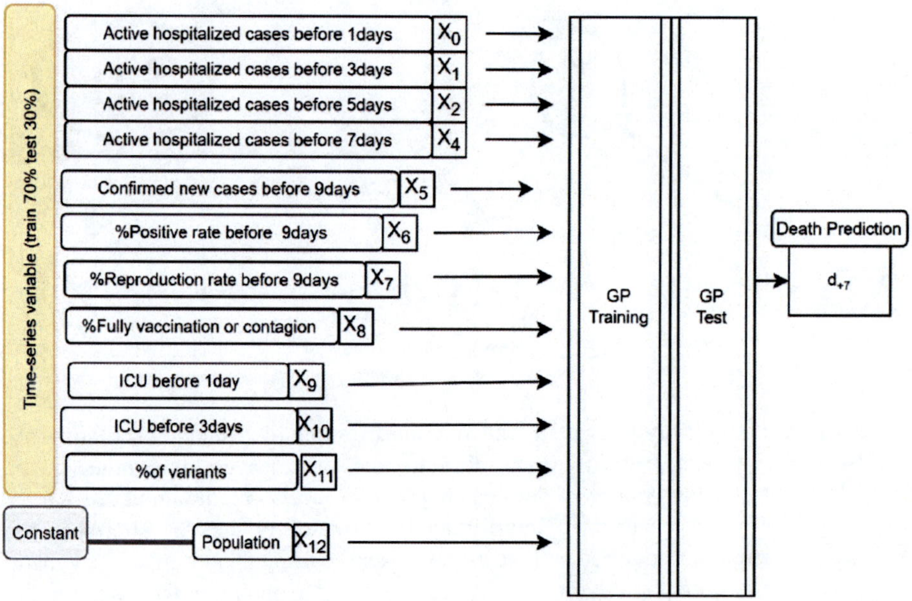

Fig. 5. Input configuration for training GP

With data from the literature [9], it is assumed that the number of deaths in the near future mostly depends on the number of:

- ICU – (input X_9) and X_{10}) but this data is not always available (for example: for Hungary the data is missing)
- Active hospitalized cases (X_1, X_2, X_3 and X_4)
- New confirmed cases (X_5), this input is less relevant in the short prediction of 1 week.

In order to adjust the behavior of the algorithm, it is necessary to define the hyper-parameter rules or at least their range (Fig. 5 and Table 1).

Table 1. Hyper parameters

	Parameter for Ncc and Ah		Parameter for ICU and Nd	
	min	max	min	max
Population	175	450	250	650
Generation	150	350	250	400
Tournament size	21	40	18	50
Crossover probability	0.9	0.96	0.9	0.96
Subtree mutation probability	0	0.09	0	0.09
Hoist mutation probability	0.01	0.08	0.01	0.04
Point mutation probability	0	0.09	0	0.09
Sum of probability	0.9	1	0.9	1
Constant range	-2000	3000	-200	1000
The minimal initial tree depth	4	8	4	8
The maximal initial tree depth	11	11	9	14
Stopping criteria coefficient	0,1		0,01	
Parsimony coefficient	0.0001	0.01	0.001	0.1
Metrics	RMSE		RMSE	
Function set	add, sub, mul, div, sqrt, log, abs, neg, inv, max, min, sin, cos, tan			

Model quality estimation:

The obtained model is evaluated using the coefficient of determination R2. It is a measure of how well the statistical model fits the investigated data. It is the proportion of variance in the dependent variable that is explained by the model. Equation:

$$R^2 = 1 - \frac{SSR}{SST} \tag{1}$$

where: SSR is a Sum of Squared Regression (variation explained by the model), and SST is the Sum of Squared Total (total variation in data) [13].

3 Results and Discussion

Figure 6 shows the real number of deceased cases (blue curve) and the estimation of the deceased cases (orange curve) for a period of 2 weeks. Both the curves are similar in nature, where the number of deceased cases (blue curve) has more fluctuation than the estimation (orange curve), thus the algorithm precisely estimates the number of deceased cases with R2 = 0.9537.

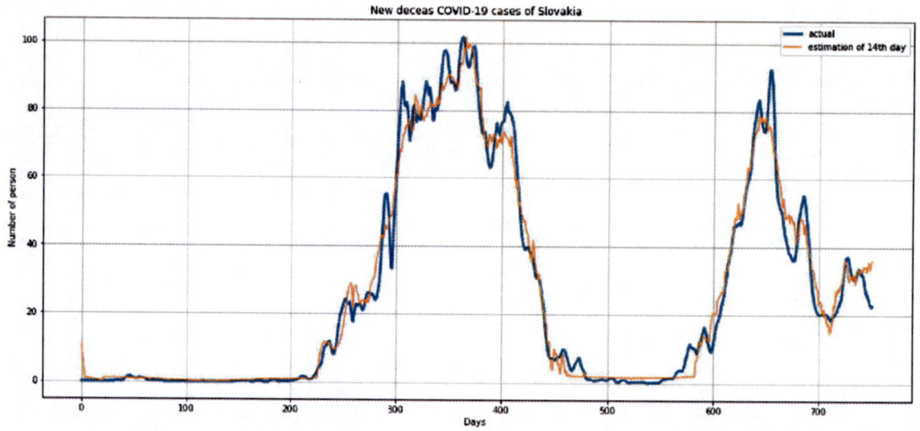

Fig. 6. Estimation of new deceased cases for Slovakia 2nd week

By displaying all predictions of one country for the 1st, 2nd, 3rd, and 4th week on one graph, we get Fig. 7, where on the bottom we have the entire time of 763 days. On the upper part of the image, the enlarged predictions on the 275th and 415th day are displayed.

The R2 score of the prediction shown in Fig. 7 is shown in Table 2, other scores are shown in Table 3, 4 and 5.

The best result is achieved with the model for deceased cases, where all models have an R2 score greater than 0.919. The country model with the best result is for Czechia with a score greater than 0.975. The worse result has the model for long term prediction (2th–3th week) for hospitalization cases, with average score of 0.93. The average R2 of all model for 1st, 2nd, 3rd and 4th week are $R2_{1st} = 0.98$, $R2_{2nd} = 0.966$, $R2_{3rd} = 0.958$ and $R2_{4th} = 0.923$ respectively.

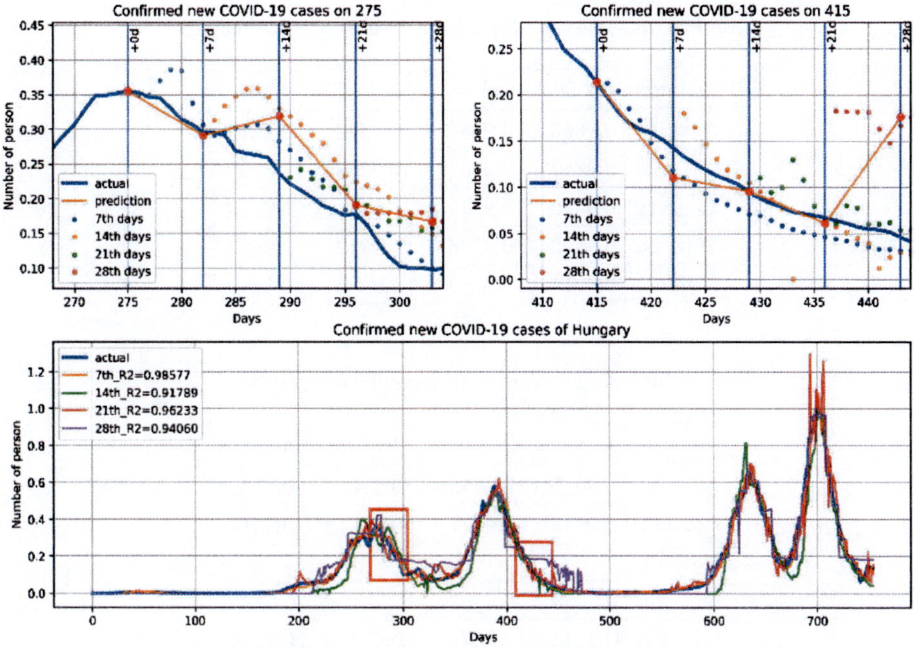

Fig. 7. Estimation of new confirmed cases for Hungary 1st, 2nd, 3rd, and 4th week

Table 2. R2 score of new confirmed cases prediction models

Country	1st week	2nd week	3rd week	4th week
Austria	0.9834	0.9561	0.9578	0.9660
Czechia	0.9671	0.9422	0.9372	0.8355
Hungary	0.9858	0.9179	0.9623	0.9406
Slovakia	0.9559	0.9751	0.9651	0.9359
Slovenia	0.9898	0.9805	0.9792	0.9249

Figure 8 shows the correlation between past NCC and ND values and as it can be seen, the correlation after 1–2 weeks increases (only Slovakia has two maxima, but the other maximum is probably an error due to the proximity of the two variants) and due to that, a delay was added to the previous diagram. Austria and Slovenia have a low correlation (0.2–0.4), and other countries: Czechia, Hungary, and Slovakia have a medium correlation (0.4–0.6).

The importance of the input variables of the model is analyzed on the obtained mathematical expression using the Sobol sensitivity analysis package from sib.lib for python that shows which input variables influence the output of the mathematical model the most. This indicates which input variables have the greatest impact on the output of the obtained mathematical model, and the sensitivity of the model (Fig. 1) to omicron,

Table 3. R2 score of hospitalized cases prediction models

Country	1st week	2nd week	3rd week	4th week
Austria	0.9866	0.9656	0.9554	0.8613
Czechia	0.9867	0.9765	0.9258	0.8797
Hungary	0.9826	0.9779	0.9586	0.9298
Slovakia	0.9898	0.9838	0.9812	0.9622
Slovenia	0.9914	0.9792	0.9686	0.8486

Table 4. R2 score of ICU prediction models

Country	1st week	2nd week	3rd week	4th week
Austria	0.9652	0.9555	0.9305	0.9283
Czechia	0.9939	0.9771	0.95512	0.8816
Hungary	-	-	-	-
Slovakia	0.9858	0.9757	0.961	0.9584
Slovenia	0.9843	0.9541	0.9622	0.9579

Table 5. R2 score of deceased cases prediction models

Country	1st week	2nd week	3rd week	4th week
Austria	0.9745	0.9622	0.9421	0.9191
Czechia	0.9863	0.9756	0.9806	0.9753
Hungary	0.9804	0.9865	0.9827	0.9685
Slovakia	0.9674	0.9537	0.9279	0.9333
Slovenia	0.9763	0.9527	0.9717	0.9314

delta, alpha, and other variants. The obtained results for the aforementioned variants (other variants include all variants at the beginning of the pandemic until the moment they began to be tested and separated) are as follows: mean Sobol index for other variant percentage is 0.48534, followed by delta (B.1.617.2) with 0.04045 and alpha (B.1.1.7) with 0.00057 while omicron does not affect the output and its Sobol index is 0. The Sobol index depending on the number of newly infected is shown in Fig. 9.

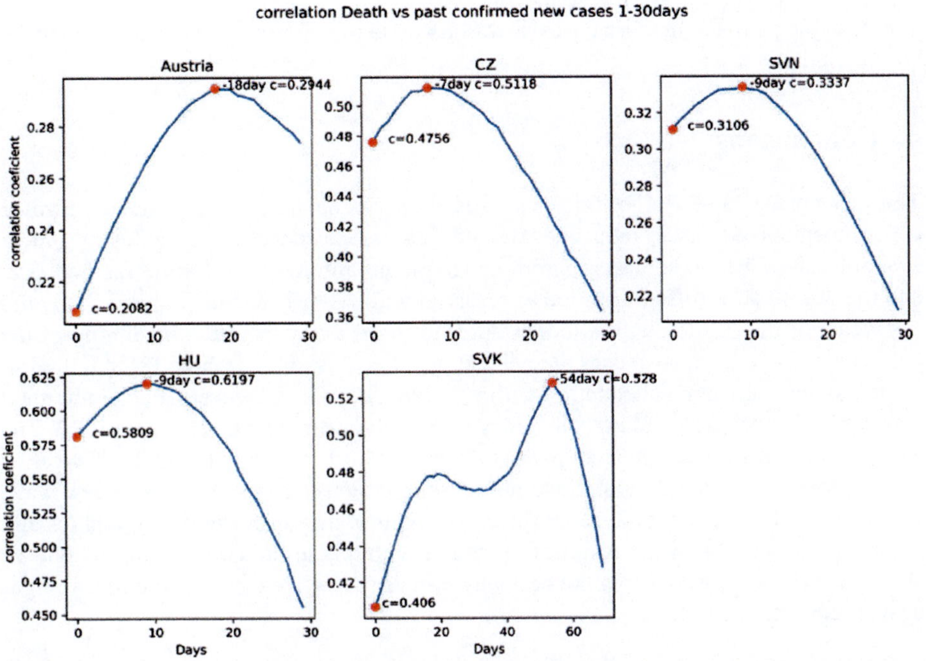

Fig. 8. Pearson's correlation between deceased cases and new confirmed cases on past time-series

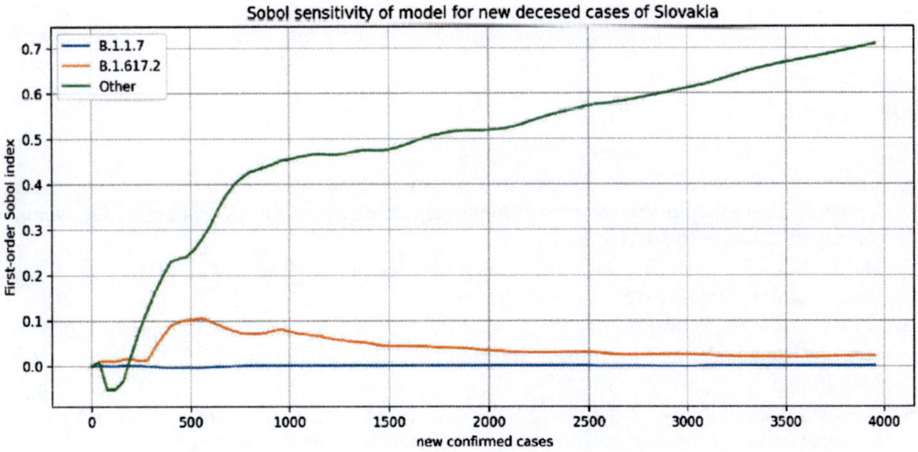

Fig. 9. Sobol sensitivity of estimation model for new deceased cases for Slovakia

The result is an understandable omicron variant that does not affect output because it has lower mortality, but it is strange for delta and alpha because they have higher mortality. The results give the highest correlation with other variants probably due to excess mortality in the beginning when states were not ready and they didn't know how to approach the patients. In these cases, it was assumed that mortality was more associated with the other variant.

4 Conclusions

This paper presents a GP algorithm used to develop a model for estimating new confirmed cases, hospitalized cases, intensive care unit cases, and deceased cases using online available data. The model was programmed to predict future values for the 1st, 2nd, 3rd, and the 4th week. It differs from other research with GP because the input is the entire two years of the pandemic, the model does not need a day from the beginning of the pandemic as input, and also uses the percentage of each variant as input [8]. All models result from a mathematical equation with variables X0, X1, X2 and etc. Each individual country model has archived high accuracy with high Coefficient determination R2. The average R2 drops for longer term predict, from 0.98 in 1st week to 0.923 in 4th week.

The virus is constantly mutating, and more infectious variants result in a new wave of infection. Spread is not the same for every country, they differ in demographics and other factors, and this is the result of modeling with this limited data and time, but we should use the tools we have in the best way because these are the best solutions to fight against the virus.

Acknowledgment. This research is supported by the project that has received funding from the European Union's Horizon 2020 research and innovation programmes under grant agreement No 952603 (SGABU project). This article reflects only the author's view. The Commission is not responsible for any use that may be made of the information it contains.

References

1. Coronavirus disease (COVID-19) Weekly Epidemiological Updates and Monthly Operational Updates. [Online]. https://www.who.int/emergencies/diseases/novel-coronavirus-2019/situation-reports. Accessed 30 Mar 2022
2. Hu, B., Guo, H., Zhou, P., Shi, Z.L.: Characteristics of SARS-CoV-2 and COVID-19. Nat. Rev. Microbiol. **19**, 141–154 (2021)
3. COVID-19 Data Repository by the Center for Systems Science and Engineering (CSSE) at Johns Hopkins University. https://github.com/CSSEGISandData/COVID-19. Accessed 30 Mar 2022
4. D'Angelo, G., Pilla, R., Tascini, C., Rampone, S.: A proposal for distinguishing between bacterial and viral meningitis using genetic programming and decision trees. Soft. Comput. **23**(22), 11775–11791 (2019). https://doi.org/10.1007/s00500-018-03729-y
5. Ain, Q.U., Xue, B., Al-Sahaf, H., Zhang, M.: Genetic programming for feature selection and feature construction in skin cancer image classification. In: Geng, X., Kang, B.-H. (eds.) PRICAI 2018. LNCS (LNAI), vol. 11012, pp. 732–745. Springer, Cham (2018). https://doi.org/10.1007/978-3-319-97304-3_56
6. Ain, Q.U., Al-Sahaf, H., Xue, B., Zhang, M.: A multi-tree genetic programming representation for melanoma detection using local and global features. In: Mitrovic, T., Xue, B., Li, X. (eds.) AI 2018. LNCS (LNAI), vol. 11320, pp. 111–123. Springer, Cham (2018). https://doi.org/10.1007/978-3-030-03991-2_12

7. Tan, M.S., Tan, J.W., Chang, S.W., Yap, H.J., Kareem, S.A., Zain, R.B.: A genetic programming approach to oral cancer prognosis. PeerJ **4**, e2482 (2016)
8. Anđelić, N., Baressi Šegota, S., Lorencin, I., Mrzljak, V., Car, Z.: Estimation of COVID-19 epidemic curves using genetic programming algorithm. Health Inform. J. **27**(1) (2021)
9. Wang, F., et al.: The timeline and risk factors of clinical progression of COVID-19 in Shenzhen China. J. Transl. Med. **18**(1), 1–11 (2020)
10. GISAID, [Online]. https://www.gisaid.org/hcov19-variants/. Accessed 30 Mar 2022
11. De Jong, K.: Learning with genetic algorithms: An overview. Mach. Learn. **3**(2), 121–138 (1988)
12. Poli, R., Langdon, W.B., McPhee, N.: A field guide to genetic programming (2009)
13. Turney, S.: Coefficient of determination (R^2) | Calculation & Interpretation. Scribbr. 14 September 2022. [Online]. https://www.scribbr.com/statistics/coefficient-of-determination/. 3 May 2022

Liver Tracking for Intraoperative Augmented Reality Navigation System

Lazar Dašić(✉)

Bioengineering Research and Development Center (BioIRC), Prvoslava Stojanovića 6, 34000 Kragujevac, Serbia
lazar.dasic@kg.ac.rs

Abstract. Liver cancer is one of the main causes of cancer-related deaths in the world. To treat this deadly disease, open liver surgery is still the preferred method. Due to the complexity of the liver structure, the addition of augmented reality (AR) navigation would provide the surgeon with a necessary 3D anatomy of the patient's liver. In this study, we propose an AR navigation system where the main focus is on providing automatic 3D model registration during the operative process. The difficulty in the process of registration, comes from the fact that the liver is not a static organ due to the deformation and movement of the soft tissue. We propose a method that finds liver movement by constantly tracking a patient's organ during operation, using liver features detected by Shi-Tomasi Feature Detector. To be able to track only liver movement and deformation, it was important first to segment only the organ itself. Liver segmentation was performed using the image segmentation method based on HSV color space.

Keywords: augmented reality · liver segmentation · open liver surgery · tracking system

1 Introduction

In the period between 2018 and 2020, liver cancer advanced from having the third-highest mortality rate amongst all cancers to the second-highest [1]. Liver metastases, as well as primary liver cancers (hepatocellular carcinoma - HCC) are the main contributors to these statistics. HCC is more commonly found in males than in females and is the third leading cause worldwide for all cancer deaths [2]. Regions of the world with the highest liver cancer rates are the same areas with the highest exposure to hepatitis B virus (HBV) and hepatitis C virus (HCV): East Asia, Western and Middle Africa [3]. In the Western cultures, the majority of liver cancer instances are caused by alcohol-induced cirrhosis and obesity-related fatty liver [4].

Even though there are different therapy approaches available, such as chemotherapy and immunotherapy, surgery is still considered as the best method for liver cancer treatment, especially for patients with early-stage HCC. Generally, resection showed good results in patients with sufficient liver-healthy tissue and surgically feasible tumor locations [5]. For these types of patients, the survival rate after resection is between 40% and

N. Filipovic (Ed.): AAI 2022, LNNS 659, pp. 332–341, 2023.
https://doi.org/10.1007/978-3-031-29717-5_20

70% [6]. Surgical methods have evolved over time and one of those advancements in the surgical field is certainly laparoscopic surgery. However, laparoscopy is a highly skilled procedure that requires an expert surgeon and special equipment. Perhaps, the biggest difficulty with laparoscopic surgery is the lack of tactile feedback. Tactile feedback is important for finding the location of tumors and for orienting the surgeon during the operating procedure. These are the reasons why globally open liver surgery still remains the most common resection method for liver cancer treatment. A proper surgical resection involves the complete removal of tumors while preserving the surrounding healthy tissues, as well as blood vessels and a biliary tree. Still, due to the complexity of the liver, proper surgical resection is quite a challenge, requiring a high level of expertise and extensive preoperative planning.

1.1 Virtual and Augmented Reality in Medicine

The fundamental technology of virtual reality (VR) is the creation of an artificial environment that can be experienced through the use of virtualized (computer-simulated) graphics where an individual can interact with the events in the virtual three-dimensional (3D) world [7]. It works by surveying the existing environment and modeling objects accordingly to mimic a real-world scenario. On the other hand, augmented reality (AR) represents the real-time direct or indirect view of a physical, real-world environment that has been enhanced/augmented by adding virtual computer-generated information to it [8]. While the VR environment is completely synthetic and it separates the user from the real world [9], the AR system combines virtual objects with reality in real time.

Virtual and augmented reality found usage in a wide variety of applications, where one of the fields with the highest utilization of these technologies is medicine. In healthcare, AR/VR are used for surgery, training and education, diagnostics, rehabilitation, etc.

Augmented reality has been especially useful in assisting surgeons during operative procedures where additional information is augmented during surgery. AR is usually used in the following surgical procedures:

1. Neurosurgery – working with the brain requires utmost focus and precision. By superimposing 3D structures of the brain into the operative view, neurosurgeons have computer-aided assistance for brain tumor dissection.
2. Soft-tissue surgery – developing an AR navigation system for soft-tissue surgery proved to be difficult to shift and deform due to the nature of these organs. The augmented 3D model needs to be able to follow changes in the real organ during surgery. The main focus has been put on developing navigation systems for liver surgery.
3. Orthopedic surgery - AR is most often used as a guide to properly align the placement of implants in hip and knee replacements.

1.2 Augmented Reality for Liver Resection

Traditionally, the planning stage in liver surgery consisted of the analysis of the patient's liver magnetic resonance imaging (MRI) or computed tomography (CT) scans, but due to the liver's complex anatomy, assessments based on simple MRI or CT scans are not adequate. In the last two decades, the preoperative phase consists of combining aforementioned imaging data in a way that gives three-dimensional visualization of liver anatomical structure. Several studies showed that livers resection that used 3D models in preoperative planning had better resection margins and oncological outcomes [10, 11]. Knowledge obtained in this phase is transferred into the operative field, because liver resection requires a reliable intraoperative navigation system. Over the past decades, the main navigation tool has been ultrasonography, but this approach is heavily dependent on the operator's radiological skills. Recent advancements in medicine focus on using a 3D model from the preoperative phase as a way of intraoperative navigation. This addition of augmented reality navigation provides the surgeon with a necessary 3D anatomy of the patient's liver. AR-assisted surgery is a surgical tool utilizing technology that superimposes a computer-generated enhanced image on a surgeon's view of the operative field, thus providing a composite view [12].

The development of AR navigation systems for liver resection is an interesting topic with a lot of advancements in the last couple of years. Most of the research is done for the laparoscopic resection navigation system, where the goal is to provide guidance of surgical tools [13, 14]. Besides laparoscopy and open liver resection, AR intraoperative navigation systems have been developed for robotic surgery as well. Buchs et al. [15] superimposed a 3D model that allowed for the visualization of the target tumor, relative to the tip of the robotic instrument, for an assessment of the distance between the tumor and the tool for the realization of safe resection margins. Clements et al. [16] based their solution on saliant anatomical features, using only an endoscopic camera. Gavaghan et al. [17] used a device to project virtual information directly on the liver surface, however, this method is not applicable in clinical practice due to organ deformations. With the main difficulty working with 3D models of soft-tissue organs being the fact that they are constantly deforming, there has been active research on trying to overcome registration errors on organ deformations during breathing [18]. Pelanis et al. [19] proposed a solution that relies on a robotic C -arm to perform registration to preoperative CT/MRI image data and allows for intraoperative updates during resection using fluoroscopic images. Their solution achieves overall median accuracy of 4.44 mm with a maximum error of 9.75 mm over the four subjects they tested. Zhang et al. [20] used the Go-ICP method to register the preoperative 3D model and intraoperative video image and if the effect of this automatic registration was not satisfactory, their system provides a manual registration function.

2 Materials and Methods

The main steps in using an AR-based intraoperative navigation system consist of:

1. Acquisition of preoperative imaging data (MRI, CT) of patient's liver – for years MRI and CT scans have been used for diagnostics of various liver diseases. These methods produce highly detailed images that give experts a high level of accuracy in the diagnosis of several liver conditions. By using MRI/CT medical experts have a clear view of the liver's structures, as well as atypical growths.

2. Imaging data preprocessing and 3D liver model creation – medical imaging data collected in the preoperative phase could be processed with various software tools to create 3D models of the patient's liver. This model can be used to make the liver structure even more comprehensive prior to resection, or it can be used as a guidance during the surgery itself.
3. 3D model registration during operative process - the third and most difficult step, that is registration, consists of correct direct matching between operative view and produced 3D model.

Finite elements (FE) mesh 3D model, that consists of livers tissue, blood vessels, bile ducts and tumor tissue, was obtained by segmenting preoperative DICOM images from CT scans. Figure 1 shows the 3D model which was created. Special attention should be brought on the red lumps of tissues that represent tumors that need to be removed.

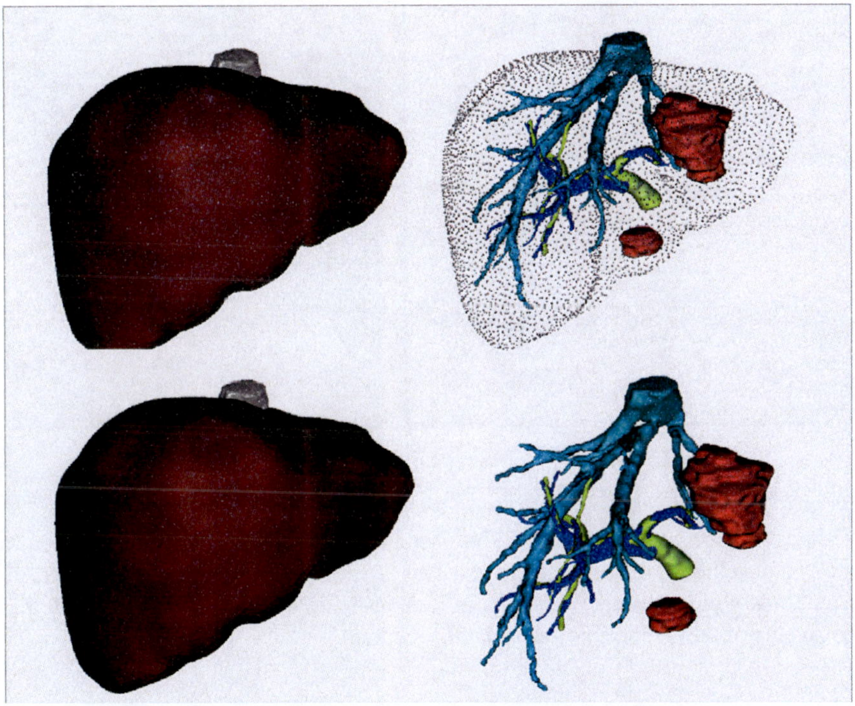

Fig. 1. Model created by 3D reconstruction

This 3D model should be overlayed over the actual organ in the navigation system. However, the process of registration comes with a lot of challenges. These complications during the registration process come from the fact that the 3D liver model is a static snapshot, while the liver itself is not due to constant deformation and movement of the soft tissue. The main causes of this deformation are heartbeat, breathing, tissue dissection and surgeon positioning, all of which alter the anatomy and the position of the liver [21]. The aforementioned drawbacks are limiting factors for the clinical usage of intraoperative navigation systems for liver resection, unlike the intraoperative navigation systems for orthopedic procedures and neurosurgery.

2.1 Manual Registration

The registration process can be manual, semi-automatic and completely automated. In the case of manual registration, the generated 3D model is shown on the monitor and manually positioned and scaled in a way that properly aligns the model and a real liver. In Fig. 2, the result of this approach is shown.

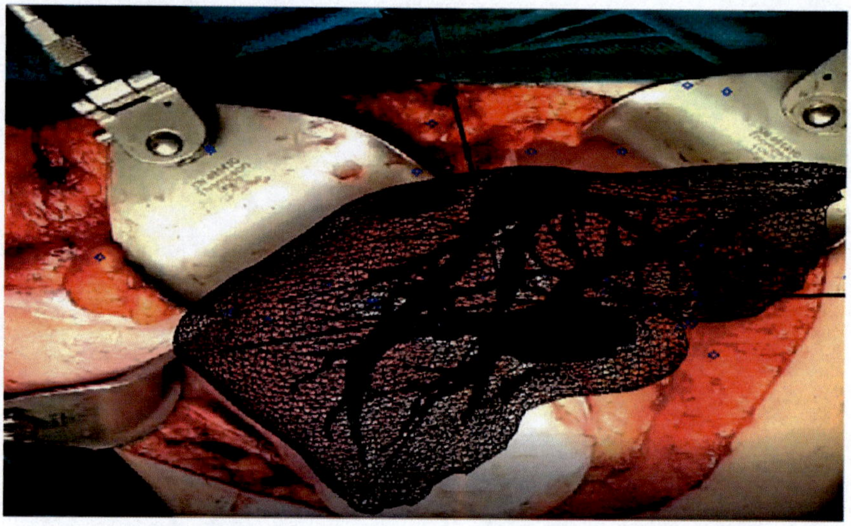

Fig. 2. AR navigation system based on manual registration

This navigation system required the user to constantly track liver movement and manually adjust the 3D model accordingly. This is a demanding task due to continual liver deformations, which ultimately results in human-made errors. Since surgery demands utmost precision, this margin of error is unacceptable.

2.2 Semi-automatic Registration

The semi-automatic registration approach requires the usage of landmark structures (markers) to be present on the 3D model and the actual liver. The AR navigation system is then superimposed over the operative field using these markers for registration. Due to deformations of the liver, markers are automatically tracked, so that the 3D model can follow the movements of the real organ. This semi-automatic method has been shown to be more precise than the manual method. The method used in the development of earlier iterations of the intraoperative navigation system, relied on the usage of binary markers that were used to define a plane in space that represents the external boundary conditions of the organ. The in-house algorithm developed for the purposes of this navigation system performs the deformation of the 3D model of the organs based on the parameters obtained from the detected markers. At the beginning of the operation, the surgeon sews two markers on the left and right lobes of the liver. Then they manually bring and position the virtual 3D organ to the real position. After this initial step, the finite elements method deforms the virtual organ to fit the shape of the real one. During the intervention, the system adjusts the 3D model using the input data from the markers if it comes from organ movement (e.g., due to patient breathing). This adjustable 3D model is displayed on monitors in the operating room (Fig. 3).

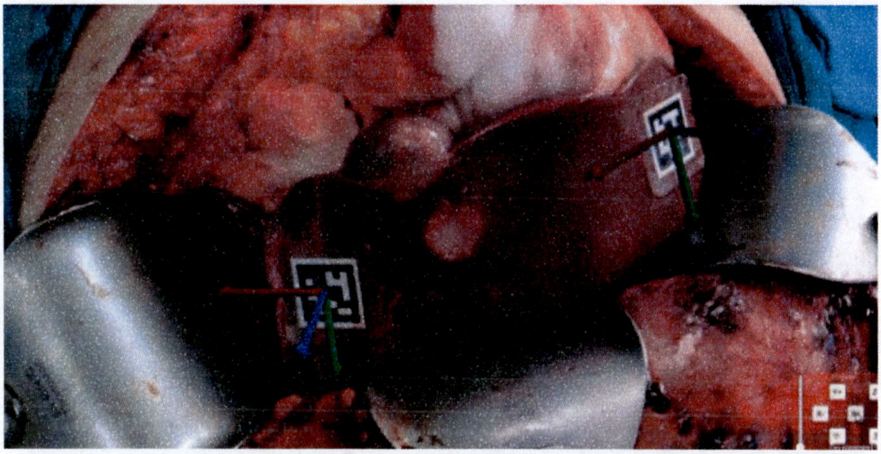

Fig. 3. Navigation system based on usage of binary markers

2.3 Automatic Registration and Liver Tracking

The semi-automatic method with usage of markers that provide information about liver position gives generally good results, but it still requires additional steps pre-resection (sewing the markers, calibrating camera etc.) that increases the duration of the operative procedure. Also, AR navigation systems based on markers have problems when the view of the markers that are placed on the organ is blocked. These situations often happen due to the movement of surgeon's hands and equipment, as well as raising and lowering of organs into abdominal cavity. Fully automatic registration is subject to ongoing research with its difficulty in obtaining reliable and real-time automatic registration in liver tissue surgery.

In order to track the liver in real time without markers, Shi-Tomasi Feature Detector was used [22]. This detector is used to extract the strongest N corners from the images, where N represents the number of wanted features. Since the main focus is on the number of features tracked, rather than their quality, N was set to 5000. To implement Shi-Tomasi Feature Detector, we used a common library in the field of computer vision, OpenCV implemented in the C++ programming language. This feature detector did a great job and discovered various features to track on the liver, but due to the width of the camera's field of view, the surrounding area that is not a region of interest was also tracked. To prevent this behavior liver area needs to be segmented before the feature detector is applied.

There are numerous ways to perform segmentation of the region of interest, but the most simple and robust way, in our case, was to perform image segmentation using HSV color space. Using the trackbar system, the user selects minimum and maximum allowed values for each HSV component (Hue, Saturation, Value). All of the pixels whose components are in a range between minimum and maximum allowed values are considered regions of interest. This trackbar system with the resulting segmentation is shown in Fig. 4.

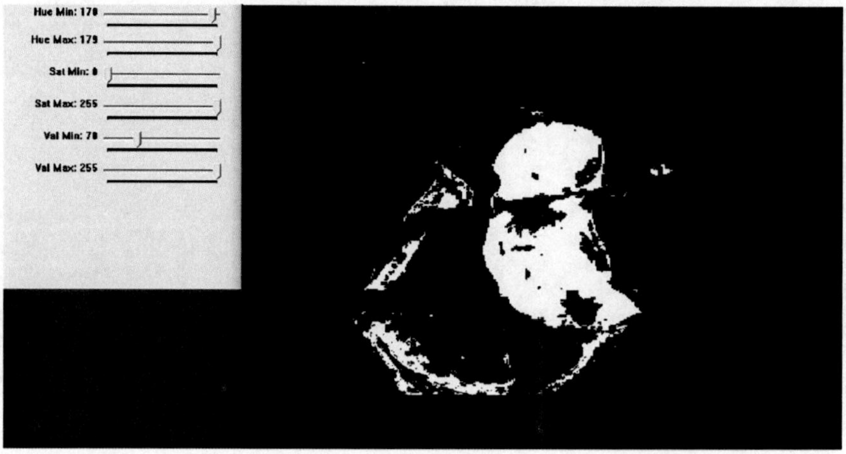

Fig. 4. HSV trackbar control system with the resulting segmentation

Looking at the result of segmentation in Fig. 4, it is clear that the segmentation was not perfect due to the fact that some pixels outside of the region of interest have the same color as the liver itself (abdominal tissue). To try and reduce this noise in the image, an operation of erosion was performed. Unfortunately, erosion also affects the region of interest and reduces the area that needs to be tracked. To try and recover some of the lost region of interest, erosion was followed by the dilation technique. In order to control the amount of applied erosion and dilation, a system of trackbars was used in a similar fashion as the HSV color segmentation. This system, with the resulting segmentation, is shown in Fig. 5.

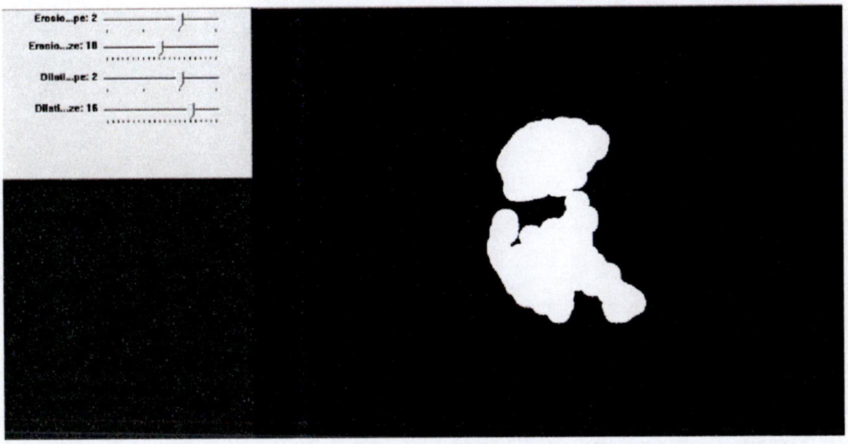

Fig. 5. Erosion and dilation control system with the resulting liver segmentation

Following the successful segmentation of the liver as the region of interest, Shi-Tomasi can be applied without worrying that the detected corner is going to be outside of the liver.

3 Results

Figure 6 shows a clear difference between the application of feature detector without and with liver segmentation. Figure 6b shows that the feature detector found adequate number of features to track so that navigation system can properly follow the intraoperative liver movement and deformations. However, due to lightning and slight color variation in some areas of the liver, it is clear that the liver is not completely segmented, which results in the detector not being able to find features in those areas of the liver.

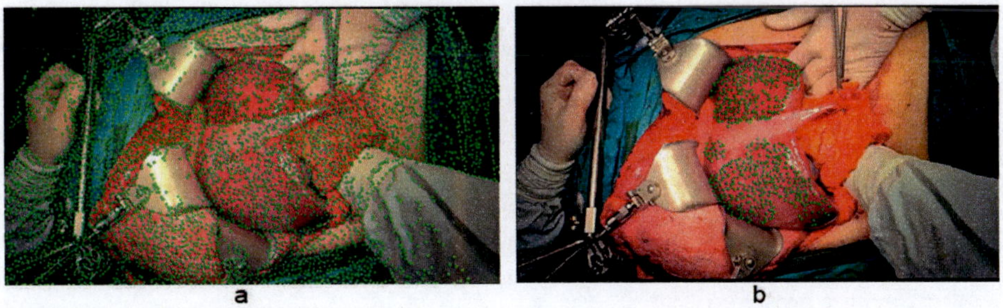

Fig. 6. Results of feature detector without liver segmentation (a), results of feature detector with liver segmentation (b)

The F1 score between the whole liver and segmentation presented in Fig. 6 is 91.38%, while the F1 scores for the rest of the patients, as well as the mean for the whole dataset, are shown in Table 1.

Table 1. Per patient and mean F1 scores

Patients	F1 scores
Patient 1	91.38%
Patient 2	88.43%
Patient 3	92.17%
Patient 4	87.67%
Mean	89.91%

Still, comparing this approach with the semi-automatic method where only two markers are being used and followed, the number of detected features is significantly higher. This makes the process of 3D model adaptation to liver deformations much more robust. Now, even if the view of the liver is being partially obstruct there is still a plethora of tracked features available to keep position of 3D liver model overlapped with actual organ.

4 Conclusion

In this paper we presented a liver tracking method that is, in further research, going to be used in an implementation of a markerless automatic model registration. In the future project development, the tracked features are going to be used as a way to bind the 3D model with the patient's liver. Using this approach, our goal is to develop a reliable, easy-to-use augmented reality intraoperative navigation system that is robust to liver deformations.

Acknowledgements. The research was funded by Serbian Ministry of Education, Science, and Technological Development, grant [451-03-68/2022-14/200107 (Faculty of Engineering, University of Kragujevac)]. This paper is supported by the project that has received funding from the European Union's Horizon 2020 research and innovation programme under grant agreement No 755320 (TAXINOMISIS project).

References

1. Cao, W., Chen, H.D., Yu, Y.W., Li, N., Chen, W.Q.: Changing profiles of cancer burden worldwide and in China: a secondary analysis of the global cancer statistics 2020. Chin. Med. J. **134**(7), 783–791 (2021)
2. Jemal, A., Bray, F., Center, M.M., Ferlay, J., Ward, E., Forman, D.: Global cancer statistics. CA: Can. J. Clin. **61**(2), 69–90 (2011)
3. Parkin, M.D.: The global health burden of infection-associated cancers in the year 2002. Int. J. Cancer **118**(12), 3030–3044 (2006)
4. Jemal, A., Center, M.M., DeSantis, C., Ward, E.M.: Global patterns of cancer incidence and mortality rates and TrendsGlobal patterns of cancer. Cancer Epidemiol. Biomark. Prev. **19**(8), 1893–1907 (2010)

5. Kubota, K., et al.: Measurement of liver volume and hepatic functional reserve as a guide to decision-making in resectional surgery for hepatic tumors. Hepatology **26**(5), 1176–1181 (1997)
6. Nathan, H., Schulick, R.D., Choti, M.A., Pawlik, T.M.: Predictors of survival after resection of early hepatocellular carcinoma. Ann. Surg. **249**(5), 799–805 (2009)
7. Hageman, A.: Virtual reality. Nursing **24**(3), 3–3 (2018). https://doi.org/10.1007/s41193-018-0032-6
8. Carmigniani, J., Furht, B.: Augmented reality: an overview. In: Handbook of Augmented Reality, pp. 3–46 (2011)
9. Azuma, R.T.: A survey of augmented reality. Presence: Teleoperators Virtual Environ. **6**(4), 355–385 (1997)
10. Okuda, Y., et al.: Usefulness of operative planning based on 3-dimensional CT cholangiography for biliary malignancies. Surgery **158**(5), 1261–1271 (2015)
11. Li, P., et al.: Preoperative three-dimensional versus two-dimensional evaluation in assessment of patients undergoing major liver resection for hepatocellular carcinoma: a propensity score matching study. Ann. Transl. Med. **8**(5), 182 (2020)
12. Quero, G., et al.: Virtual and augmented reality in oncologic liver surgery. Surg. Oncol. Clin. **28**(1), 31–44 (2019)
13. Kang, X., et al.: Stereoscopic augmented reality for laparoscopic surgery. Surg. Endosc. **28**(7), 2227–2235 (2014). https://doi.org/10.1007/s00464-014-3433-x
14. Zhang, P., et al.: Real-time navigation for laparoscopic hepatectomy using image fusion of preoperative 3D surgical plan and intraoperative indocyanine green fluorescence imaging. Surg. Endosc. **34**(8), 3449–3459 (2019). https://doi.org/10.1007/s00464-019-07121-1
15. Buchs, N.C., et al.: Augmented environments for the targeting of hepatic lesions during image-guided robotic liver surgery. J. Surg. Res. **184**(2), 825–831 (2013)
16. Clements, L.W., Chapman, W.C., Dawant, B.M., Galloway Jr. R.L., Miga, M.I.: Robust surface registration using salient anatomical features for image-guided liver surgery: algorithm and validation. Med. Phys. **35**(6)Part1, 2528–2540 (2008)
17. Gavaghan, K.A., Anderegg, S., Peterhans, M., Oliveira-Santos, T., Weber, S.: Augmented reality image overlay projection for image guided open liver ablation of metastatic liver cancer. In: Workshop on Augmented Environments for Computer-Assisted Interventions, Berlin, (2011)
18. Haouchine, N., Dequidt, J., Berger, M.-O., Cotin, S.: Deformation-based augmented reality for hepatic surgery. Stud. Health Technol. Inf. **184**, 182–188 (2013)
19. Pelanis, E., et al.: Evaluation of a novel navigation platform for laparoscopic liver surgery with organ deformation compensation using injected fiducials. Med. Image Anal. **69**, 101946 (2021)
20. Zhang, W., et al.: Augmented reality navigation for stereoscopic laparoscopic anatomical hepatectomy of primary liver cancer: preliminary experience. Front. Oncol. **11**, 663236 (2021)
21. Zijlmans, M., Langø, T., Hofstad, E.F., Van Swol, C.F., Rethy, A.: Navigated laparoscopy–liver shift and deformation due to pneumoperitoneum in an animal model. Minim. Invasive Ther. Allied Technol. **21**(3), 241–248 (2021)
22. Shi, J.: Good features to track. In: Proceedings of IEEE Conference on Computer Vision and Pattern Recognition (1994)

Intelligent Drug Delivery Systems

Ana Mirić[✉] and Nevena Milivojević

Institute for Information Technologies, University of Kragujevac, Jovana Cvijića bb, 34000 Kragujevac, Serbia
{ana.miric,nevena.milivojevic}@uni.kg.ac.rs

Abstract. For the last few decades, scientists have been working on the development of advanced drug delivery systems, which involve the use of carriers of medically active substances for precise and effective therapy accompanied by a reduction in the occurrence of side effects. The principle of controlled delivery of drugs is based on the control over the place where the drug will be released, the moment of the start of its release, the time interval during which it will be released, the amount of drug that will be released over time, by modifying the characteristics of the carrier, mostly depend on the properties of the carrier. When developing a system for modern drug delivery, it is of great importance to create an optimal material design, but also to predict how the material interacts with cells and tissue. Modern drug delivery systems include 3D printed tablets, patches, liposomes or nanoparticles. Also, novel technologies represent small devices with personalized drug administration where it is possible to combine different principles of release (constant, linear, pulsatile) or several different drugs together. The development of computer methods has made it possible to simulate the change in drug concentration over time on a computer, thus saving time on testing and materials. In order for computer methods to be as accurate as possible and correspond to the real system, it is necessary to create adequate computer models. These models are very useful tools in this field because with artificial intelligence we can predict the release of a drug of a certain concentration in time.

Keywords: intelligent systems · drug delivery · controlled release · mathematical models · artificial intelligence

1 Introduction

For the last few decades, scientists have been working on the development of advanced drug delivery systems, which involve the use of carriers of medically active substances for precise and effective therapy accompanied by a reduction in the occurrence of side effects. By designing the properties of the carriers, it is possible to control the place of drug release, the moment of its release, the time interval during which the drug will be released, as well as the amount of drug that will be released. This increases the accuracy and control of the treatment, thus increasing the efficiency, shortening the duration of the treatment and decreasing or completely eliminating the possibility of side effects [1].

The ideal way of modern drug delivery is to release in a precise period of time and action at a specific place in the body. The main advantages of such systems are

N. Filipovic (Ed.): AAI 2022, LNNS 659, pp. 342–366, 2023.
https://doi.org/10.1007/978-3-031-29717-5_21

targeted local drug delivery, control of drug concentration released over time, lower drug concentration required for treatment, minimization or complete absence of side effects, improvement of treatment efficiency [2].

The goal of modern medicine is an individual approach to each patient and the adaptation of therapy to his needs in order to achieve the best therapeutic outcome. In order to achieve this, it is necessary to adjust the dose of the drug. Adjusting the dose to each individual patient can be achieved by using macro, micro and nanocarriers, which would be used to introduce the medicinal substance into the body, as well as with mathematical models that contribute to the accurate determination of individual dose [3].

Contemporary research in this area aims to change the pharmacokinetics and specificity of the drug by designing different drug carriers and medical devices. By increasing the bioavailability and shortening the half-time elimination, the duration of the drug's lasting effect increases, thus improving the therapeutic outcome [4].

2 Principles of Controlled Drug Delivery

The traditional way of treating diseases, which is used in practice today, involves several stages. The first step is based on the empiric administration of broad-spectrum of drugs before identifying the illness. The drug can be administered orally or intravenously so that the concentration of the drug in the blood plasma is around the therapeutic concentration [5].

For example, the empiric method of treatment with antibiotics has a large number of disadvantages. By applying high doses of antibiotics, the level in the blood plasma becomes significantly higher than in the place where it should work, and the antibiotic accumulates in other regions of the body. There is also a problem with the imprecise time interval during which the drug is administered, which is usually estimated intuitively. In case of inadequate therapy, the disease turns into a chronic form that can be treated for years [6]. This type of treatment first of all implies the possibility of side effects due to a long-time interval during which intravenous catheters are applied (with intravenous administration of antibiotics), extremely increases the possibility of side effects of antibiotics due to their accumulation in other organs, the appearance of resistant species, a decrease in immunity, and then the necessity of long-term hospitalization, which entails both personal and economic disadvantages [5].

Contrary to the traditional method of treatment, in the last couple of decades a more advanced drug delivery system has been developed, which is based on the application of carriers of medically active substances. The principle of controlled delivery of drugs is based on control over the place where the drug will be released, the moment of the start of its release, the time interval during which it will be released, the amount of drug that will be released over time, by modifying the characteristics of the carrier, mostly depend on the properties of the carrier. By precisely designing its properties, it is possible to adapt the drug delivery to the specific needs of individual types of therapeutic, diagnostic or preventive needs. In this way, an increase in the precision and control of the treatment is achieved, which consequently enables the improvement of efficiency, shortens the time of concealment of the treatment and potentially alleviates or completely eliminates the occurrence of side effects [7].

Drug delivery using carriers has been developed in several directions:

- Controlled drug delivery,
- Delayed (initiated) delivery of medicines,
- Targeted drug delivery.

2.1 Controlled Delivery

Controlled drug delivery represents the dependence of the release kinetics on the gradual change in the properties of the carrier used, an example is the degradation of biodegradable polymers in physiological conditions. In this way, control is achieved over the amount of drug that will be released, as well as over the interval during which the drug will be released. Controlled drug delivery can also be of a local nature and it involves the implantation of a carrier with an encapsulated drug in the place where it should manifest its effect. One example of the application of drug carrier implants is the local intramuscular delivery of antidiabetic drugs [8]. Another method for controlled delivery is programmed-pulsatile drug release. Such systems are suitable for drugs or hormones that need to be released in a precise pattern. Pulsatile release improves the effectiveness of therapy because they follow biological and physiological conditions, and with their help overcomes the disadvantages of conventional therapy such as first-pass metabolism through the liver or the use of drugs at night [9].

2.2 Delayed or Initiated Delivery

Initiated drug delivery is characterized by control over the moment at which drug release will occur. For this type of delivery, carriers that are used have the ability to change certain properties under the influence of exogenous or endogenous factors. At the moment when there is a change in one of the parameters characteristic of the pathological state of the environment, the properties of the carrier material will be changed, which enables the drug to be released. For this purpose, polymer structures are most often used, which have the ability to change conditions under the influence of the mentioned environmental parameters. An example is the transition from a tubular to a planar polymer structure, which can be initiated by a deviation from the average temperature of a living organism when drug release begins [2].

2.3 Targeted Delivery

In the case of targeted drug delivery, the carrier, which in this case is also called a vector, moves to the place where the drug should manifest its effect. This means that with this approach control is achieved over the place where the drug will be released. The way in which this type of control over the drug carrier is achieved most often refers to the precise modification of the surface characteristics of the carrier, which involves the binding of ligands capable of specifically binding to the appropriate, targeted group of receptors. An alternative way of driving the carrier with the drug to the target site refers to the possibility of manipulation using an external field (electrical, magnetic, ultrasonic). In that case, the vector must have characteristics that make it sensitive to external influences in order to achieve its movement, as well as the possibility of visualization that facilitates its direction. This method of drug delivery is currently attracting a lot of attention precisely because of the possibility of manipulation at the cellular or subcellular level. In this way, greater specificity of treatment is achieved. Increased specificity of treatment is particularly important in the case of antibiotics and cytostatics [10].

Today, there is an extremely large number of different drug carriers developed according to the needs of the potential application. Their constant improvement follows the upward trend in the number of requirements that define the suitability of modern materials for biomedical applications. In general, the drug carriers which have been developed so far can be divided into three groups:

- Synthetic (polymers, nanoparticles, composites),
- Natural (proteins, peptides, enzymes),
- Cell carriers (macrophages, erythrocytes, bacteria, viruses) [2].

2.4 Endogenous and Exogenous Stimuli for Drug Release

Stimuli for the controlled release of the drug can be from the internal or from the external environment. Changes in pH value, redox reactions, ATP, H_2S, are some of the applications of internal triggers for different stimulation of drug release, while temperature, pressure, radiation, magnetic or electric field are examples of exogenous stimuli [11].

The temperature has so far been investigated as a stimulus for drug release. Here, materials that are sensitive to heat are of the greatest importance in the sense that they change their aggregate state when applying temperature. Changes from gel to a solution or vice versa occur at different temperature thresholds depending on the targeted creation of the structure of the controlled drug release system. There are many examples developed so far, such as hydrogels, composite gels, liposomes, nanoparticles, etc. [12].

Light as a trigger represents a group of exogenous stimuli that can be used if there are photo-sensitive functional groups on the material for manufacturing the system for the controlled release of drugs. These motifs that absorb light of a specific wavelength (from 2500 to 380 nm) can be, for example, derivatives of azobenzene, nitrobenzene. In this way, the drug can be released at a targeted place on the body, especially one that is difficult to reach. Also, pulsatile drug release can be achieved by applying light for a period of time, or darkness, and so on in a circle. Wavelengths below 700 nm do not penetrate deeper than 1 cm into the tissue due to interaction with melanin, hemoglobin and myoglobin chromophores, while wavelengths above 900 nm are absorbed by water molecules. It follows that wavelengths between these two are the most suitable stimulus solution for drug release. A large amount of research deals with the use of gold for these purposes because it has the ability to absorb different wavelengths depending on its shape [9, 13].

Magnetic field as a trigger: The incorporation of atoms such as iron, cobalt, nickel, into materials for controlled drug release can be used for magnetic field stimulation. Such systems can cause drug release by causing hyperthermia or deformation of the material after exposure to a magnetic field. Such materials are also used for magnetically guiding the drug carriers to the target site [13].

By applying and removing the electric field, controlled drug release can be achieved, and it is especially suitable for pulsatile release because it can be precisely controlled. One example of the application of an electric field is iontophoresis, which is used to increase the passage of drug molecules through membranes (it is limited to drugs of ionic nature, small size and low molecular mass) [14].

The ultrasound field as a stimulus is most often used to control the passage of medicine through biological barriers such as skin and blood vessels. A good feature of this method is that the exact depth of drug penetration can be controlled. An ultrasonic trigger can achieve drug release by increasing the temperature during field absorption or by increasing or oscillating gases in the target medium (known as cavitation) [15].

Metal materials have recently been combined with chemotherapeutics because they increase the success of therapy due to the controlled release of drugs. They play an important role due to their properties such as high atomic number, photoelectric absorption and production of hydroxyl radicals. These features make them suitable triggers for radio-sensitive controlled drug release [16].

Nanotechnology-based drug carriers can be functionalized with stimuli-responsive motifs to achieve controlled drug release. On the surface of nanoparticles, there can be motifs such as hydrogen/hydrazone bonds or ionizing functional groups for pH-dependent activation, bound substrates for individual enzymes as motifs for enzyme-sensitive release, or disulfide bonds as sites sensitive to redox changes [17].

3 Modern Drug Delivery Systems

3.1 Tablets

Tablets are solid dosage forms of drugs, administered orally, and resorbed in the oral cavity, stomach or intestines. Most often, they are round or oval in shape, flat or convex surfaces, rounded edges. They can be coated with a thin film (film tablets) or a sugar membrane (drags), in order to cover the unpleasant smell and taste, to protect the medicinal substance from the influence of external factors or stomach acid, and due to the resorption of the drug in certain parts of the digestive tract. They can be multi-layered if they contain several active substances which are not mutually compatible [18, 19].

Tablets used for controlled release of medicinal substances can be:

- Hydrophilic,
- Hydrophobic [18].

In order for the active substance to be released from the hydrophilic tablet, it is necessary for the tablet to come into contact with water. The release mechanism is controlled based on the properties of the polymer and takes place in four phases. When the tablet comes into contact with water, the polymer breaks down through the hydration process creating a gel layer and then the initial release of the drug from the inner layer of the tablet occurs. Water penetrates the tablet, increasing the gel layer, through which the active substance diffuses. The outer layer of the tablet is completely hydrolyzed, and water continues to penetrate into the core of the tablet. The soluble active substance is primarily released by diffusion through the gel layer, and the insoluble active substance by degradation of the tablet [19].

Hydrophobic tablets can be:

- Lipid
- Plastic

In the case of hydrophobic tablets, the drug is released by wetting, hydrolysis or dissolution of lipid components under the influence of enzymes and changes in the pH value in the gastro-intestinal tract. Fatty acids are more soluble in an alkaline solution ($pH > 8$) [20].

3.2 Patches

Transdermal drug delivery systems, or transdermal patches, are pharmaceutical preparations of various sizes that contain one or more medicinal substances and are intended for application to intact skin [21]. After absorption through the skin, medicinal substances reach the systemic circulation where they achieve their effect. Transdermal patches have an outer covering layer, which is impermeable to the medicinal substance and water, its role is to protect the patch during application and a protective layer that is removed before applying the patch [22].

Depending on the mechanism by which the release of the medicinal substance is controlled, patches can be divided into matrix patches and reservoir-type patches with a membrane [23].

Matrix-type transdermal patches can have a single-layer or multi-layer matrix, solid or semi-solid consistency, the speed of diffusion of the medicinal substance through the skin depends on its composition and structure. They have a simple design, they are thinner than patches with a reservoir, which is why they are more acceptable to patients. However, the choice of auxiliary substances is extremely complicated, especially adhesive substances, because this layer must, in addition to sticking the patch to the skin, also ensure the controlled release of the medicinal substance [22, 23].

Another type of patch has a reservoir of semi-solid consistency that has a porous membrane on one side that controls the release of the drug. The rate of drug release depends on the porosity, permeability and thickness of the membrane [23].

A novel type of transdermal patch is the microreservoir patch, which contains a medicinal substance that is dissolved in a hydrophilic polymer that is dispersed in a suitable hydrophobic polymer. The solubility of the medicinal substance is significantly lower in the hydrophobic polymer, which enables a high concentration gradient for diffusion through the skin [24].

In addition to the medicinal substance, the composition of transdermal patches includes auxiliary substances such as: solubilizers, stabilizers, substances that modify the rate of release of the medicinal substance, and substances that improve percutaneous absorption [25].

Transdermal patches are widely used today because of their advantages. These systems enable:

- Better compliance - they are more acceptable for the patient due to a more comfortable dosing regimen;
- Achieving a constant therapeutic concentration of the drug in the blood;
- Bypassing the passage of the drug through the gastrointestinal tract, which reduces the risk of side effects;
- Discontinue the use of the drug as needed, by removing the patch [26].

Examples of patches that can be found on the market:

- Scopolamine (hyoscine) transdermal patches - used to treat motion sickness;
- Nitroglycerin transdermal patches - used to treat angina pectoris;
- Clonidine transdermal patches - used in hypertension therapy;
- Transdermal patches with fentanyl - used in pain therapy;
- Estradiol patches - contraceptive therapy;
- Nicotine patch for smoking cessation [23].

3.3 Liposomes

Liposomes are spherical dispersion systems for drug transport. These are microparticles consisting of an amphiphilic layer of lipids that surrounds a water core. A liposome can be surrounded by one or more layers of lipids, depending on their use. The lipophilic membrane can be double-layered, where the polar groups face the interior of the vesicle and the outer surface of the liposome. Various ligands (specific molecules, antibodies, receptors, opsonins) can be attached to their surface, which drives the liposome to bind to the desired site. The number of lipid bilayers surrounding the water space in the interior can be one (unilamellar) or more (multilamellar). Depending on the structure, the size of liposomes ranges from 20 nm for small unilamellar to 4000 nm for large multilamellar liposomes [27].

The importance of lipids as drug carriers is reflected in the fact that they can transport both hydrophilic and hydrophobic components. Hydrophilic drugs are placed in the lipid core where the aqueous solution is, while hydrophobic drugs are dissolved in the lipid membrane. The lipid layer on the surface connects to the cell membrane (which is also composed of lipids), and thus the drug is transported from the liposome into the cell. There is another type of liposome that transports the drug into the cell by diffusion. These liposomes are made in such a way that the pH value at which the drug will be charged is determined. There are already a large number of drugs on the market that are placed in a liposomal transport system. Liposomes can also be used to transport specific DNA sequences [28].

3.4 Nanoparticles

Nanotechnologies represent a new type of technologies that are based on materials of the order of the size of small molecules, as well as instrumentation that is capable of characterizing such materials. They can be manipulated at will independently of the usual spontaneous chemical reactions between atoms and molecules, and they can be built in classical structures based on the principles of self-organization, quantum and intelligent nanomaterials and systems [29].

From the point of view of nanostructures, we distinguish three types: nanocomposites, nanostructurally designed surfaces and nanocomponents where the nanoparticles embedded in the material are fixed or free. Free nanoparticles can be simple or complex compounds where nanoparticles of individual elements are covered with another substance (covered nanoparticles or "core-shell" nanoparticles). A powder or liquid containing nanoparticles is almost never monodisperse and contains some distribution, which complicates the analysis even more because larger nanoparticles have different properties compared to smaller ones. Nanoparticles also show a tendency to aggregate, and aggregates have different properties compared to individual nanoparticles [30, 31].

In order for a material to be a nanomaterial, it is necessary that its growth be limited at the nanometer level by one, two and/or three dimensions. The division of nanomaterials based on size was proposed by Pokropivny and Skorokhod. They are divided into:

- zero-dimensional (0D, all dimensions less than 100 nm),
- one-dimensional (1D, at least one dimension larger than 100 nm),
- two-dimensional (2D)
- three-dimensional (3D) [32, 33].

There are different morphological forms of 1D nanostructures, the most studied of which are: nanofibers, nanowires, nanobelts, nanorings, nanorods, and nanotubes [34].

Based on the chemical composition, nanomaterials are divided into organic and inorganic. Organic materials include carbon-containing materials. Among the most important are nanodiamond, fullerene C60, carbon nanotubes and nanofibers. Carbon nanomaterials can be in the form of hollow spheres, ellipsoids or tubes. Spherical and ellipsoidal are called fullerenes. Inorganic nanomaterials are particles based on metal oxides: zinc oxide, iron oxide, titanium dioxide and metals: gold, silver, iron and cobalt [35, 36].

The physical and chemical properties of the material change significantly when the number of atoms that make up the material is significantly reduced, even in nano size, for example, gold is no longer yellow and can have green, red, blue and other colors [37].

4 Materials for Controlled Drug Delivery

When developing a system for modern drug delivery, it is of great importance to create an optimal material design, but also to predict how the material interacts with cells and tissue. This material property is called biocompatibility. Biocompatibility can be defined as the property of a material to provide an effect after application, without causing an unwanted response of the organism. For the manufacture of the drug carriers, synthetic materials are generally used, which to a certain extent can cause the body's immune response. It is necessary to optimize the level of biocompatibility of the material used [38].

Polymeric materials were the first to be used as drug carriers. Polymeric materials were the first to be licensed for practical use and commercial sale. According to their origin, they are classified into natural, organic synthetic, inorganic synthetic and semi-synthetic materials. And based on the possibility of degradation in physiological conditions into biodegradable and non-biodegradable [39].

Non-biodegradable polymers are known "reservoirs" of drugs that are released from them by desorption from the surface or diffusion from the container. Porosity, specific surface and surface roughness of the material are of great importance for this method of release. The first polymers tested for controlled drug delivery were poly (ethylene vinyl acetate - PEVA), silicones (poly (dimethyl - siloxane) - PDMS and poly (methyl methacrylate - PMMA) [40, 41]. These materials showed good properties during *in vivo* application. For various types of polymers, the tonicity associated with the polymerization process was observed. After this process, residual monomers can inactivate the drug and cause necrosis of the surrounding tissue [41].

Biodegradable polymers overcome the difficulties associated with the use of PMMA and other non-biodegradable polymers. These polymers have a greater capacity to encapsulate the drug which is gradually released until the entire amount of the drug is released. The most commonly used synthetic degradable polymer is poly (lactide - co - glycolide) – PLGA. The release of the drug depends on the degradation of the polymer, so by modifying its properties it is possible to achieve the release of the drug in different time intervals. A good knowledge of the characteristics of the material is necessary for the proper selection of biodegradable polymer material. The most important characteristics of adequate material are:

1. Possibility of controlled degradation under physiological conditions,
2. Change in surface characteristics during the degradation process,
3. Non-toxicity of degradation products [42].

Natural biodegradable polymers are extremely important from the point of view of controlled drug delivery because they are non-toxic, have excellent surface characteristics and a very high degree of biocompatibility. Some of the main representatives of this group of polymers are chitosan, gelatin, collagen. Chitosan is a biopolymer that has bacteriostatic activity by itself. Collagen is the most abundant biopolymer of the organic part of bone tissue and its degradation enables a high degree of control over the released drug [43].

4.1 Poly (Lactide - Co - Glycolide)

Poly (lactide - co - glycolide) - PLGA is a copolymer of glycolide with l - lactide or dl - lactide, obtained by a copolymerization process, whereby a polymer is obtained that retains the properties of these two homopolymers [44]. The degradation of this polymer largely depends on the ratio of glycolide and lactide monomer units [45]. In order for the polymer to have an amorphous structure and have all the characteristics needed to be used as a carrier for the controlled release of drugs, it needs to contain 25% and 70% polyglycolic acid. The increase in the number of these monomers increases the hydrophilicity of the polymer [44]. In addition to the share of monomer units, the process of polymer degradation also depends on other factors such as: molar mass, chain length, shape and size of particles, mass share of the drug in the polymer, medium [45, 46].

The process of degradation actually involves the process of hydrolysis. In the presence of water, at an elevated temperature, polymer chains break, resulting in acidic products, polyalcohols and polyketones. Degradation products migrate from the volume of particles to the external environment where they are neutralized. At the same time, basic components from the environment enter the particle volume where they neutralize the acidic components. Since the diffusion processes are relatively slow, the rate of formation of acidic products is related to the rate of their neutralization, which results in an increase in acidity. A decrease in the pH value of the environment causes tissue irritation and increases the possibility of activating the immune response. The final degradation products are lactic and glycolic acid, which are non-toxic metabolic products that are eliminated from the body through the Krebs cycle [47–49].

4.2 Gelatin

Gelatin is a polypeptide composed of 20 types of amino acids connected by peptide bonds, which is obtained by thermal denaturation or partial hydrolysis of collagen. In addition to being used in the food industry, gelatin is also used in the pharmaceutical industry and in the fields of drug delivery, tissue engineering and regenerative medicine [50].

Depending on the source of collagen and the method of obtaining it, there are several types of gelatin:

1. Pork and beef gelatin - gelatin similar to human,
2. Fish gelatin - this type of gelatin has a significantly lower melting and gelling temperature, lower thermal stability, but higher viscosity compared to gelatin obtained from mammals [50].

Gelatin dissolves in the void, and the rate of dissolution increases with increasing temperature. It is amphoteric in nature, that is, it can act as both an acid and a base thanks to the presence of acidic functional groups and alkaline amino acids. The behavior of gelatin solutions is influenced by several factors, such as temperature, pH value, concentration and method of preparation [50].

The most important characteristic of gelatin is its ability to gel at low temperatures. During cooling, hydrogen bonds, electrostatic and hydrophobic interactions between polypeptide chains are formed, resulting in a hydrogel. These bonds are broken by heating, which is why this hydrogel is thermoreversible. This feature is very important for 3D bioprinting technology because it allows for controlled printing. The melting temperature of gelatin is around 28 °C [50].

4.3 Collagen

Collagen is a fibrous protein whose most important role is to maintain the integrity of connective tissues. There are about 16 types of collagen in the human body, of which types I, II and III are the most abundant. These three types make up 80–90% of the collagen in the human body [51]. Collagen is a natural material with low immunogenicity, which makes it suitable for the production of drug delivery systems [52]. As a drug carrier, collagen has found its application in ophthalmology, in the treatment of cancer, in tissue engineering, and it is also used to make dressings for wounds and burns [53]. Collagen is obtained by extraction from skin (veal, beef and pork), beef and human placenta, and from bones. It is very important that animal skin is young, because it contains active collagen that has the ability to regenerate. Collagen denaturation occurs at a temperature between 36 and 37 °C [52].

5 Fabrication of Drug Delivery Systems

5.1 Conventional Tablet Manufacturing

Hydrophilic tablets are made by tableting a mixture of active ingredients and hydrophilic materials. Some of the materials used in making these tablets are sodium carboxymethyl cellulose, polyethylene oxide, polyvinyl pyrrolidone, polyvinyl acetate [54]. There are numerous factors that influence the formulation of the matrix for making tablets, namely:

- Hydration capacity - it is necessary that the polymer has a high hydration capacity, otherwise there will be premature release of the medicinal substance and disintegration of the tablet;
- Amount of hydrophilic matrix - tablet disintegration time increases with an increasing amount of matrix. Polymers that are less soluble are used in higher concentrations;
- Size of polymer particles - reducing the size of polymer particles enables rapid hydration and gel formation;
- Polymer viscosity - if the viscosity of the polymer is high, the release of the medicinal substance is slower;
- Solubility of matrix components - water-soluble components increase the rate of drug release; however, the rate can be reduced by adding soluble components due to an increase in gel viscosity;
- Influence of pH value – pH value affects the viscosity of the gel. A decrease in the pH value decreases the rate of release of the medicinal substance. For example, gelatin forms a gel of high viscosity in an acidic environment and thus reduces the rate of drug release;
- Size and shape of the tablet [55].

Lipid (hydrophobic) tablets are made by incorporating the active substance into fat- and wax-based granules using spray-thickening, mixing-thickening in an aqueous solution, with or without the addition of substances that regulate surface tension, and the spray-drying technique [56].

Plastic (hydrophobic) tablets are made by direct pressing of the active substance with a plastic matrix, whereby the plastic material is granulated in order to obtain the desired particle size. There are three approaches to mixing the active substance and the plastic material:

- Solid active substance and plastic material in powder are mixed with a solution of the same plastic mass or some other binding material in an organic solvent, after which the mixture is granulated;
- The active substance is dissolved in the plastic material using an organic solvent and granulated with the evaporation of the solvent;
- The active substance and the plastic matrix are granulated using latex or pseudolatex as a granulation liquid [56].

5.2 Tablet Manufacturing Using 3D Printing

3D printing, also called additive manufacturing, was first proposed by engineer Charles Hull in the early 1980s. 3D printing is a manufacturing process in which materials are deposited layer by layer to form an entity. Based on a pre-designed 3D digital model, it accumulates printed layers to complete the construction of the 3D object. 3D printing is extremely flexible, allowing local control of material composition and microstructure. Compared to traditional processes, 3D printing has great advantages in the production of highly complex and customized products, so it is more economical and time-saving [57]. Currently, 3D printing technologies applied in the field of pharmaceutical manufacturing mainly include fused deposition modeling (FDM), stereolithography (SLA), and extrusion printing, etc. [58]. 3D printing can be used to make molds for pill production or to print pills directly using drug powders as raw materials. Researchers or doctors can use computer-aided design to create instructions for the print path of the nozzle. With this instruction, the printer nozzle lays down ink containing a binder and powder of active pharmaceutical ingredients layer by layer to print a 3D product. 3D printing has enormous potential in personalized medicine. 3D printing has the ability to precisely microcontrol and can obtain different release profiles by controlling the external shape and internal structure of tablets [57].

According to the modality of bioprinting, bioprinters can be classified into three main groups, which include bioprinters based on: extrusion, droplet or laser. An extrusion bioprinter uses a mechanical or pneumatic system to deposit cells in filament form, while a droplet bioprinter uses a thermal, piezo, or acoustic mechanism to deposit droplets of cell suspension with high throughput, while laser bioprinters use laser energy to deposit cells from the donor to the recipient, without the need for nozzle [59].

5.2.1 Extrusion-Based Bioprinting

The most widely applied method is extrusion-based bioprinting, which is suitable for a large number of biocompatible materials. Biomaterials are applied through the nozzle opening in the form of a continuous filament, using an extrusion system, which can be pneumatic, piston or screw. By depositing the material on the substrate or the previous layer, the desired 3D shapes are obtained. This method is suitable for temperature-sensitive materials and materials whose viscosity is in the range of 30 to 6 x 107 MPa. Structures with higher viscosity retain their shape more easily. The basic unit of production in this type of bioprinting is fiber [59]. This technique enables the precise deposition of materials and the production of complex structures. Drug carriers obtained by extrusion bioprinting are usually soft, due to their high-water content [60, 61].

5.2.2 Inkjet Bioprinting

Inkjet bioprinting enables the precise deposition of cells and biomaterials, using some of the benefits of 2D inkjet printing to create 3D structures. Inkjet bioprinters use thermal or piezoelectric energy to deposit solution droplets into desired shapes, and consist of one or more ink chambers with multiple nozzles and corresponding piezoelectric or heating components. Due to chemical cross-linking, many natural materials often change their chemical properties. In addition, some cross-linking mechanisms induce a decrease in cell viability and functionality, which is a major drawback of inkjet printing [60, 62].

5.2.3 Stereolithography (SLA)

Photo drying-based bioprinting applies the curing of photosensitive polymers to form a structure under precisely controlled light. Stereolithography or SLA uses photopolymers to harden resins or liquid materials. It uses a scanning laser to create layer by layer, in a vat of light-cured resin. In the case of this technology, drugs can be incorporated into a polymer network to produce carriers loaded with active ingredients. This technology is the one that best enables the combination of different drugs in the same 3D container. Post-processing includes the removal of excess resin and the curing process under UV light [63].

5.2.4 Fused Deposition Modeling (FDM)

An FDM printer is essentially a robotic glue gun; the extruder either passes over the stationary platform, or the platform moves under the stationary extruder. The software "cuts" the objects into layers, and the coordinates are transferred to the printer. Materials by definition must be thermoplastic. A common material is the biodegradable polymer polylactic acid. These or similar materials have been used as key components of materials used for bioprinting. Building complex geometries usually requires the laying of support structures that can either be formed from the same material, or from a different material placed by a different extruder – which, for example, can extrude a water-soluble support material. The accuracy will depend on the speed of the extruder, as well as on the material flow and the size of each step [64] (Fig. 1).

Fig. 1. Example of 3D printed tablets using the Inkjet bioprinting method (taken from Miric et al. [86])

5.3 Electrospinning

The process of electrospinning technology represents obtaining nanofibers with a defined geometry optimal for the incorporation of drug molecules. It is a technique that uses electrical force to drive the process of knitting and fiber formation. In order to obtain nanofibers with a defined diameter and geometry, the electrospinning process is optimized in relation to the electric field, the chemical composition of the electrospinning solution and the type of polymer. The cross-linking of nanofibers adjusts the biodegradability of the resulting support. The characterization of these nanofibers can be done by electron microscopy (SEM and TEM). Electrospinning represents an approach with great potential in modern medicine for the delivery of drugs over a long period of time. They have already been investigated for use in the treatment of skin wounds or in cancer therapy [65, 66].

Carriers can be made from different biomaterials such as alginate, chitosan, hyaluronic acid, etc. Some natural materials, such as polysaccharides, have a high surface tension which reduces their ability to make fibers through electrospinning technology. Such materials can be chemically modified to overcome these defects and form cross-linked fibers. Synthetic materials that can be used are poly (lactic acid) (PLA), poly (glycolic acid) (PGA), poly (D, L-lactic-co-glycolic acid) (PLGA), poly (ε-caprolactone) (PCL) [67].

5.4 Production of Nanoparticles

There are different methods used for the synthesis of nanoparticles, depending on the nature of the materials from which they are obtained, their size, or use. The synthesis of metal nanocomposites includes spray pyrolysis, liquid infiltration, rapid solidification process, high energy ball milling, chemical vapor deposition, physical vapor deposition, chemical processes – sol-gel and colloidal. The synthesis of ceramic nanocomposites includes a powder process, a precursor polymer process, and a sol-gel process. Finally, the production of polymer nanocomposites includes intercalation, *in situ* intercalative polymerization, melt intercalation, template synthesis, blending, *in situ* polymerization, and the sol–gel process. Some other methods by which some types of nanoparticles can be produced are hydrothermal synthesis, inert gas condensation, ion scattering, microemulsions, pulsed laser ablation, spark discharge, or biological synthesis [68, 69].

5.5 Production of Liposomes

Liposomes are produced by the sonification of phospholipids in water. By using low frequencies, concentric multi-layered liposomes with a size of 100 nm to 10 μm are obtained, while high frequencies produce single-layered liposomes with a diameter of 20 to 100 nm. There are also newer methods for producing liposomes such as extrusion.

Liposomes can be obtained using natural, semi-synthetic or synthetic phospholipids. More recently, they are obtained from sphingolipids, which have a very similar lipid composition to the surface layers of the skin [28, 70].

6 Selection of Medicine

Immediately before starting the optimization of the encapsulation process, it is necessary to have a good knowledge of the characteristics of the drug that are of key importance for encapsulation and release control. The most important features are:

- Solubility of the active substance,
- Molecular mass,
- Concentration of the active substance,
- Hydrophilic and hydrophobic additives,
- Particle size [71].

Drug solubility is an extremely important factor that affects the encapsulation process, but also the drug release process. Different solubility conditions different encapsulation possibilities. It refers to the different possibility of achieving a certain level of encapsulation efficiency and overcoming the problem of adsorption of the drug to the outer surface of the carrier, which leads to the effect of a sudden release of the drug. Solubility also affects release kinetics. This property is related to the diffusion coefficient. The solubility depends on whether the drug will be released faster or slower in a certain environment [38].

The Food and Drug Administration classifies drugs based on permeability and solubility into 4 basic groups:

1. Medicines of high permeability and solubility,
2. Medicines with high permeability and low solubility,
3. Medicines of low permeability and high solubility,
4. Medicines of low permeability and solubility [72].

The use of hydrophobic drugs is quite limited in clinical practice due to their poor solubility in water, and therefore poor bioavailability. Such drugs are released very slowly or incompletely after a long release period [73].

The molecular weight of the drug molecule significantly affects the release process. The effect of a sudden release of the drug is specific for drugs of small molecular weight and for peptides and proteins, but it manifests itself differently. In the case of molecules of low molecular weight, this effect is primarily related to their high mobility due to greater solubility or greater ability to pass through very small pores on the support before starting the degradation process. In the case of high molecular mass molecules, the effect of sudden release is related to incomplete encapsulation due to the adsorption of macromolecule chains on the surface of the carrier [45].

The constant release rate of the active substance can be increased by increasing the concentration of the active substance, adding hydrophilic or hydrophobic additives (polyethylene glycol, sugars, electrolytes, oils) that increase the solubility of poorly soluble active substances, increasing the size of the carrier particles, and decreasing the size of the drug particles [73].

Particular attention should be paid to whether the selected drug exhibits pharmacological activity in the form in which it is found. Many drugs show their effect only after certain metabolic processes, which are simply called bioactivation. This kind of drug is called a prodrug. Designing prodrugs improves the pharmacokinetic profile of drugs, increases drug stability in *in vivo* and *in vitro* conditions, and modifies drug solubility [71].

For example, when treating infectious diseases, the selection of antibiotics is based on two very important criteria: the spectrum of antibacterial action and the potential of the drug to cause an allergic reaction. Penicillin antibiotics have a very wide range of effects, but due to frequent allergic reactions, which can occur at any age, they are not suitable for encapsulation inside the carrier [71]. Aminoglycoside antibiotics were initially chosen for this purpose, because of their bactericidal effect, but also because of their satisfactory allergic profile. The best-known representatives of this group, suitable for encapsulation, are gentamicin and tobramycin. One of the more serious drawbacks observed in the encapsulation of gentamicin and tobramycin is the appearance of resistant species [74]. Lincosamide antibiotics are a very attractive group of antibiotics that are suitable for encapsulation within carriers for the treatment of bone tissue infections. This group of drugs does not have a wide range of effects, but representatives of this group are designed to act against all strains of aerobic cocci, which are the most common causes of such infections [75]. After the encapsulation of this type of antibiotic, systems are obtained that are primarily intended for local application through the skin, for peridental application, and for the construction of implants and bone fillers [76].

7 Smart Devices for Controlled Drug Delivery

The previous text described different drug carriers and technologies for their manufacture, as well as different drug release profiles. These methods can achieve one way of drug release per type of carrier. New technologies with adaptive release profiles are already in trials and in use. They represent small devices with personalized drug administration where it is possible to combine different principles of release (constant, linear, pulsatile) or several different drugs together [9].

7.1 Microchips

Microchips are devices with multiple small reservoirs that carry drugs for controlled release. These reservoirs can be limited by polymer membranes of different molecular weight and thus adjust their degradation rate depending on the need. Such devices can be implanted in the patient's body through surgery and constantly release the required amount of medicine. Also, these microchips can have metal membranes instead of polymer ones, and with the command via an external stimulus, the current can be released as needed. Such devices, for example, can be applied for diseases like cancer, osteoporosis or multiple sclerosis, which some of the authors of such research are planning in the future [9, 77].

7.2 Ophthalmological Devices

New delivery systems that transport drugs into ophthalmology more safely and efficiently are also being investigated. In the eye, there are multiple barriers to tissue penetration and action, including blinking, lacrimation, mucin. Most ophthalmic formulations are short-lived on the surface of the eye, and only a small fraction of the dose is available for penetration into the eye. Microelectromechanical ocular implants that have micro-reservoirs can sustain drug release upon initiation of an external stimulus. Examples include drug delivery using sonophoresis (i.e., ultrasound), iontophoresis (application of low voltage current), lipid-based delivery systems, and polymers that deform using magnets to release the drug [78].

7.3 Transdermal Devices

Transdermal patch devices are promising on-demand drug delivery systems through the skin. They are highly acceptable by patients because they are non-invasive, patients can use them alone, reduce gastrointestinal side effects, and overall improve treatment success. Such patches contain a dissolved drug in a reservoir behind an impermeable membrane, made of natural or synthetic polymer or synthetic elastomers, which controls the rate of drug release. Until now, there are already several such systems approved for use for hormonal therapy, for some sedatives, antihypertensive, for anti-nausea drugs or for anti-smoking drugs. The application and effect of some of them can last up to 7 days [79, 80].

7.4 Medicine Delivery Pumps

Recent advances in closed drug delivery systems promise automated and continuous pump regulation in the near future. Such devices are already in use with patients with diabetes. When delivering insulin, a big problem is the daily administration of the drug, both because of the drug dose and the time regimen of administration, and because of the method of application, which involves subcutaneous injections. All these requirements have been overcome by the use of a smart device such as an insulin pump. These devices mimic the physiological rhythm of insulin secretion, basal and postprandial, thus contributing to better disease regulation, better compliance, and less frequent occurrence of adverse events such as hyperglycemia and hypoglycemia. By using these pumps, it is possible to monitor the level of insulin and glucose in the blood, as well as the effect of therapy in real-time [81].

8 Mathematical Models and Artificial Intelligence

Analyzing the change in drug concentration over time is a well-known process, but it requires a lot of manual work and testing on samples of different concentrations. The development of computer methods has made it possible to simulate these analyzes on a computer, thus saving time on testing and materials. In order for computer methods to be as accurate as possible and correspond to the real system, it is necessary to create adequate computer models. Numerous examples of analyzes and simulations based on diffusion effects can be found in the literature. These mathematical models find application in the pharmaceutical industry as well, enabling scientists and laboratory workers to reach results faster and understand the way the drug diffuses and spreads through the medium [82–84].

One way to translate a real model into an adequate computer model is to use genetic algorithms. Genetic algorithms (GA) are a type of parallel heuristic search method. Genetic algorithms belong to a larger class of evolutionary algorithms (EA) that generate solutions to optimization problems using techniques inspired by natural evolution, such as inheritance, mutation, selection, and crossing-over. The goal of genetic algorithms is curve fitting to find functional coefficients (i.e., parameter values) that minimize the total error in collecting the data points under consideration. The genetic algorithm terminates either when the maximum number of generations has been produced or a satisfactory level for the population has been reached [84–86].

A GA approach can be applied that fits the concentration-time curves and calculates the squared error of overlapping the real models with the newly created GA model. For this purpose, specific GA algorithms and software were developed. This method showed a very good correlation between the real curves and the estimated ones. In both cases, the coefficient of determination is very close to 1 (0.995 and 0.997) [86].

Example 1. As shown in the example by authors Mirić et al., the real concentration of the drug is determined by measuring the absorption, at the wavelength of 210 nm at which this drug shows absorption. A standard curve was obtained by a dilution series (0.1; 0.15; 0.2; and 1 mg/ml), and the 1 mg/ml results were obtained from this curve and used to fit the GA methods. Figure 2 shows a comparison of the actual measured changes in drug concentration over time in relation to the estimated results obtained by GA methods. Based on these results, we can conclude that the GA method has a very good potential for evaluation and provides an opportunity to evaluate the drug release of some other concentrations [86].

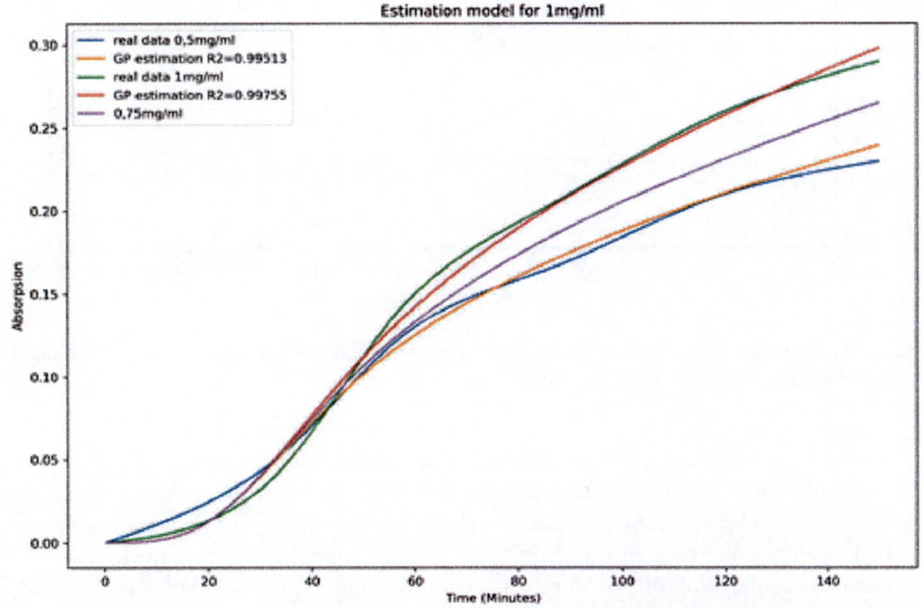

Fig. 2. Standard curve versus GA solution (taken from Mirić et al. [86])

In this way, it is possible to achieve control over the concentration of the released drug, which ensures that its concentration is optimal at the target site. It is expected that the application of such modern drug delivery systems would overcome the shortcomings of traditional treatment [86, 87].

Example 2. There are already developed computer models that simulate the release of drugs from carrier systems obtained by electrospinning methodology. Depending on the properties of the material, the mechanism of drug release such as degradation, erosion or diffusion, environmental conditions, characteristics of drug molecules, complex models are developed with high accuracy and applicable in the development of drug delivery systems. These models are very useful tools in this field because with artificial intelligence we can predict the release of a drug of a certain concentration in time from a different network of fibers [88].

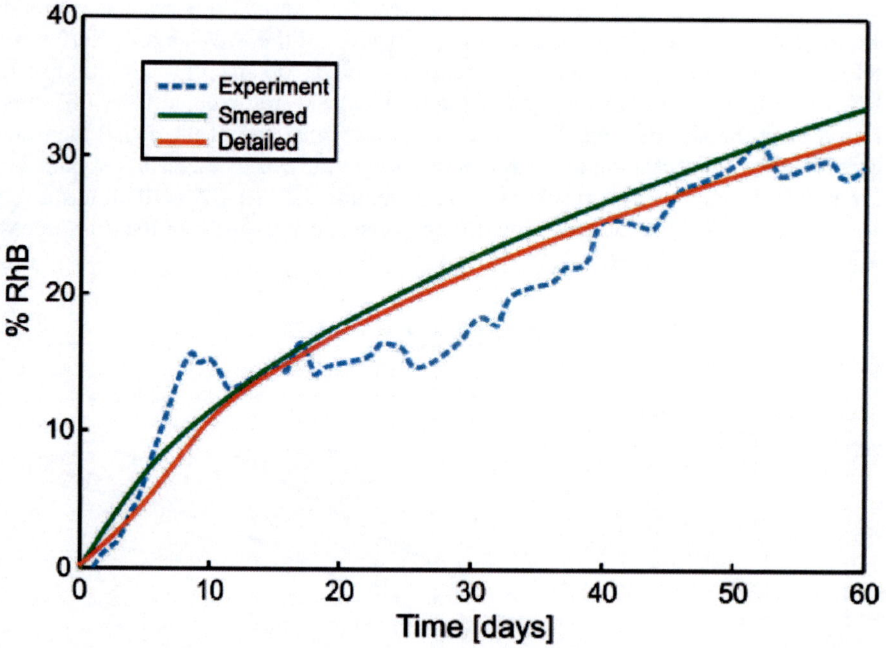

Fig. 3. Experimental curve and computational results obtained using true (detailed) and smeared models of PLGA nanofibers (taken from Milosevic et al. [88]).

Figure 3 shows a comparison of numerical and experimental results of drug (in this case is Rhodamine B - RhB) diffusion from PLGA fibers. It can be seen that there are slight differences between the two models; therefore, the concept of smeared modeling can be used to predict drug transport from drug-impregnated nanofibers [88].

To simulate the drug diffusion process from nanofibers, two variables are needed, the drug concentration that is released per unit of time and the time for which that concentration is released. The goal of the artificial neural network is to predict those two variables, using the input parameters. Two groups of data can be used as input parameters. The first group are material parameters, such as: type of polymer, its density, hydrophilicity, amount of medicinal substance used, concentration and temperature of the solution in the core and shell. Another group of data that is used as an input are electrospinning parameters such as: speed of pulling the core into the sheath, applied voltage, flow rate, distance of the collector from the capillary tube, temperature and ambient humidity. These parameters significantly affect the thickness of the fiber, thus affecting the drug diffusion process as well [89, 90].

9 Conclusion

The concept of controlled delivery of drugs today implies a new, more advanced way of administering medicines, the aim of which is to improve the efficiency and quality of treatment. The advantages of this method of treatment compared to the traditional one includes the control over the concentration of the released drug, as well as the target site of action. Achieving this, in the same amount of time it prevents accumulation of drug in parts of the body where it is not used, which potentially reduces the occurrence of side effects. It is expected that the application of this concept in the treatment of acute or chronic diseases would be able to overcome most of the shortcomings associated with traditional treatment.

Computational models created using genetic algorithms and artificial intelligence methods show very good estimation of drug release concentrations over time and have huge potential for future *in silico* testing. These methods could greatly simplify the tablet modeling process and, with the use of 3D printing, electrospinning, liposome and nanoparticle production, and open a new path to targeted therapy.

References

1. Dash, A.K., Cudworth II, G.C.: Therapeutic applications of implantable drug delivery systems. J. Pharmacol. Toxicol. Methods **40**(1), 1–12 (1998). https://doi.org/10.1016/s1056-8719(98)00027-6
2. Kopecek, J.: Smart and genetically engineered biomaterials and drug delivery systems. Eur. J. Pharm. Sci. **20**(1), 1–16 (2003). https://doi.org/10.1016/s0928-0987(03)00164-7
3. Redekop, W.K., Mladsi, D.: The faces of personalized medicine: a framework for understanding its meaning and scope. Value Health **16**(6 Suppl.), S4– S9 (2013). https://doi.org/10.1016/j.jval.2013.06.005
4. Stanojević, G., Medarević, D., Adamov, I., Pešić, N., Kovačević, J., Ibrić, S.: Tailoring atomoxetine release rate from DLP 3D-printed tablets using artificial neural networks: influence of tablet thickness and drug loading. Molecules **26**(1), 111 (2020). https://doi.org/10.3390/molecules26010111
5. Tan, J.P., et al.: Hierarchical supermolecular structures for sustained drug release. Small **5**(13), 1504–1507 (2009). https://doi.org/10.1002/smll.200801756
6. Briones, E., Colino, C.I., Lanao, J.M.: Delivery systems to increase the selectivity of antibiotics in phagocytic cells. J. Control Release **125**(3), 210–227 (2008). https://doi.org/10.1016/j.jconrel.2007.10.027
7. Tian, Y., et al.: A series of naphthalimide derivatives as intra and extracellular pH sensors. Biomaterials **31**(29), 7411–7422 (2010). https://doi.org/10.1016/j.biomaterials.2010.06.023

8. Alymani, N.A., Smith, M.D., Williams, D.J., Petty, R.D.: Predictive biomarkers for personalised anti-cancer drug use: discovery to clinical implementation. Eur. J. Cancer **46**(5), 869–879 (2010). https://doi.org/10.1016/j.ejca.2010.01.001
9. Davoodi, P., et al.: Drug delivery systems for programmed and on-demand release. Adv. Drug Deliv. Rev. **132**, 104–138 (2018). https://doi.org/10.1016/j.addr.2018.07.002
10. Biondi, M., Ungaro, F., Quaglia, F., Netti, P.A.: Controlled drug delivery in tissue engineering. Adv. Drug Deliv. Rev. **60**(2), 229–242 (2008). https://doi.org/10.1016/j.addr.2007.08.038
11. Akbar, M.U., Badar, M., Zaheer, M.: Programmable drug release from a dual-stimuli responsive magnetic metal-organic framework. ACS Omega **7**(36), 32588–32598 (2022). https://doi.org/10.1021/acsomega.2c04144
12. Parveen, F., et al.: Investigation of eutectic mixtures of fatty acids as a novel construct for temperature-responsive drug delivery. Int. J. Nanomed. **17**, 2413–2434 (2022). https://doi.org/10.2147/IJN.S359664
13. Song, Y., Li, Y., Xu, Q., Liu, Z.: Mesoporous silica nanoparticles for stimuli-responsive controlled drug delivery: advances, challenges, and outlook. Int. J. Nanomed. **12**, 87–110 (2016). https://doi.org/10.2147/IJN.S117495
14. Ge, J., Neofytou, E., Cahill III, T.J., Beygui, R.E., Zare, R.N.: Drug release from electric-field-responsive nanoparticles. ACS Nano **6**(1), 227–233 (2012). https://doi.org/10.1021/nn203430m
15. Ten Hagen, T.L.M., et al.: Drug transport kinetics of intravascular triggered drug delivery systems. Commun. Biol. **4**(1), 920 (2021) https://doi.org/10.1038/s42003-021-02428-z
16. Afereydoon, S., et al.: Multifunctional PEGylated niosomal nanoparticle-loaded herbal drugs as a novel nano-radiosensitizer and stimuli-sensitive nanocarrier for synergistic cancer therapy. Front. Bioeng. Biotechnol. **10**, 917368 (2022). https://doi.org/10.3389/fbioe.2022.917368
17. Fatfat, Z., Fatfat, M., Gali-Muhtasib, H.: Micelles as potential drug delivery systems for colorectal cancer treatment. World J. Gastroenterol. **28**(25), 2867–2880 (2022). https://doi.org/10.3748/wjg.v28.i25.2867
18. Gupta, H., Bhandari, D., Sharma, A.: Recent trends in oral drug delivery: a review. Recent Pat. Drug Deliv. Formul. **3**(2), 162–173 (2009). https://doi.org/10.2174/187221109788452267
19. Maver, U., Milojević, M., Štos, J., Adrenšek, S., Planinšek, O.: Matrix tablets for controlled release of drugs incorporated using capillary absorption. AAPS PharmSciTech **20**(2), 91 (2019). https://doi.org/10.1208/s12249-019-1303-5
20. Abdelkader, H., Youssef Abdalla, O., Salem, H.: Formulation of controlled-release baclofen matrix tablets II: influence of some hydrophobic excipients on the release rate and in vitro evaluation. AAPS PharmSciTech **9**(2), 675–683 (2008). https://doi.org/10.1208/s12249-008-9094-0
21. Al Hanbali, O.A., et al.: Transdermal patches: design and current approaches to painless drug delivery. Acta Pharm. **69**(2), 197–215 (2019). https://doi.org/10.2478/acph-2019-0016
22. Cilurzo, F., Gennari, C.G., Minghetti, P.: Adhesive properties: a critical issue in transdermal patch development. Expert Opin. Drug Deliv. **9**(1), 33–45 (2012). https://doi.org/10.1517/17425247.2012.637107
23. Pastore, M.N., Kalia, Y.N., Horstmann, M., Roberts, M.S.: Transdermal patches: history, development and pharmacology. Br. J. Pharmacol. **172**(9), 2179–2209 (2015). https://doi.org/10.1111/bph.13059
24. Banerjee, S., Chattopadhyay, P., Ghosh, A., Datta, P., Veer, V.: Aspect of adhesives in transdermal drug delivery systems. Int. J. Adhes. Adhes. **50**, 70–84 (2014). https://doi.org/10.1016/j.ijadhadh.2014.01.001
25. Wokovich, A.M., Prodduturi, S., Doub, W.H., Hussain, A.S., Buhse, L.F.: Transdermal drug delivery system (TDDS) adhesion as a critical safety, efficacy and quality attribute. Eur. J. Pharm. Biopharm. **64**(1), 1–8 (2006). https://doi.org/10.1016/j.ejpb.2006.03.009

26. Alexander, A., et al.: Approaches for breaking the barriers of drug permeation through transdermal drug delivery. J Control Release **164**(1), 26–40 (2012). https://doi.org/10.1016/j.jconrel.2012.09.017
27. Daniels, R., Knie, U.: Galenics of dermal products–vehicles, properties and drug release. J. Dtsch. Dermatol. Ges. **5**(5), 367–383 (2007). https://doi.org/10.1111/j.1610-0387.2007.06321.x
28. Basu, S.C., Basu, M.: Liposome Methods and Protocols, vol. 199. Springer, Charm (2008)
29. Koruga, Đ: Nanotehnologije u medicini i kozmetici. Arhiv za Farmaciju **56**(2), 164–177 (2006)
30. Utell, M.J., Frampton, M.W.: Acute health effects of ambient air pollution: the ultrafine particle hypothesis. J. Aerosol. Med. **13**(4), 355–359 (2000). https://doi.org/10.1089/jam.2000.13.355
31. Nikolić, S., et al.: Orally administered fluorescent nanosized polystyrene particles affect cell viability, hormonal and inflammatory profile, and behavior in treated mice. Environ. Pollut. **305**, 119206 (2022). https://doi.org/10.1016/j.envpol.2022.119206
32. Pokropivny, V.V., Skorokhod, V.V.: Classification of nanostructures by dimensionality and concept of surface forms engineering in nanomaterial science. Mater. Sci. Eng. C **27**(5–8), 990–993 (2007). https://doi.org/10.1016/j.msec.2006.09.023
33. Wu, B., Wu, X., Liu, S., Wang, Z., Chen, L.: Size-dependent effects of polystyrene microplastics on cytotoxicity and efflux pump inhibition in human Caco-2 cells. Chemosphere **221**, 333–341 (2019). https://doi.org/10.1016/j.chemosphere.2019.01.056
34. Tiwari, P.M., Bawage, S.S., Singh, S.R.: Gold nanoparticles and their applications in photomedicine, diagnosis and therapy. Appl. Nanosci. Photomed., 249–266 (2015). https://doi.org/10.1533/9781908818782.249
35. Partha, R., Conyers, J.L.: Biomedical applications of functionalized fullerene-based nanomaterials. Int. J. Nanomed. **4**, 261–275 (2009)
36. Stern, S.T., McNeil, S.E.: Nanotechnology safety concerns revisited. Toxicol. Sci. **10**(1), 4–21 (2007). https://doi.org/10.1093/toxsci/kfm169
37. Conde, J., et al.: Revisiting 30 years of biofunctionalization and surface chemistry of inorganic nanoparticles for nanomedicine. Front. Chem. **2**, 48 (2014). https://doi.org/10.3389/fchem.2014.00048
38. Lewis, G., Janna, S., Bhattaram, A.: Influence of the method of blending an antibiotic powder with an acrylic bone cement powder on physical, mechanical, and thermal properties of the cured cement. Biomaterials **26**(20), 4317–4325 (2005). https://doi.org/10.1016/j.biomaterials.2004.11.003
39. Kumari, A., Yadav, S.K., Yadav, S.C.: Biodegradable polymeric nanoparticles based drug delivery systems. Colloids Surf B Biointerfaces **75**(1), 1–18 (2010). https://doi.org/10.1016/j.colsurfb.2009.09.001
40. Khang, G., Choi, H.S., Rhee, J.M.: Controlled release of gentamicin sulphate from poly (3-hydroxybuttyrateco-3-hydroxyvalerate) wafers for the treatment of osteomyelitis. Korea Polym. J. **8**, 253–262 (2000)
41. Noel, S.P., Courtney, H., Bumgardner, J.D., Haggard, W.O.: Chitosan films: a potential local drug delivery system for antibiotics. Clin. Orthop. Relat. Res. **466**(6), 1377–1382 (2008). https://doi.org/10.1007/s11999-008-0228-1
42. Aoki, K., Saito, N.: Biodegradable polymers as drug delivery systems for bone regeneration. Pharmaceutics **12**(2), 95 (2020). https://doi.org/10.3390/pharmaceutics12020095
43. Kanellakopoulou, K., et al.: Lactic acid polymers as biodegradable carriers of fluoroquinolones: an in vitro study. Antimicrob. Agents Chemother. **43**(3), 714–716 (1999). https://doi.org/10.1128/AAC.43.3.714

44. Adamović, D., Ristić, B., Živić, F.: Review of existing biomaterials – method of material selection for specific applications in orthopedics. Fakultet inženjenjrskih nauka, Univezitet u Kragujevcu (2018)
45. Martins, V.C., Goissis, G., Ribeiro, A.C., Marcantônio Jr., E., Bet, M.R.: The controlled release of antibiotic by hydroxyapatite: anionic collagen composites. Artif. Organs. **22**(3), 215–21 (1998). https://doi.org/10.1046/j.1525-1594.1998.06004.x
46. Vey, E., et al.: Degradation mechanism of poly (lactic-co-glycolic) acid block copolymer cast films in phosphate buffer solution. Polym. Degrad. Stab. **93**(10), 1869–1876 (2008). https://doi.org/10.1016/j.polymdegradstab.2008.07.018
47. Jung, J.H., Ree, M., Kim, H.: Acid- and base-catalyzed hydrolyses of aliphatic polycarbonates and polyesters. Catal. Today **115**(1–4), 283–287 (2006). https://doi.org/10.1016/j.cattod.2006.02.060
48. Gunatillake, P.A., Adhikari, R.: Biodegradable synthetic polymers for tissue engineering. Eur. Cell Mater. **5**., 1–16 (2003). https://doi.org/10.22203/ecm.v005a01
49. Stevanović, M., et al.: Comparison of hydroxyapatite/poly (lactide-co-glycolide) and hydroxyapatite/polyethyleneimine composite scaffolds in bone regeneration of swine mandibular critical size defects: vivo study. Molecules **27**(5), 1694 (2022). https://doi.org/10.3390/molecules27051694
50. Wang, X., et al.: Gelatin-based hydrogels for organ 3D bioprinting. Polymers **9**(9), 401 (2017). https://doi.org/10.3390/polym9090401
51. Li, J., Wu, C., Chu, P.K., Gelinsky, M.: 3D printing of hydrogels: Rational design strategies and emerging biomedical applications. Mater. Sci. Eng. Rep. **140**, 100543 (2020). https://doi.org/10.1016/j.mser.2020.100543
52. Yannas, I.V., Burke, J.F.: Design of an artificial skin. I. Basic design principles. J. Biomed. Mater. Res. **14**(1), 65–81 (1980). https://doi.org/10.1002/jbm.820140108
53. Fu Lu, M.Z., Thies, C.: Collagen-Based Drug Delivery Devices. Polymers for Controlled Drug Delivery, CRC Press, Boca Raton, FL, pp. 149–161 (1991)
54. Nardi-Ricart, A., et al.: Formulation of sustained release hydrophilic matrix tablets of tolcapone with the application of sedem diagram: influence of tolcapone's particle size on sustained release. Pharmaceutics **12**(7), 674 (2020). https://doi.org/10.3390/pharmaceutics12070674
55. Bejugam, N.K., Parish, H.J., Shankar, G.N.: Influence of formulation factors on tablet formulations with liquid permeation enhancer using factorial design. AAPS PharmSciTech **10**(4), 1437–1443 (2009). https://doi.org/10.1208/s12249-009-9345-8
56. Nguyen, T.T., Hwang, K.M., Kim, S.H., Park, E.S.: Development of novel bilayer gastroretentive tablets based on hydrophobic polymers. Int. J. Pharm. **574**, 118865 (2020). https://doi.org/10.1016/j.ijpharm.2019.118865
57. Ozbolat, I.T., Moncal, K.K., Gudapati, H.: Evaluation of bioprinter technologies. Addit. Manuf. **13**, 179–200 (2017). https://doi.org/10.1016/j.addma.2016.10.003
58. Webb, B., Doyle, B.J.: Parameter optimization for 3D bioprinting of hydrogels. Bioprinting **8**, 8–12 (2017). https://doi.org/10.1016/j.bprint.2017.09.001
59. He, Y., Gu, Z., Xie, M., Fu, J., Lin, H.: Why choose 3D bioprinting? Part II: methods and bioprinters. Bio Des. Manuf. **3**(1), 1–4 (2020). https://doi.org/10.1007/s42242-020-00064-w
60. Sears, N.A., Seshadri, D.R., Dhavalikar, P.S., Cosgriff-Hernandez, E.: A review of three-dimensional printing in tissue engineering. Tissue Eng. Part B Rev. **22**(4), 298–310 (2016). https://doi.org/10.1089/ten.TEB.2015.0464
61. Kang, H.W., Lee, S.J., Ko, I.K., Kengla, C., Yoo, J.J., Atala, A.: A 3D bioprinting system to produce human-scale tissue constructs with structural integrity. Nat. Biotechnol. **34**(3), 312–319 (2016). https://doi.org/10.1038/nbt.3413
62. Murphy, S.V., Atala, A.: 3D bioprinting of tissues and organs. Nat. Biotechnol. **32**(8), 773–785 (2014). https://doi.org/10.1038/nbt.2958

63. He, Y., et al.: Research on the printability of hydrogels in 3D bioprinting. Sci. Rep. **6**, 29977 (2016). https://doi.org/10.1038/srep29977
64. Melocchi, A., et al.: 3D printing by fused deposition modeling of single- and multi-compartment hollow systems for oral delivery - a review. Int. J. Pharm. **579**, 119155 (2020). https://doi.org/10.1016/j.ijpharm.2020.119155
65. Živanović, M.N.: Use of electrospinning to enhance the versatility of drug delivery. In: Lai, W.F. (eds.) Systemic Delivery Technologies in Anti-Aging Medicine: Methods and Applications. Healthy Ageing and Longevity, vol. 13, pp. 347–364. Springer, Cham (2020). https://doi.org/10.1007/978-3-030-54490-4_14. ISBN 978-3-030-54489-8
66. Wang, T., Yang, L., Xie, Y., Cheng, S., Xiong, M., Luo, X.: An injectable hydrogel/staple fiber composite for sustained release of CA4P and doxorubicin for combined chemotherapy of xenografted breast tumor in mice. Nan Fang Yi Ke Da Xue Xue Bao. **42**(5), 625-632 (2022). https://doi.org/10.12122/j.issn.1673-4254.2022.05.01
67. Miloševic, M., et al.: Preparation and modeling of three-layered PCL/PLGA/PCL fibrous scaffolds for prolonged drug release. Sci Rep. **10**(1), 11126 (2020). https://doi.org/10.1038/s41598-020-68117-9
68. Gu, J., Wensing, M., Uhde, E., Salthammer, T.: Characterization of particulate and gaseous pollutants emitted during operation of a desktop 3D printer. Environ. Int. **123**, 476–485 (2019). https://doi.org/10.1016/j.envint.2018.12.014
69. Hayashi, Y., Inoue, M., Takizawa, H., Suganuma, K.: Nanoparticle Fabrication. Nanopackaging, 109–120 (2008). https://doi.org/10.1007/978-0-387-47325-3_6
70. Yamada, H., Yamana, K., Kawasaki, R., Yasuhara, K., Ikeda, A.: Cyclodextrin-induced release of drug-entrapping liposomes associated with the solation of liposome gels. RSC Adv. **12**(34), 22202–22209 (2022). https://doi.org/10.1039/d2ra03837d
71. Price, J.S., Tencer, A.F., Arm, D.M., Bohach, G.A.: Controlled release of antibiotics from coated orthopedic implants. J. Biomed. Mater. Res. **30**(3), 281–286 (1996). https://doi.org/10.1002/(SICI)1097-4636(199603)30:3%3c281:AID-JBM2%3e3.0.CO,2-M
72. Food, D.: Administration, guidance for industry: waiver of in vivo bioavailability and bioequivalence studies for immediate-release solid oral dosage forms based on a biopharmaceutics classification system. In: Food and Drug Administration, Rockville, MD (2000)
73. Vimalson, D.C., Parimalakrishnan, S., Jeganathan, N.S., Anbazhagan, S.: Techniques to enhance solubility of hydrophobic drugs: an overview. Asian J. Pharm. Sci. **10**(2), S67–S75 (2016)
74. Ambrose, C.G., et al.: Antibiotic microspheres: preliminary testing for potential treatment of osteomyelitis. Clin. Orthop. Relat. Res. **415**, 279–285 (2003). https://doi.org/10.1097/01.blo.0000093920.26658.ae
75. Lewis, R.E., et al.: Evaluation of low-dose, extended-interval clindamycin regimens against Staphylococcus aureus and Streptococcus pneumoniae using a dynamic in vitro model of infection. Antimicrob. Agents Chemother. **43**(8), 2005–2009 (1999). https://doi.org/10.1128/AAC.43.8.2005
76. Virto, M.R., Elorza, B., Torrado, S., Elorza Mde, L., Frutos, G.: Improvement of gentamicin poly (D, L-lactic-co-glycolic acid) microspheres for treatment of osteomyelitis induced by orthopedic procedures. Biomaterials **28**(5), 877–85 (2007). https://doi.org/10.1016/j.biomaterials.2006.09.045
77. Stajić, D., Živanović, S., Mirić, A., Sekulić, M., Đonović, N.: Prevalence of risk factors among women with osteoporosis. Serbian J. Exp. Clin. Res. **18**(3), 239–243 (2017). https://doi.org/10.1515/sjecr-2016-0080
78. Perez, V.L., Wirostko, B., Korenfeld, M., From, S., Raizman, M.: Ophthalmic drug delivery using iontophoresis: recent clinical applications. J. Ocul. Pharmacol. Ther. **36**(2), 75–87 (2020). https://doi.org/10.1089/jop.2019.0034

79. Bird, D., Ravindra, N.M.: Transdermal drug delivery and patches—an overview. Med. Devices Sens. **3**(6) (2020). https://doi.org/10.1002/mds3.10069
80. Smaoui, M.R., Lafi, A.: Leeno: type 1 diabetes management training environment using smart algorithms. PLoS ONE **17**(9), e0274534 (2022). https://doi.org/10.1371/journal.pone.0274534
81. Keyu, G., et al.: Comparing the effectiveness of continuous subcutaneous insulin infusion with multiple daily insulin injection for patients with type 1 diabetes mellitus evaluated by retrospective continuous glucose monitoring: a real-world data analysis. Front. Public Health **10**, 990281 (2022). https://doi.org/10.3389/fpubh.2022.990281
82. Demetriades, M., et al.: Interrogating and quantifying in vitro cancer drug pharmacodynamics via agent-based and Bayesian monte Carlo modelling. Pharmaceutics **14**(4), 749 (2022). https://doi.org/10.3390/pharmaceutics14040749
83. Galdi, I., Lamberti, G.: Drug release from matrix systems: analysis by finite element methods. Heat Mass Transfer. **48**, 519–528 (2012). https://doi.org/10.1007/s00231-011-0900-y
84. Filipović, N., Živanović, M.N.: Use of numerical simulation in carrier characterization and optimization. In: Lai, WF. (eds.) Systemic Delivery Technologies in Anti-Aging Medicine: Methods and Applications. Healthy Ageing and Longevity, vol. 13, pp. 435–446. Springer, Cham (2020). https://doi.org/10.1007/978-3-030-54490-4_18
85. Hadjianfar, M., Semnani, D., Varshosaz, J., Mohammadi, S., Rezazadeh Tehrani, S.P.: 5FU-loaded PCL/Chitosan/Fe3O4 core-shell nanofibers structure: an approach to multi-mode anti-cancer system. Adv. Pharm. Bull. **12**(3), 568–582 (2022). https://doi.org/10.34172/apb.2022.060
86. Mirić, A., et al.: Controlled drug release from a 3D printed tablet. In: 1st Serbian International Conference on Applied Artificial Intelligence, Kragujevac, Serbia, p 86, 19–20 May 2022. ISBN: 978-86-81037-71-3
87. Yokoi, K., et al.: Liposomal doxorubicin extravasation controlled by phenotype-specific transport properties of tumor microenvironment and vascular barrier. J. Control Release **217**, 293–299 (2015). https://doi.org/10.1016/j.jconrel.2015.09.044
88. Milošević, M., et al.: A computational model for drug release from PLGA implant. Materials **11**(12), 2416 (2018). https://doi.org/10.3390/ma11122416
89. Maleki, M., Amani-Tehran, M., Latifi, M., Mathur, S.: Drug release profile in core–shell nanofibrous structures: a study on Peppas equation and artificial neural network modeling. Comput Meth. Prog. Biomed. **113**(1), 92–100 (2014). https://doi.org/10.1016/j.cmpb.2013.09.003
90. Musulin, J., et al.: Application of artificial intelligence-based regression methods in the problem of COVID-19 spread prediction: a systematic review. Int. J. Environ. Res. Public Health **18**(8), 4287 (2021). https://doi.org/10.3390/ijerph18084287

Cost Effectiveness Analysis of Real and in Silico Clinical Trials for Stent Deployment Based on Decision Tree

Marija Gačić[1,2](✉)

[1] Institute for Information Technologies, University of Kragujevac, Jovana Cvijića bb Street, 34000 Kragujevac, Serbia
marija.gacic@kg.ac.rs

[2] Bioengineering Research and Development Centre (BioIRC), 6 Prvoslava Stojanovića Street, 34000 Kragujevac, Serbia

Abstract. The global coronary and peripheral stent market size was valued at 5.91 billion USD in 2019 and is projected to reach 8.08 billion USD by 2027 as new and innovative devices are being invented and developed rapidly. In this process of developing new models of stents, one of the key phases is clinical testing on patients. The aim of in silico medicine is to Reduce, Refine and Replace (3R concept) real clinical trials with an aim to decrease costs and time needed to perform a clinical study. Within the InSilc project (funded by H2020 programme, GA 777119) the platform for designing, developing and assessing stents was developed. The platform consists of several modules, some of which can be used as standalone modules. Characteristics of Mechanical and Deployment Module are presented in this chapter. Pricing strategies for different scenarios are described in order to prove the effectiveness and benefits of in-silico testing. The workflow begins from the Mechanical module, and continues to the prediction of the stenting outcomes for different virtual anatomies and different modules: 3D reconstruction and plaque characterization tool, Deployment Module, Fluid dynamics Module, Drug-delivery Module, Degradation Module and Myocardial Perfusion Module. Cost-effectiveness analysis is done using decision tree for in silico and real clinical trials for coronary stent deployment.

Keywords: coronary and peripheral artery · in silico clinical trials · pricing strategy · stenting outcome · virtual anatomy · stent deployment · cost effectiveness analysis · decision tree

1 Introduction

Real clinical trials require approval by a regulatory authority and an ethics committee review of the pre-clinical regulatory submission [1]. The basic assumption is that a data base collected in the clinical study is relatively small but a representative selection of subjects and the researchers have to generalise the results so they could be applicable to the larger patient population. If the sample is too constrained or poorly selected,

N. Filipovic (Ed.): AAI 2022, LNNS 659, pp. 367–382, 2023.
https://doi.org/10.1007/978-3-031-29717-5_22

the broad applicability of the results is hindered. This is not only a statistical concern, but also an ethical and medical one [2]. Within the InSilc project (2017–2021) an in silico clinical trial platform was developed for designing, developing and assessing drug-eluting bioresorbable vascular scaffolds (BVS), by building on the comprehensive biological and biomedical knowledge and advanced modelling approaches, to simulate their implantation performance in the individual cardiovascular physiology. The platform is also applicable on the other models of stents, such as BMS, DES, peripheral stents etc. [3].

Testing of new models of vascular stents, scaffolds and balloons in real clinical trials is time-consuming, expensive and highly inconvenient for patients included in the study. Therefore, the intention is to replace, reduce and refine real clinical studies with in silico clinical studies and in silico testing of innovative models of stents in order to decrease the costs and the time required to perform real clinical studies. In this chapter, analyse cost effectiveness of using in silico clinical trials for stent deployment was compared with real clinical trials.

Mechanical Modeling module assists in reducing the required number of real mechanical tests and the associated costs. In brief, the module provides the ability of the following mechanical tests to be simulated *in silico*: Simulated use – Pushability, Torquability, Trackability, Recoil, Crush resistance, Flex/kink, Longitudinal tensile strength, Crush resistance with parallel plates, Local Compression, Radial Force, Foreshortening, Dog Boning, Three-point bending, Inflation and Radial Fatigue tests.

These scenarios are designed for prediction of the stenting outcomes for different virtual anatomies and different modules as 3D reconstruction and plaque characterization tool, Deployment Module, Fluid dynamics Module, Drug-delivery Module, Degradation Module and Myocardial Perfusion Module.

In-silico simulation of the stenting procedure consists of the following steps that should be repeated for each device (stent or balloon): (i) device positioning, (ii) balloon inflation and, (iii) stent deployment.

2 Stent Market

2.1 Coronary Stents

Coronary stents have revolutionised the treatment and prognosis for patients with cardiovascular disease by lowering the risk of restenosis and providing improved long-term clinical results. However, the development and introduction of these devices to the market is slow and expensive. As a result, the time to reach the market of the innovative model is too long (3 to 7 years are required on average for the stent industry to bring these devices to the market), while the costs are high (throughout the different phases of the development process between 31 and 94 million EUR). The InSilc platform is an integral part of the product definition (prototype development), product development (development of product design) and product confirmation (design verification and validation) phases of the development lifecycle. In brief, InSilc optimizes the product development process by reducing:

- the number of design iteration cycles, and the cost of the associated testing, required to optimize the stents,
- the risk of device redesign, and the significant costs associated with it, particularly in the later phases of the product development cycle,
- the return-on-investment threshold below which the development of the stents would not be cost-effective,
- time to the market release.

According to the Global Balloon-Expandable Stents Market Industry Trends and Forecast to 2028 (report by Data Bridge Market Research) [4], the demand for balloon-expandable stents has significantly increased over the past years and the market is estimated to grow at formidable rate. Some of the main factors driving the growth of the market are increasing prevalence of vascular diseases, preference for minimally invasive procedures, rising geriatric population, increased research activities focusing on improving the balloon-expandable stent technology.

Currently, some of the most important players in the balloon-expandable stents market are Medtronic, Abbott, Boston Scientific Corporation, Biotronik, B. Braun Melsungen AG, Terumo Corporation, MicroPort Scientific Corporation, Meril Life Sciences Pvt. Ltd, STENTYS SA, Vascular Concept., BD, W. L. Gore & Associates, Inc., ENDOLOGIX, etc. [5].

2.2 Peripheral Stents Market

Peripheral stents are small tubular metal scaffolds that are used for deployment in the peripheral vessels to treat narrowing or blockage within peripheral arteries or veins, in order to provide more blood flow. The market report by Future Market Insights "Peripheral Vascular Stents Market: Global Industry Analysis 2013–2017 and Opportunity Assessment 2018–2028" [6] indicates that the global peripheral vascular stents market is anticipated to growth at a compound annual growth rate (CAGR) of 6.6% over the forecast period 2018–2028 and reach 5,324.1 million USD by the end of 2028. The self-expanding stents segment is expected to grow at above 6.6% CAGR over the forecast period 2018–2028. However, high cost of peripheral vascular stents and related procedures are stated to be the main restraint for the market growth. The negative impact of the COVID-19 pandemic is expected to reduce the demand for these stents. On the other hand, introduction of new and state-of-the-art technologies and growing demand for minimally invasive procedures, accompanied with the launching of the last generation of stent devices (drug-eluting stents and bio-based stents) present key opportunities for the market expansion. This market includes the self-expanding stents and the balloon expanding stents. The self-expanding stents segment has a major share in the market, with nearly 51% of the global revenue in 2019.

North America dominated the global peripheral vascular stents market in 2021, with nearly 48.5% market share, partly due to the penetration of newly approved drug eluting stents in the U.S. and is expected stay in the leading position in the forecast period. Western Europe is anticipated to be the second large market, with the U.K., Italy, France and Germany showing high growth rates. Asia Pacific is anticipated to grow at a higher CAGR over the forecast period due to the increasing number of patients, availability of cost-effective surgeries and rise in the percentage of elderly population in the region. The main factors driving the demand for peripheral vascular stents are high prevalence of peripheral vascular disease across the globe, the increase of the aging population, the increased use of peripheral stents in angioplasty and conventional open surgery.

The main market players are Medtronic Plc., Abbott Vascular Inc., Dickinson and Co., Cordis Corporation, Cardinal Health, Inc., B. Braun Melsungen AG, MicroPort Scientific Corporation, Boston Scientific Corp., W.L. Gore & Associates Inc., and Cook Medical Inc. [7].

2.3 InSilc Platform

In-silico medicine is still in the emerging stage, but this market is extremely promising, and it shows the potential to reduce the number of traditional clinical trials. The high potential for market penetration of in-silico solutions, such as the InSilc platform, is based on the overall megatrends that digital transformation will bring to healthcare.

In-silico medicine, which involves the adoption of predictive computer models, promises to drastically accelerate and amplify medical device and pharmaceutical innovation process. At the same time, it will slow down the unsustainable rise in the healthcare costs [8].

The InSilc platform is based on the extension of existing multidisciplinary and multi-scale models for simulating the drug-eluting BVS mechanical behaviour, the deployment and degradation, the fluid dynamics in the micro- and macroscale, and the myocardial perfusion, for predicting the drug-eluting BVS and vascular wall interaction in the short- and medium/long term.

The developed InSilc platform consists of different simulation modules/tools - some of which can be considered as stand-alone modules and, therefore, can be used separately if there is such demand from the targeted users. These modules integrated in the InSilc platform are: Mechanical Modelling Module, 3D reconstruction and plaque characterization tool, Deployment Module, Fluid Dynamics Module, Drug Delivery Module, Degradation Module, Myocardial Perfusion Module, Virtual Population Physiology and Virtual Population database (Fig. 1). These tools are applicable all types of coronary and peripheral stents, such as Bare Metal Stents (BMS), Drug-eluting Stents (DES) and Bioresorbable Vascular Stents (BVS). This is a great advantage of InSilc which enables the penetration of the InSilc platform and modules to a wide range of market and interested stakeholders [9]. Drug-coated balloon simulation and optimization system for the improved treatment of peripheral artery disease has been considered in DECODE project [10]. Cost effectiveness for the real and in silico clinical trial performed on the BMS, DES and BVS stents using the in silico cloud platform solution was analysed.

Decision analysis and cost effectiveness analysis are quantitative techniques that provide a systematic approach to integrating evidence within the context of a specific decision problem. Some steps in decision analysis can be defined: 1) define the decision problem (including specifying the decision-maker and the ultimate goal or objective of the decision); 2) identify all the decision alternatives; 3) list all the possible outcomes of each decision alternative; 4) define the relevant time horizon; 5) map out the sequence of events leading from the initial decision to the relevant outcomes including chance events and secondary decisions; 6) quantify uncertainty: determine the probability of

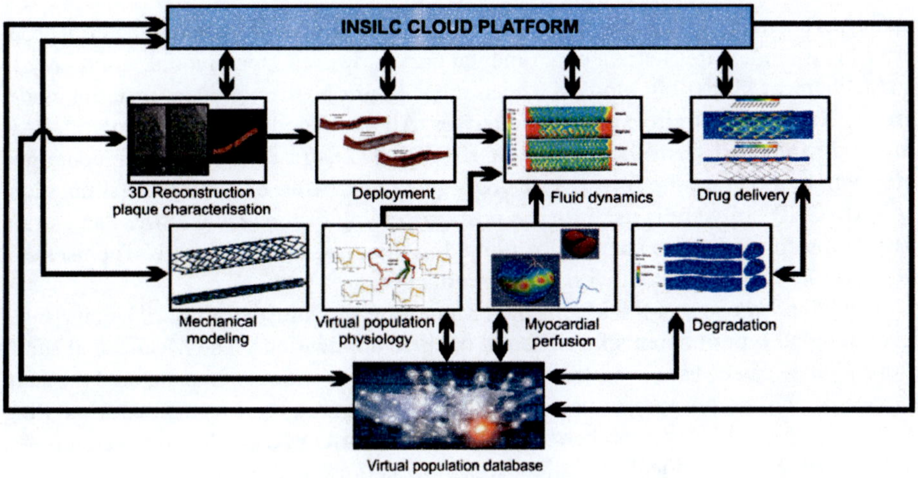

Fig. 1. InSilc cloud platform.

each chance outcome; 7) quantify values: assign a value to each outcome; 8) calculate the expected value of each decision alternative [11]. The process of explicitly quantifying the uncertainty and values involved in a decision problem provides valuable insight into the key issues and controversies inherent in the decision.

3 Mechanical Modelling Module

The Stent industry follows standard mechanical stent testing in the entire process of stent evaluation, i.e., according to ISO test standards. Mechanical testing is often considered overly time-consuming and expensive, and requires many cycles/iterations, while in some cases a total stent redesign is required or even the examined stent design is abandoned. The Mechanical Modelling module assists in reducing the required number of real mechanical tests and the relevant costs. In brief, the module provides the ability of the following mechanical tests to be simulated *in silico*: Simulated use – Pushability, Torquability, Trackability, Recoil, Crush resistance, Flex/kink, Longitudinal tensile strength, Crush resistance with parallel plates, Local Compression, Radial Force, Foreshortening, Dog Boning, Three-point bending, Inflation and Radial Fatigue test. The risk of fatigue failure is also predicted using fatigue criteria for metal stents with polymer.

The entire process of the Mechanical Module development includes the design, set up and implementation of several finite element simulations performed with the advanced and beyond the state-of-the-art in-house solver PAK developed by BIOIRC [12]. The solver is able to perform simulation of nonlinear material and geometry problems, nonlinear contact problems, dynamics and statics with residual stress and strain analysis. The process that is followed, in general, includes the following steps: (i) creation of the 3D stent geometry (in case this is not available directly in a 3D format from the manufacturer), (ii) mesh generation, (ii) application of appropriate boundary conditions (depending on the test a variety of boundary conditions are applied). Nonlinear material model has been developed in the finite element solver PAK for prescribing material property from uniaxial stress-strain experimental curves. It is an Open module used only in the Mechanical Modelling Module. The user can prescribe the nonlinear material model directly from stress-strain experimental curves. All open modules for the program PAK can be downloaded from the website of BIOIRC [12]. Based on the aforementioned information, the input file for solver PAK is generated. After executing the simulation, the PAK-CAD software is used for the post-processing of the results. Stress and strain distribution for each time step can be plotted in the PAK-CAD software. The user can prescribe the material model with stress-strain curves.

The intended users of the Mechanical Modelling Module are, primarily, companies developing all types of stents for different medical applications. The Mechanical Modelling Module can be used for different types of stents, regardless of the material the stent is made of and the purpose of use/application. In addition, the Mechanical Modelling Module can be used by researchers and students for further research and education. An overview of the Mechanical Modelling workflow is presented in Fig. 2.

An estimated price for all mechanical tests for one case is approximately 5.000 EUR. More details on the different pricing options are described in Sect. 3. The researchers and the academic community can use the Mechanical Modelling Module as Open Source through GNL, GPL license. The price for academic users will be negotiable, since they need limited performances of the workflow. It will depend on the number of mechanical tests, mesh density, size of the stent model, specific output results for stress, strain, displacement, using pre-stress conditions, nonlinear contact problem etc.

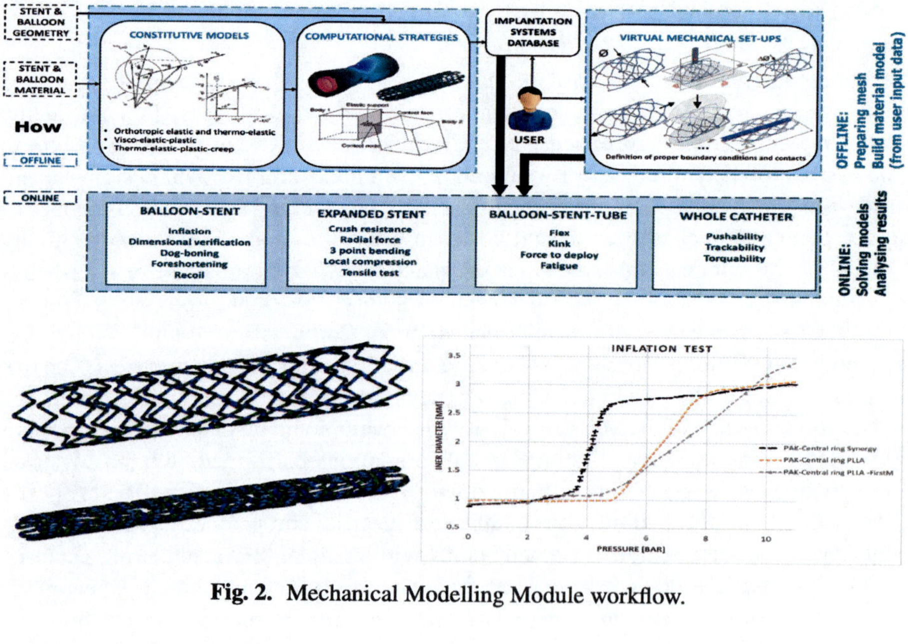

Fig. 2. Mechanical Modelling Module workflow.

3.1 Deployment Module

The Deployment Module requires detailed information about the delivery systems to be simulated to create reliable and realistic virtual FE models of the devices involved in the stenting procedure. In silico simulations of the stenting procedure consists of the following steps, to be repeated for each device (stent or balloon): (i) device positioning, (ii) balloon inflation and, (iii) stent deployment. Most of the computational steps are automatized and this allows a significant reduction in preparing and performing the simulations. In turn, this allowed a reduction of the process to be sustained by the users of the Deployment Module.

Thanks to the realistic predictions of in silico simulations of stenting, the Deployment Module can be exploited by stent companies to refine, reduce or even replace in vitro tests, animal studies and clinical trials. Moreover, the Deployment Module can compare the efficacy of different devices in a specific anatomy. Finally, academic institutions and research centres as well as Regulatory agencies could benefit from the Deployment module simulations (Fig. 3).

The interaction between Deployment Module and other modules has allowed starting useful scientific collaborations that will positively affect the research activities of different groups and will produce papers of interest to the community involved in the development of treatments for cardiovascular diseases.

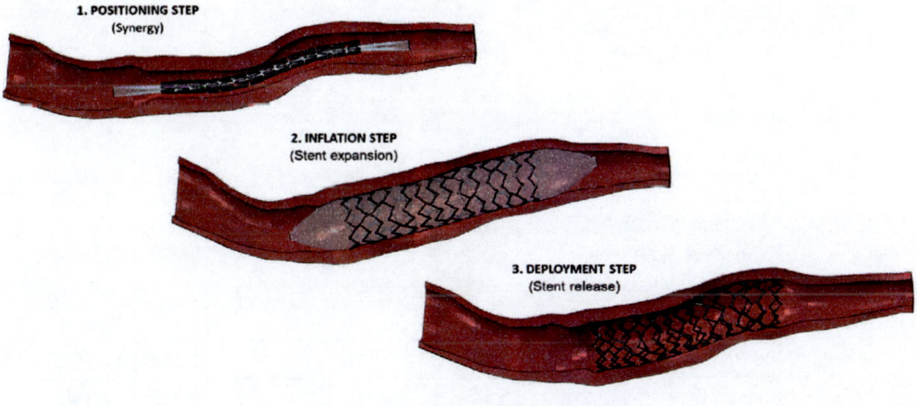

Fig. 3. Stent deployment process.

4 Cost Effectiveness for the InSilc Platform

4.1 Mechanical Modelling Module Pricing Strategy

The pricing strategy for mechanical in-silico module is presented in this section. Starting with the Mechanical Modelling Module, 16 tests that are considered to be included in the bench testing are presented in Table 1, divided into 10 groups of tests. The price for the setup of the model is considered to be 1.000 EUR. The column "Price" (A) includes the prices per simulation. Column B represents the number of tests that should be performed in order to obtain a result (12 is maximum number of tests). Column C is the price per test with some minor changes in the conditions. Finally, in column "Cumulative price", the final price for each test is presented.

Table 1. Mechanical Modelling Module pricing strategy

Test ID	Group ID	Test	Price [€]	# Tested configurations	Out of which, *New* tested configurations	Price	Cumulative Price
			A	**B**	**C**	**D_1**[a]	**$E = \Sigma D_i$**
1	1	Radial Force/Compression	400	3	3	2.400	2.400
2	2	Three-Point Bending	400	5	2	2.200	4.600
3	3	Longitudinal Tensile Strength	300	3	0	900	5.500
4	4	Profile/Diameter	400	12	7	5.500	11.000
5	4	Foreshortening					
6	4	Dog Boning					
7	4	Stent-Free Surface Area					
8	4	Inflation					
9	4	Recoil					
10	5	Crush Resistance/Crush Resistance with Parallel Plates	200	3	0	600	11.600
11	6	Local Compression	200	3	0	600	12.200
12	7	Flex/Kink	500	4	0	2.000	14.200
13	8	Pushability	600	5	1	3.100	17.300
14	8	Trackability					
15	9	Torquability	400	3	0	1.200	18.500
16	10	Radial Fatigue	600	1	1	700	19.200
		TOTAL		**12 different configurations**			

[a] $D_1 = 1{,}000€ + A * B + (C-1) * 100€$ $D_{2\text{–}16} = A * B + C * 100€$

From Table 1 it can be seen that total expenses for in silico mechanical testing for 10 groups of tests amount to around 19.200 EUR. On the other side, for similar in vitro mechanical testing the total price for 12 different configurations of stents is at least

80.000 EUR. Therefore, it can be concluded that in silico mechanical testing is almost 4 times cheaper than real mechanical testing. Moreover, the time needed for the execution of in silico testing is at least 10 times shorter than it is the case with similar in vitro mechanical testing.

4.2 Scenario 1 – Pre-clinical Testing Assessment

All stent manufacturers have the obligation to perform standard mechanical stent testing according to ISO standards. The objective of Scenario 1 is to simulate in silico all the tests required by the ISO. In this example, the aforementioned tests, in silico, so as to compare the performance of two stents (with difference in the design and performances) have been perfomed. The cost and time required is presented in Table 2.

Table 2. Costs and time to perform Scenario 1

Mechanical Module	Cost €	Actual Cost €	Time (in days)	Actual Time (in days)
Radial	800	6.000	1	6
Inflation	800	10.000	1	42
Three-point bending	800	6.000	1	6
Crush	400	6.000	1	6
Local Compression	400	6.000	1	6
Longitudinal Tensile Strength	600	6.000	1	6
Kinking	1.000	12.000	1	6
Flex	1.000	12.000	1	6
Total	**5.800**	**64.000**	**8**	**84**

*Actual cost/time: concerns the testing of 10–15 minimum samples per test

As it can be seen from Table 2, the total price for 8 mechanical tests in silico is 5.800 EUR. On the other hand, in vitro real mechanical tests are approximately 64.000 EUR. Time calculated in days for in vitro mechanical tests is 84 days and for numerical simulation only 8 days. If all 8 in silico tests are running in parallel, they can be performed within 1 day.

4.3 Scenario 2 - Design New Stents

Scenario 2 aims to predict the stenting outcomes for a virtual anatomy, when parameters such as design or material change in a specific stent. In this example, the following modules/tools are included: 3D reconstruction and plaque characterization tool, Deployment Module, Fluid dynamics Module, Drug-delivery Module, Degradation Module and Myocardial Perfusion Module. Relevant modules, their prices and time required are presented in Table.

Table 3. Costs and time required to perform Scenario 2

Tool/Module used in Scenario 2	3D reconstruction Tool	Deployment Module	Fluid dynamics Module	Drug-delivery Module	Degradation Module	Myocardial perfusion module	Total €
Cost €	200	2.000	1.000	3.800	1.000	600	**8.600**
Actual cost of clinical trial	**Metallic stent** clinical study cost per stent: **11.788 €** **BVS** clinical study cost per stent: **14.400 €**						
Time (in days)	2	3	2	7	2	2	18
Time required to execute a clinical study	Up to 2 years (enrolment + 9–12 months FU) for first-in-man study						

As can be observed from Table 3, the total price for 3D reconstruction and plaque characterization tool + Deployment Module + Fluid dynamics Module + Drug-delivery Module + Degradation Module + Myocardial Perfusion Module is around 8.600 EUR and it will take around 18 days. On the other hand, the price for real clinical study for Metallic per stent is 11.788 EUR and for BVS per stent is 14.400 EUR and the time required to execute it is up to 2 years.

4.4 Scenario 3 - Compare Existing Stents

The aim of this scenario is to compare two stents using the same virtual anatomy/anatomies, available in the Virtual vessel database. In this example, the following modules/tools are included: 3D reconstruction and plaque characterization tool, Deployment Module, Fluid dynamics Module, Drug-delivery Module and Degradation Module. The cost and time for performing this scenario is presented in Table.

As can be observed from Table 4, the total price for in silico simulation using two stents in the same virtual anatomies is 8.000 EUR and it will take 18 days. Duration of a clinical trial depends upon the study design. The price for a real clinical study for Metallic per stent is 11.788 EUR and for BVS per stent is 14.400 EUR. If we consider including the whole pipeline up to degradation, the duration can be up to 5 years (patient enrolment + follow-up for degradation). If we limit to restenosis, 2–3 years are needed (enrolment + 12 months Follow-Up).

Table 4. Cost and time required to perform Scenario 3

Tool/Module used in Scenario 3	3D reconstruction Tool	Deployment Module	Fluid dynamics Module	Degradation Module	Drug-delivery Module	Total €
Cost €	200	2.000	1.000[a]	1.000	3.800	**8.000**
Actual cost of clinical trial		**Metallic stent** clinical study cost per stent: **11.788 €** **BVS** clinical study cost per stent: **14.400 €**				
Time (in days)	3	3	2	3	7	**18**
Days required to execute a clinical study		Duration of a clinical trial depends upon the study design 1) If we consider including the whole pipeline up to degradation, the duration can be **up to 5 years** (patient enrolment + follow-up for degradation) 2) If we limit to restenosis, **2–3 years** are needed (enrolment + 12 months FU)				

[a]It only includes macro-scale, two simulations.

4.5 Scenario 4 - Compare Anatomy Configurations and Patient Conditions

Scenario 4 aims to predict the stenting procedure, for a specific stent, considering different virtual anatomies. In this example, the following modules/tools are included: 3D reconstruction and plaque characterization tool, Deployment Module, Fluid dynamics Module and Drug-delivery Module. The cost and time required to perform Scenario 4 are presented in Table.

Table 5. Cost and time required to perform Scenario 4

Tool/Module used in Scenario 4	3D reconstruction Tool	Deployment Module	Fluid dynamics Module	Drug-delivery Module	Total €
Cost €	400	2.000	1.000	3.800	**7.200**
Actual cost of clinical trial	**Metallic stent** clinical study cost per stent: **11.788 €** **BVS** clinical study cost per stent: **14.400 €**				
Time (in days)	6	3	2	7	**18**
Days required to execute a clinical study	If we limit to restenosis, **2–3 years** are needed (enrolment + 12 months FU)				

As can be observed from Table 5, the total price for in silico simulation using one stent in the different virtual anatomies is 7.200 EUR and it will take 18 days. Duration of a clinical trial depends upon the study design. The price for a real clinical study for Metallic per stent is 11.788 EUR and for BVS per stent is 14.400 EUR. If it is limited to restenosis, it is estimated to take 2–3 years.

4.6 Scenario 5 - Compare Different Revascularization Procedures

Scenario 5 aims to predict the stenting outcome, for the selected virtual anatomy and the considered stent, when different implantation procedures are simulated. In this example, the following modules/tools are included: 3D reconstruction and plaque characterization tool, Deployment Module, Fluid dynamics Module, and Myocardial Perfusion Module. The cost and time required to perform Scenario 5 are presented in Table 6.

Table 6. Cost and time required to perform Scenario 5

Tool/Module used in Scenario 5	3D reconstruction Tool	Deployment Module	Fluid dynamics Module	Myocardial perfusion module	Total€
Cost €	200	2.000	1.000	600	**3.800**
Actual cost of clinical trial	**Metallic stent** clinical study cost per stent: **11788 €** **BVS** clinical study cost per stent: **14400 €**				
Time (in days)	2	3	2	2	9
Time required to execute a clinical study	At least 5 years				

As can be observed from Table 6, the total price for in silico simulation used to predict the stenting outcome, for the selected virtual anatomy and the considered stent, when different implantation procedure is around 3.800 EUR and it will take 9 days. Duration of a clinical trial depends upon the study design. The price for a real clinical study for Metallic per stent is 11.788 EUR and for BVS per stent is 14.400 EUR, while the execution time is at least 5 years.

5 Surface Reconstruction Based on Stent Deployment Simulation and 3D Imaging Data

Stent deployment simulations are done using the finite element method (FEM). Geometry for the artery is taken from 3D image data from a real patient. The starting point for this reconstruction technique is the preoperative 3D reconstruction of the coronary artery. This data is combined with a FEM model of the stent, crimped over a balloon. This model is virtually placed in the 3D reconstructed lumen of the artery, and subsequently deployed by virtually inflating the balloon. Finally, the pressure is released, and the balloon is removed. Pressure inside the balloon, shape and properties of the stent and the vessel wall eventually determine the postprocedural shape of the stent and the vessel wall. This technique is rather costly from a computational viewpoint, and required advanced knowledge regarding FEM procedures. It is therefore only used by a few groups, mainly in studies in idealized geometries (2). In this project, the approach described above was used in one of the scenarios, comparing two different stent designs (Fig. 4).

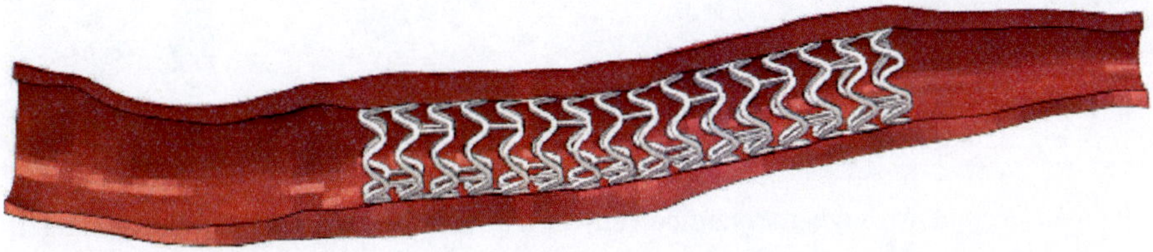

Fig. 4. 3D reconstruction of the surface of the scaffold using stent deployment simulations.

Stent malapposition, or incomplete stent apposition, is a morphological description defined by the lack of contact between at least one stent strut and the underlying intimal surface of the arterial wall in a segment not overlying a side branch [13]. Stent malapposition with stent-artery distance in mm has been presented in Fig. 5.

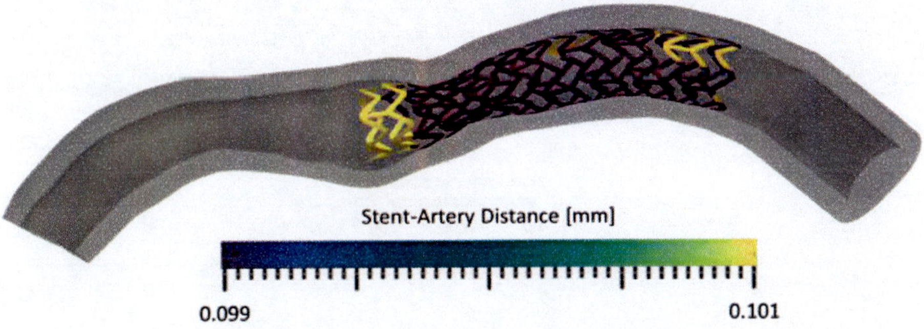

Fig. 5. Stent malapposition (stent-artery distance in mm).

Stent flexibility can influence clinical outcome, especially in complex multiple stenosis. The rigid stent can impose mechanical stress on the artery at the stent edges and alter both arterial geometry and blood flow dynamics in artery [14]. Mechanical stresses inside the arterial wall have been presented in Fig. 6.

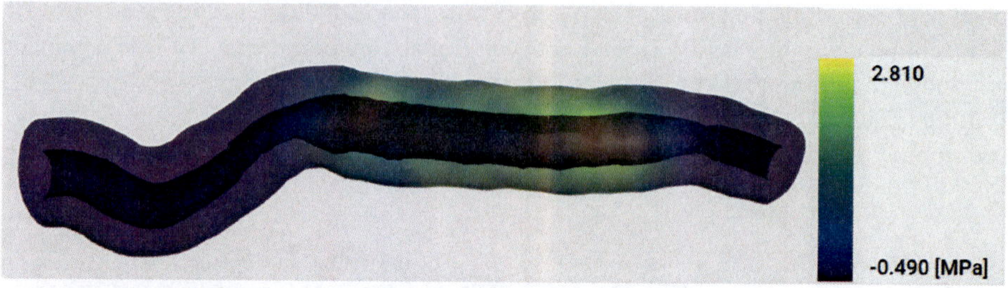

Fig. 6. Artery wall injury (stresses at max inflation).

It is important to have mechanical stress distribution inside the stent during the process of stent deployment. This distribution from the Deployment Module has been presented in Fig. 7.

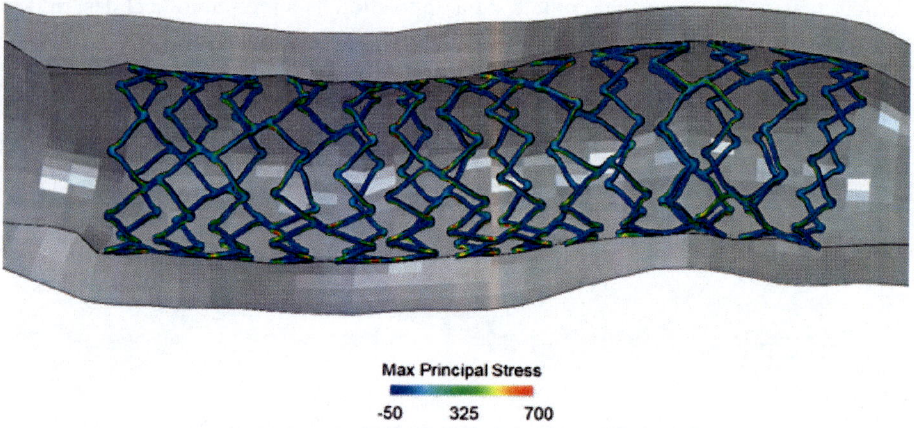

Fig. 7. Mechanical stress in the stent expressed in MPa.

6 Cost Decision Tree Calculation

A *decision node*, typically represented by a square, is a point where several alternatives are possible. A *chance node*, typically represented by a circle (blue color in Fig. 8), is a point in a decision tree where chance determines which event will occur. The sum of probabilities for all branches emanating from a chance node must equal 1.0 or 100%, because one of the events must occur.

The cost decision tree for in silico and real stent deployment has been presented in Fig. 8. Accordingly, there firstly two main decisions are real or in silico stent deployment. We took into account Bare Metal Stents (BMS), Drug-eluting Stents (DES) and Bioresorbable Vascular Stents (BVS) stents. The average value for success is 87.7% while failure amounts to 12.3%. Failure goes again to stent procedure 93.7% while for

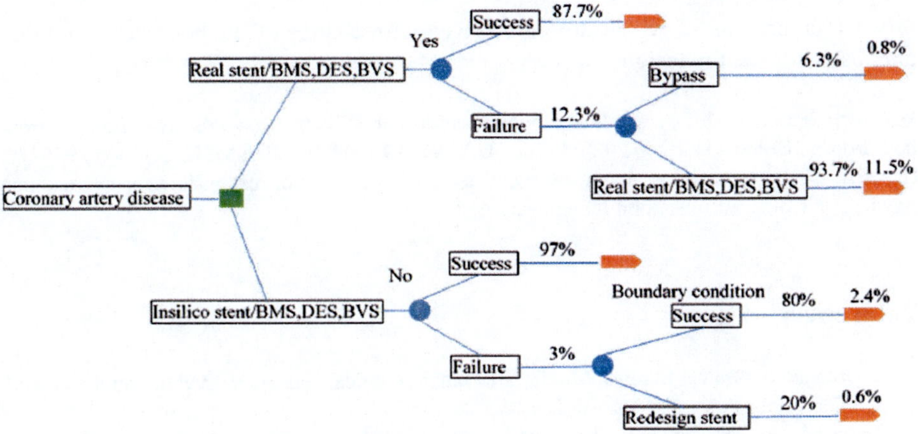

Fig. 8. Cost decision tree for in silico and real stent (BMS, DES, BVS) trial.

bypass procedure it is 6.3%. On the other side, in silico trials give only 3% failure where 80% are solved with change of the boundary condition and 20% goes again to redesign of the stent.

7 Conclusion

In this study, the cost effectiveness of the in silico clinical trials versus real clinical trials and benefits for industry and patients was presented [15, 16]. Also, the pricing strategies for in-silico stent deployment and cost effectiveness analysis with decision tree for Bare Metal Stents (BMS), Drug-eluting Stents (DES) and Bioresorbable Vascular Stents (BVS) stents deployment in the coronary arteries were presented. The InSilc platform for in-silico stent deployment which was developed within the InSilc project (H2020 GA 777119) was presented and all its modules: Mechanical Modelling Module, 3D reconstruction and plaque characterization tool, Deployment Module, Fluid Dynamics Module, Drug Delivery Module, Degradation Module, Myocardial Perfusion Module, Virtual Population Physiology and Virtual Population database. Stent market for coronary and peripheral arteries has been described and all its potential for the future development and research. Mechanical and Deployment Module are described in more detail, since they have the highest TRL (technology readiness level), and they are already validated. Pricing strategies, designed by InSilc consortium, for different scenarios are described and the conclusion is that InSilc cloud platform has high competitive advantage comparing to real clinical trials regarding costs and time required. These scenarios are crucial for prediction of the stenting outcomes for different virtual anatomies and different modules, such as 3D reconstruction and plaque characterization tool, Deployment Module, Fluid dynamics Module, Drug-delivery Module, Degradation Module and Myocardial Perfusion Module.

After the completion of the InSilc project, the work in the cloud platform has been continued within the DECODE project and the conclusion is that in silico trials have only 3% failure, out of which 80% are solved with change of the boundary condition and 20% with stent redesign.

Acknowledgment. This research is supported by the projects that have received funding from the European Union's Horizon 2020: InSilc - GA 777119 and DECODE MSCA - GA No 956470. This article reflects only the author's view. The Commission is not responsible for any use that may be made of the information it contains.

References

1. Learn About Clinical Studies, March 2019. https://clinicaltrials.gov/ct2/about-studies/learn#ClinicalTrials. Accessed 15 Aug 2022
2. Properzi, F., Taylor, K., Steedman, M.: Intelligent drug discovery, 7 November 2019. https://www2.deloitte.com/us/en/insights/industry/life-sciences/artificial-intelligence-biopharma-intelligent-drug-discovery.html. Accessed 19 Aug 2022
3. H2020 InSilc project: In-silico trials for drug-eluting BVS design, development and evaluation. https://cordis.europa.eu/project/id/777119. Accessed 19 Aug 2022
4. Global Balloon-expandable Stents Market – Industry Trends and Forecast to 2028, February 2021. https://www.databridgemarketresearch.com/reports/global-balloon-expandable-stents-market. Accessed 12 Aug 2022
5. Press Release, February 2021. https://www.databridgemarketresearch.com/press-release. Accessed 21 Aug 2022
6. Peripheral Vascular Stents Market, July 2021. https://www.futuremarketinsights.com/reports/peripheral-vascular-stents-market. Accessed 5 Aug 2022
7. Peripheral Vascular Stents Market By Growth, Demand & Opportunities and Forecast to 2028, 7 May 2021. https://www.pharmiweb.com/press-release/2021-05-07/peripheral-vascular-stents-market-by-growth-demand-opportunities-and-forecast-to-2028-medtronic. Accessed 22 Aug 2022
8. PHARMA.AI. https://insilico.com/platform. Accessed 5 Aug 2022
9. H2020 DECODE ITN project: Drug-coated balloon simulation and optimization system for the improved treatment of peripheral artery disease, 12 August 2022. https://www.decodeitn.eu/
10. Fotiadis, D., Filipović, N.: In silico trials for drug-eluting BVS design, development and evaluation. Proj. Repos. J. **8** (2021)
11. Sox, H.C., Higgins, M.C., Owens, D.K.: Medical Decision Making, 2nd edn. Wiley-Blackwell (2013)
12. Software PAK. http://www.bioirc.ac.rs/index.php/software/5-pak. Accessed 22 Aug 2022
13. Hong, M., et al.: Late stent malapposition after drug-eluting stent implantation: an intravascular ultrasound analysis with long-term follow-up. Circulation (2006)
14. Saito, N., Mori, Y., Komatsu, T.: Influence of stent flexibility on artery wall stress and wall shear stress in bifurcation lesions, 2 November 2020
15. Miković, R., et al.: The influence of social capital on knowledge management maturity of non-profit organizations – predictive modelling based on a multilevel analysis. IEEE Access J., 1–15 (2019)
16. Stefanović, M., et al.: An assessment of maintenance performance indicators using the fuzzy sets approach and genetic algorithms. Part B J. Eng. Manuf. **231**, 15–27 (2015)

Author Index

A
Aleksić, Aleksandar 132
Anić, Milos 223

B
Benolić, Leo 320
Bielińska-Dusza, Edyta 198
Blagojević, Anđela 271
Bolbotinović, Željko Z. 75
Božanić, Darko I. 151

C
Car, Zlatan 121, 320
Cardillo, Franco Alberto 58
Comić, Tijana Z. 75

D
Dakić, Dragan 45
Dašić, Lazar 332
Đukić, Tijana 223

E
Elsayed, Mahmoud 186

F
Filipović, Nenad 186, 320

G
Gačić, Marija 367
Geroski, Tijana 271

H
Hamerska, Monika 198

I
Ilić, Đorđe 306

K
Kotowicz, Magdalena 198

L
Lula, Paweł 90, 198

M
Medojević, Milovan M. 29
Milivojević, Nevena 342
Miljković, Boža D. 151
Milošević, Danijela 109
Milutinović, Veljko 103
Mirić, Ana 342
Mitrović, Katarina 109

N
Nestić, Snežana 132

P
Petrović, Miloš 1

S
Salom, Jakob 103
Saveljić, Slavica Mačužić 249
Stanojević, Nebojsa D. 75
Štifanić, Daniel 121
Štifanic, Jelena 121
Stojadinović, Zoran 170

T
Tadić, Danijela 132
Tešić, Duško Z. 151

V
Vasiljević Toskić, Marko M. 29
Vukićević, Arso M. 1
Vukmirović, Dragan V. 75

Z
Zembura, Marcela 90
Živković, Jelena 306
Zulijani, Ana 121

N. Filipovic (Ed.): AAI 2022, LNNS 659, p. 383, 2023.
https://doi.org/10.1007/978-3-031-29717-5